AF360899

Fundamentals of Advanced Pyrometallurgy

Fundamentals of Advanced Pyrometallurgy

Editors

Baojun Zhao
Jianliang Zhang
Xiaodong Ma

MDPI • Basel • Beijing • Wuhan • Barcelona • Belgrade • Manchester • Tokyo • Cluj • Tianjin

Editors

Baojun Zhao
Faculty of Materials
Metallurgy and Chemistry
Jiangxi University of Science
and Technology
Ganzhou
China

Jianliang Zhang
School of Metallurgical and
Ecological Engineering
University of Science and
Technology Beijing
Beijing
China

Xiaodong Ma
Sustainable Minerals Institute
The University of Queensland
Brisbane
Australia

Editorial Office
MDPI
St. Alban-Anlage 66
4052 Basel, Switzerland

This is a reprint of articles from the Special Issue published online in the open access journal *Metals* (ISSN 2075-4701) (available at: www.mdpi.com/journal/metals/special_issues/fundamentals_ pyrometallurgy).

For citation purposes, cite each article independently as indicated on the article page online and as indicated below:

LastName, A.A.; LastName, B.B.; LastName, C.C. Article Title. *Journal Name* **Year**, *Volume Number*, Page Range.

ISBN 978-3-0365-7315-1 (Hbk)
ISBN 978-3-0365-7314-4 (PDF)

Contents

Preface to "Fundamentals of Advanced Pyrometallurgy"

Pyrometallurgical technologies have the advantages of the ability to process low-grade ores, to have a high productivity, and to easily control by-products. With the depletion of minerals and the growing interest in processing of secondary raw materials, new pyrometallurgical technologies have been developed in recent years that are more efficient, economical, and environmentally friendly processes. These new technologies enabled complex primary and secondary raw materials to be processed and advanced materials to be produced more efficiently.

High-temperature reactions inside furnaces are difficult to observe directly. High-quality fundamental research provides not only a deep understanding of pyrometallurgical operations but also direct support for optimization of the processes and development of new technologies. This reprint includes experimental research and modeling simulations for the pyrometallurgical production of ferrous and nonferrous metals including cobalt, copper, lead, tungsten, iron, and steel.

The Special Issue Editors of this reprint express their gratitude to the authors for their contribution and willingness to share their research findings. The Special Issue Editors also thank the academic editors and reviewers for their time and effort, which ensured the quality and timeliness of the publication.

Baojun Zhao, Jianliang Zhang, and Xiaodong Ma
Editors

Article

Experimental Determination of Phase Equilibria in the Na$_2$O-SiO$_2$-WO$_3$ System

Baojun Zhao [1,2], Kun Su [2,*] and Xiaodong Ma [2,*]

[1] Faculty of Materials Metallurgy and Chemistry, Jiangxi University of Science and Technology, Ganzhou 341000, China; bzhao@jxust.edu.cn
[2] Sustainable Minerals Institute, University of Queensland, Brisbane 4072, Australia
* Correspondence: kun.su@uq.edu.au (K.S.); x.ma@uq.edu.au (X.M.)

Abstract: The present study investigated phase equilibria in the Na$_2$O-SiO$_2$-WO$_3$ system experimentally using high-temperature equilibration, quenching, and electron probe X-ray microanalysis (EPMA). New thermodynamic information on the Na$_2$O-SiO$_2$-WO$_3$ system was derived based on the newly obtained experimental results and data from the literature. The primary phase fields of sodium metasilicate, sodium disilicate, and tridymite were determined along with the isotherms at 1073, 1173, and 1273 K. The solubilities of WO$_3$ in SiO$_2$, Na$_2$Si$_2$O$_5$, and Na$_2$SiO$_3$, and the solubility of SiO$_2$ in Na$_2$WO$_4$ were accurately measured using EPMA. Comparisons between the existing and newly constructed phase diagram were carried out and the differences are discussed. The phase equilibrium data will be beneficial to the future development of sustainable tungsten industries and thermodynamic modelling in WO$_3$ related systems.

Keywords: phase equilibria; thermodynamics; Na$_2$O-SiO$_2$-WO$_3$ system; liquidus temperature; pyrometallurgy

Citation: Zhao, B.; Su, K.; Ma, X. Experimental Determination of Phase Equilibria in the Na$_2$O-SiO$_2$-WO$_3$ System. *Metals* **2021**, *11*, 2014. https://doi.org/10.3390/met11122014

Academic Editor: Mark E. Schlesinger

Received: 9 November 2021
Accepted: 9 December 2021
Published: 13 December 2021

Publisher's Note: MDPI stays neutral with regard to jurisdictional claims in published maps and institutional affiliations.

1. Introduction

Tungsten, a metal, is identified as one of the strategic and critical minerals by numerous countries, including The United States, Europe, Russia, Britain, and Australia due to its significant economic importance and supply instability [1–4]. Historically, the extraction of tungsten from ores relied on a series of hydrometallurgy processes, such as alkali digestion followed by solvent extraction and ion exchange, that consume large amounts of reagents, water, and energy [5–7]. Meanwhile, the difficulty of processing tungsten ores continuously raises due to the increased exploitation of primary tungsten resources, which results in a lower grade more complex in mineralogy [8]. These are key hindrances to the future of tungsten extraction processes.

Gomes et al. studied a two-phase extraction method to process tungsten minerals [9,10]. This method utilized the liquid–liquid immiscibility between sodium tungstate and silicon dioxide, which is a well-known phenomenon in glass-forming oxide systems [11]. The two-phase extraction method could light a new direction in the development of future sustainable tungsten extraction processes. Phase diagram is an essential tool to predict the crystallization path and control crystallization behaviour, which has a pivotal role in the understanding of the crystallization process [12–15]. To gain a better understanding of the two-phase extraction method, the phase diagram for the Na$_2$O-SiO$_2$-WO$_3$ system needs to be studied comprehensively to provide fundamental understandings of its thermophysical properties.

Systems containing alkali tungstate have been studied extensively. In the Na$_2$O-WO$_3$ system, two congruently melting compounds, Na$_2$WO$_4$ and Na$_2$W$_2$O$_7$, and one incongruently melting compound, Na$_2$W$_4$O$_{13}$, were identified along with two eutectic reactions [16–18]. The presence of Na$_2$W$_6$O$_{19}$ was a subject of controversy, which was reported by Sakka, Chang, and Sachdev; however, this was not found by

Caillet [16–18]. No information is available on the WO_3-SiO_2 binary system. Moreover, only limited literature relating to the ternary system Na_2O-SiO_2-WO_3 has been published. Štemprok reported a Na_2O-SiO_2-WO_3 ternary phase diagram at 1473 K, which presented a wide two-liquid immiscibility [19]. It was reported that the two-liquid field corresponds to a stable demixing of the system between silicate glass and sodium tungstate. The two liquids' immiscible region disappears in the Na_2O rich sector. Furthermore, it was reported that the presence of WO_3 reduces the liquidus temperature of sodium silicates. However, the region with higher silica contents was not investigated [19]. Various homogenization techniques were carried out in the study, including repeated firing with intermittent crushing and mixing, slow cooling, and quenching. The previous study provided initial information on the Na_2O-SiO_2-WO_3 system. However, more comprehensive studies of the system are needed due to the lack of detailed methodology, microstructure analysis, and isotherms to fill the gap.

In the present study, the phase equilibria of the Na_2O-SiO_2-WO_3 system was investigated using the equilibration/quenching/EPMA technique from 1073 to 1273 K in air. The technique used has been proven to be efficient in acquiring high temperature equilibrium information for metallurgical slags [20]. It also can resolve some uncertainties resulting from dynamic and static methods for studying molten oxides that contain volatile materials, such as sodium and silica [21]. By applying this methodology, the change of initial bulk composition caused by the evaporation of the sodium oxide during high temperature equilibration will not affect the final liquid and solid compositions. This is because the change in the initial bulk composition will only affect the proportion of liquid and solid phases when equilibrium is achieved. The determined phase equilibrium data will be beneficial to the future development of sustainable tungsten industries and for scientific interests, especially for thermodynamic modelling studies where limited thermodynamic data have been reported on WO_3-containing systems.

2. Experimental

The phase equilibrium investigation was carried out using the typical melting-holding-quenching method [22,23]. The samples were prepared by mixing the required amounts of chemicals thoroughly in an agate mortar for 30 min. Chemicals used to prepare the sample were powders of SiO_2 (99.9 wt% Sigma-Aldrich, Missouri, MO, USA), Na_2CO_3 (>99.0 wt%, Sigma-Aldrich, Missouri, MO, USA), and WO_3 (99.9 wt%, Sigma-Aldrich, Missouri, MO, USA). The Na_2CO_3 powder was dried in an oven at 473 K for at least 48 h to remove any moisture prior to sample preparation. After mixing, the mixture was pelletized and placed in a platinum envelope. Approximately 0.2 g of pelletized sample was used in each equilibration experiment.

Equilibration experiments were carried out in a vertical tube furnace with lanthanum chromate heating elements (supplied by Pyrox Thermique Matériaux, Rambouille, France). A B-type thermocouple (Pt-30% Rh/Pt-6%) was placed in an alumina sheath next to the crucible located in the hot zone of the furnace to accurately measure and monitor the temperature during the experiments. The thermocouple was periodically calibrated against a standard thermocouple (supplied by the National Measurement Institute of Australia) to ensure the temperature accuracy within ±2 K. A schematic view of the vertical furnace setup is shown in Figure 1. The platinum envelope that contained the pelletized sample was held in the furnace using a FeCrAl wire (0.5 mm diameter). Samples were inserted from the bottom of the furnace, then raised and kept in the furnace hot zone for the desired period. After equilibration, the samples were dropped directly into mineral oil to achieve rapid quenching. The mineral oil was refrigerated at 263 K (−10 °C) for three hours to ensure the retaining of glassy phase by achieving rapid quenching. Water was not selected as quenching media due to the solubilities of sodium tungstate and other sodium oxide compounds. The quenched samples were washed carefully with ethanol, dried, and mounted in epoxy resin and polished for electron probe X-ray microanalysis (EPMA).

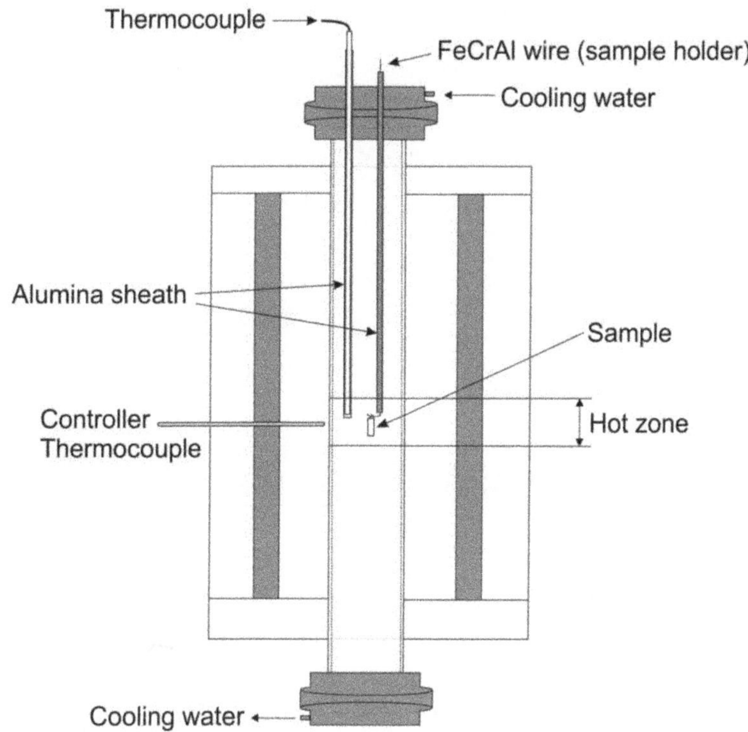

Figure 1. A schematic view of the experimental apparatus.

Carbon coater JEOL (Japan Electron Optics Ltd. * JEOL is a trademark of Japan Electron Optics Ltd., Tokyo, Japan) was used to apply a carbon coat on the surface of the polished samples prior to electron microscopic examination. JEOL JXA-8200 EPMA (Japan Electron Optics Ltd., Tokyo, Japan) with wavelength dispersive detectors was used to analyse the microstructures and phase compositions of the quenched samples. EPMA was conducted at an accelerating voltage of 15 kV and a probe current of 15 nA. The ZAF (Z is atomic number correction factor, A is absorption correction factor, and F is fluorescence correction factor) correction procedure supplied with the electron-probe software was applied. Overall, the measurements of EPMA have an average accuracy within ± 1 wt% [24]. In general, 10–20 and 5–10 measurements were performed for the liquid and solid phases, respectively, at different areas of the samples. The homogeneities of compositions in the phases indicate that the sample reached equilibrium. Only the compositions with standard deviations less than 1 wt% were used in the construction of the phase diagram.

3. Results and Discussion

3.1. Determination of Equilibration Time

At the beginning of the study, suitable equilibration times were determined to ensure that the equilibrium state was achieved in all experiments. The time needed to reach equilibrium is affected by the mass transfer in the sample. Hence, achieving equilibrium in the liquid phase with high viscosity is directly influenced by the low molecule transportation rate. Both composition and temperature can influence liquid viscosity. In general, the liquid phase with high SiO_2 content is considered as highly viscous. In the present study, we investigated the equilibration time for samples in selected primary phase fields with high SiO_2 content in the liquid phase at relatively low temperatures. The appropriate equilibration time was determined by examining the compositional homogeneity using EPMA when the equilibration time was prolonged. Samples with identical initial composition were held at different temperatures and durations. Three phases were observed from the microstructures of the quenched sample that was held for 24 h at 1073 K, indicating the 24 h was not sufficient to allow the samples to reach equilibrium. As a result, the equilibration time at 1073 K was increased to 72 and 168 h. Table 1. shows the comparisons of the samples held for 72 and 168 h at 1073 K, and 24 and 72 h at 1173 K and 1273 K respectively. It can be seen from the table that the phases present in the samples with the same initial composition and different equilibration time are close. The results of the samples in the SiO_2 primary phase field demonstrated that equilibrium was achieved within 72 h at 1073 K, and 24 h at 1173 and 1273 K.

Table 1. Comparison of the samples with same compositions and experimental temperature equilibrated for different times in selected primary phase fields.

Temperature (K)	Holding Time (h)	Phase	Composition (wt%)		
			Na_2O	SiO_2	WO_3
1073	72	Liquid	23.5	73.7	2.8
		Tridymite	0.1	99.9	n.d
	168	Liquid	23.4	73.6	3.0
		Tridymite	0.2	99.8	n.d
1173	24	Liquid	22.1	74.2	3.7
		Tridymite	0.3	99.6	0.1
	72	Liquid	22.2	74.3	3.5
		Tridymite	0.2	99.8	n.d
1273	24	Liquid	20.5	75.6	3.9
		Tridymite	0.3	99.6	0.1
	72	Liquid	20.4	75.8	3.8
		Tridymite	0.2	99.7	0.1

3.2. Description of the Ternary Phase Diagram

Over 50 experiments were conducted in the Na_2O-SiO_2-WO_3 system targeting the two-phase immiscibility region and adjacent primary phase fields at fixed temperatures of 1073, 1173, and 1273 K. Some experimental results were found in fully liquid condition. Four primary phases were identified in the composition range investigated, including Na_2SiO_3, $Na_2Si_2O_5$, SiO_2, and Na_2WO_4. The present results indicate that the phase relationships in the SiO_2-rich region appear to be more complex than that in the previous investigation [19]. Typical microstructures observed in the quenched samples are presented in Figure 2. Figure 2 shows the equilibrium of liquids with (a) sodium metasilicate (Na_2SiO_3), (b) sodium disilicate ($Na_2Si_2O_5$), (c) tridymite (SiO_2), and (d) liquid sodium tungstate (Na_2WO_4). It can be seen from Figure 2d that the sodium tungstate liquid separated from the silicate glass (liquid) and settled on the bottom of the Pt envelope, due to its high density. Furthermore, the phase boundaries in the phase diagram were identified by determining the composition of the liquid phase co-existing with the two solid/liquid phases. Figure 3 shows the microstructures where the liquid was in equilibrium with two other phases, (a) tridymite and sodium tungstate, and (b) sodium disilicate and sodium tungstate. From the microstructures of both samples, the formation of drop-like segregated sodium tungstate liquids against the silicate glass (liquid) and crystals, due to the high viscosity of the silicate glass, were observed.

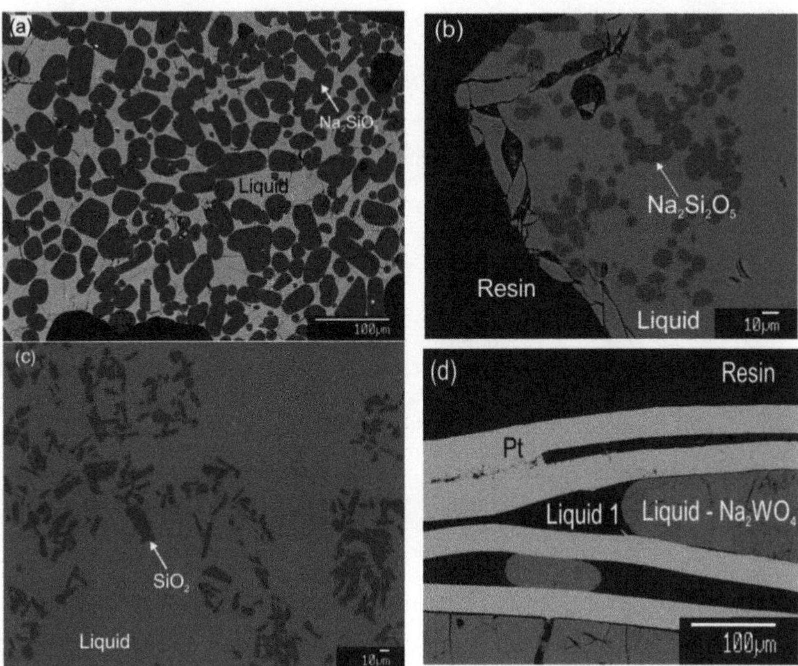

Figure 2. Typical backscattered scanning electron micrographs of quenched samples in the Na$_2$O-SiO$_2$-WO$_3$ system, (**a**) WE5—sodium metasilicate with liquid phase; (**b**) WE6—sodium disilicate with liquid phase; (**c**) WE9—tridymite with liquid phase; (**d**) WE10—sodium tungstate with liquid phase.

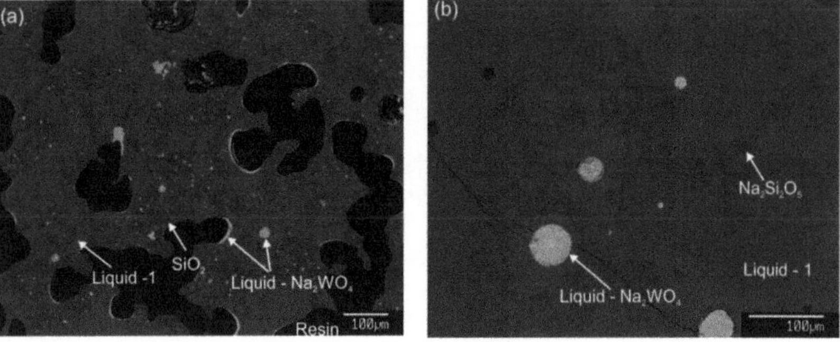

Figure 3. Typical backscattered scanning electron micrographs of the quenched samples in the Na$_2$O-SiO$_2$-WO$_3$ system showing equilibrium of liquid with (**a**) WE13—tridymite and liquid sodium tungstate, and (**b**) WE11—sodium disilicate and liquid sodium tungstate.

The compositions of the solid and liquid phases in the quenched samples analysed by EPMA that were used to construct the isotherms and phase boundaries of the ternary phase diagram are listed in Table 2. The initial compositions of the experiments shown in Table 2 are summarized in Table 3. It can be seen from Table 2 that up to 0.5 wt% WO$_3$ is present in the sodium metasilicate and tridymite. WO$_3$ in the sodium disilicate is only 0.2 wt%. SiO$_2$ in the sodium tungstate is less than 0.1 wt%. The melting points and crystal structures of the solids present in the phase diagram are summarized in Table 4. Measuring the compositions of the liquid and solid phases using the equilibration-quenching technique has one of the greatest advantages: the measured phase compositions are not affected by starting compositions and changes during equilibration. The phase diagram of Na_2O-SiO_2-WO_3 system is constructed from the experimental results, and the isotherms at 1073, 1173, and 1273 K are determined. The construction was based on the ascertained information on the phase equilibria. Information from the well-established binary Na_2O-SiO_2 system is also presented on the ternary phase diagram, and used to construct the diagram by connecting the liquidus to the Na_2O-SiO_2 binary. The hollow dots represent the liquidus obtained from the Na_2O-SiO_2 system reported by Santoso and Taskinen [21] and Zaitsev et al. [25]. In the targeting region of the phase diagram (40–100 wt% SiO_2, 40–60 wt% Na_2O, and 0–20 wt% WO_3), four primary phase fields (sodium metasilicate, sodium disilicate, sodium tungstate and tridymite) were identified at the temperature ranging 1073–1273 K. It can be seen from Figure 4 that, sodium metasilicate and tridymite occupy extraordinarily large primary phase fields in the studied area. As the SiO_2 content increased from 40 to 73.4 wt%, the sodium metasilicate gradually dissolves in the slag, and the sodium disilicate starts to precipitate. Meanwhile, the boundary of the demixing region of the two immiscible liquids enters into the low-WO_3 region. The solubility of the WO_3 in the liquid decreases until it reaches the tridymite primary phase field by continuously increasing the SiO_2 content.

Table 2. Na_2O-SiO_2-WO_3 system analysed by EPMA.

Sample Number	Temperature (K)	Phase	Composition (wt%)		
			Na_2O	SiO_2	WO_3
Sodium metasilicate primary phase field					
WE1	1073	Na_2SiO_3	50.1	49.4	0.5
		Liquid	37.5	48.8	13.7
WE2	1173	Na_2SiO_3	50.5	49.2	0.3
		Liquid	38.7	56.1	5.2
WE3	1173	Na_2SiO_3	50.6	49.2	0.2
		Liquid	38.2	54.4	7.4
WE4	1173	Na_2SiO_3	50.6	49.1	0.3
		Liquid	37.9	48.9	13.2
WE5	1273	Na_2SiO_3	50.2	49.3	0.5
		Liquid	38.9	43.0	18.1
Sodium disilicate primary phase field					
WE6	1073	$Na_2Si_2O_5$	32.8	67	0.2
		Liquid	26.4	71.1	2.5
Tridymite primary phase field					
WE7	1073	SiO_2	0.1	99.9	n.d
		Liquid	23.5	73.7	2.8
WE8	1173	SiO_2	0.3	99.6	0.1
		Liquid	22.1	74.2	3.7
WE9	1273	SiO_2	0.3	99.6	0.1
		Liquid	20.5	75.6	3.9
Sodium tungstate two immiscible liquid region					
WE10	1073	Liquid -Na_2WO_4	20.1	n.d	79.9
		Liquid -1	32.8	59.7	7.5
		Phase boundary			
WE11	1073	Liquid -1	25.4	69.5	4.1
		Liquid -Na_2WO_4	20.2	n.d	79.8
		$Na_2Si_2O_5$	32.9	67.1	0.2
WE12	1173	Liquid -1	22.1	72.3	5.6
		Liquid -Na_2WO_4	20.1	n.d	79.9
		SiO_2	0.1	99.4	0.5
WE13	1273	Liquid -1	20.5	73.5	6.0
		Liquid -Na_2WO_4	20.4	n.d	79.6
		SiO_2	0.2	99.5	0.3
Only one liquid phase exists					
WE14	1073	Liquid	28.4	66.3	5.3
WE15	1173	Liquid	36.8	53.8	9.4
WE16	1173	Liquid	24.0	71.3	4.7
WE17	1173	Liquid	26.3	68.5	5.2
WE18	1273	Liquid	40.6	48.7	10.7

Table 3. Experimental conditions for investigating the Na_2O-SiO_2-WO_3 system.

Sample Number	Temperature (K)	Composition (wt%)		
		Na_2O	SiO_2	WO_3
WE1	1073	25	70	5
WE2	1173	40	55	5
WE3	1173	38	54	8
WE4	1173	42	49	9
WE5	1273	42.5	42.5	5
WE6	1073	27	70	3
WE7	1073	17	80	3
WE8	1173	17	80	3
WE9	1273	17	80	3
WE10	1073	33	57	10
WE11	1073	27	65	8
WE12	1173	15	79	6
WE13	1273	15	79	6
WE14	1073	30	65	5
WE15	1173	35	50	15
WE16	1173	25	70	5
WE17	1173	30	65	5
WE18	1273	42	49	9

Table 4. Properties of the crystalline solids present in the ternary phase diagram.

Crystalline Solid	Structure	Melting Temperature (K)	Ref.
Na_2SiO_3	Tetragonal	1361	[26,27]
$Na_2Si_2O_5$	Tetragonal	1143	[28]
SiO_2	Tetragonal	1984	[29–31]
Na_2WO_4	Spinel	971	[32,33]

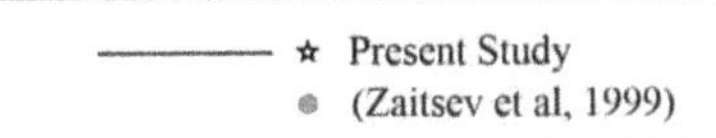

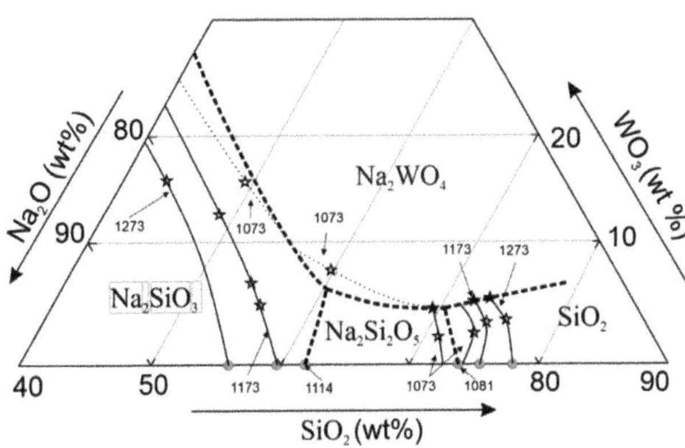

Figure 4. Newly determined isotherms at 1073 K (800 °C), 1173 K (900 °C), and 1273 K (1000 °C) for the ternary phase diagram in the Na_2O-SiO_2-WO_3 system 25; faint dotted line: estimated liquidus.

3.3. Comparisons with the Existing Ternary Phase Diagram

The data reported by Štemprok [19] are extracted and shown in the phase diagram for comparison with the present results in Figure 5. It can be seen from the figure that

Štemprok reported the same primary phase fields. However, isotherms were only reported in the sodium metasilicate primary phase field. The isotherms in the other primary phase fields were not reported in his study. It can be seen from Figure 5 that the size of the Na_2SiO_3 primary phase field determined in the present study is larger than that reported by Štemprok. In contrast, the size of the $Na_2Si_2O_5$ primary phase field determined in the present study is smaller than that reported by Štemprok. The temperature of the eutectic point joining the Na_2SiO_3, $Na_2Si_2O_5$, and Na_2WO_4 primary phase fields is below 1073 K in the present study, which is lower than that (1098 K) reported by Štemprok. The experimental points and the isotherms in the primary phase fields of SiO_2, $Na_2Si_2O_5$, and Na_2WO_4 fill the gaps not reported by Štemprok [19].

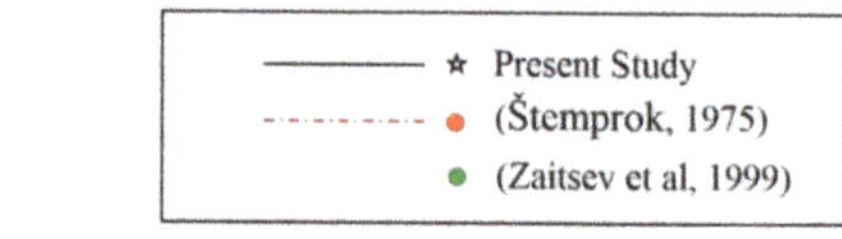

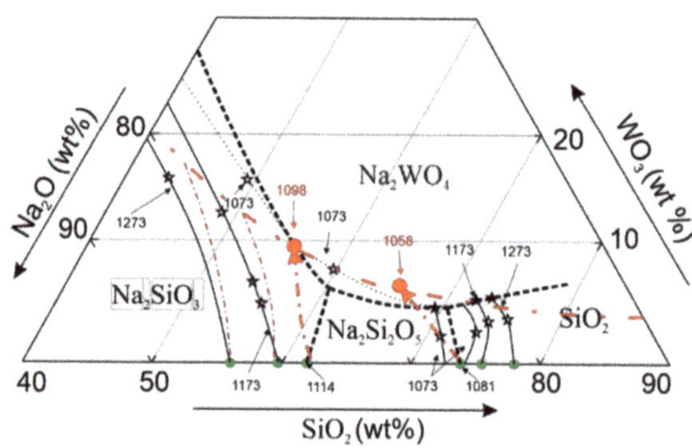

Figure 5. Comparison of the phase diagram between the present study and previous results [19,25], temperature in K.

Due to the limitations in the methodologies and techniques used in previous studies, large deviations were observed while comparing previously reported data with the present measurements. To the best of our knowledge, FactSage cannot calculate the liquidus temperatures for the Na_2O-WO_3 containing systems due to a lack of reliable experimental data. It is believed that the present study is beneficial for the further construction and optimization of thermodynamic modelling by fulfilling the FactSage database with this Na_2O-SiO_2-WO_3 ternary system.

4. Conclusions

New phase equilibria information for the Na_2O-SiO_2-WO_3 system were determined experimentally. A phase diagram targeting a high silica and low tungsten region (40–100 wt% SiO_2, 40–60 wt% Na_2O, and 0–20 wt% WO_3) was constructed with isotherms of 1073, 1173, and 1273 K. The solubilities of WO_3 in SiO_2, $Na_2Si_2O_5$, and Na_2SiO_3, and the solubility of SiO_2 in Na_2WO_4 were accurately measured using EPMA. The newly established phase diagram was compared with the existing data and phase diagram. The present study provides valuable information for the future development of the thermodynamic database and development of tungsten extraction technology.

Author Contributions: Conceptualization, B.Z.; methodology, K.S. and X.M.; validation, K.S.; formal analysis, B.Z. and K.S.; investigation, K.S.; resources, X.M.; writing—original draft preparation, K.S.;

writing—review and editing, B.Z., K.S. and X.M.; supervision, B.Z. and X.M. All authors have read and agreed to the published version of the manuscript.

Funding: This research received no external funding.

Institutional Review Board Statement: Not applicable.

Informed Consent Statement: Not applicable.

Data Availability Statement: Not applicable.

Acknowledgments: The authors would like to thank the financial support from Jiangxi University of Science and Technology through the joint laboratory on high-temperature processing. The authors acknowledge the facilities, scientific, and technical assistance of the Australian Microscopy and Microanalysis Research Facility at the Centre for Microscopy and Microanalysis, The University of Queensland.

Conflicts of Interest: The authors declare no conflict of interest.

References

1. US Department of Defense. *Strategic and Critical Materials 2013 Report on Stockpile Requirements*; Office of the Under Secretary of Defense, US Department of Defense: Washington, DC, USA, 2013; p. 189.
2. European Commission. *Report on Critical Raw Materials for the EU*; European Commission: Brussels, Belgium, 2014; pp. 1–37.
3. Skirrow, R.G.; Huston, D.L.; Mernagh, T.P.; Thorne, J.P.; Duffer, H.; Senior, A. *Critical Commodities for a High-Tech World: Australia's Potential to Supply Global Demand*; Geoscience Australia: Canberra, Australia, 2013.
4. Fortier, S.M.; Nassar, N.T.; Lederer, G.W.; Brainard, J.; Gambogi, J.; McCullough., E.A. *Draft Critical Mineral List—Summary of Methodology and Background Information—US Geological Survey Technical Input Document in Response to Secretarial Order No. 3359*; U.S. Geological Survey: Washington, DC, USA, 2018; Volume 15. [CrossRef]
5. Trasorras, J.R.L.; Wolfe, T.A.; Knabl, W.; Venezia, C.; Lemus, R.; Lassner, E.; Schubert, W.D.; Lüderitz, E.; Wolf, H.U. Tungsten, tungsten alloys, and tungsten compounds. In *Ullmann's Encyclopedia of Industrial Chemistry*; Wiley: New York, NY, USA, 2016; pp. 1–53.
6. Yih, S.; Wang, C. *Tungsten: Sources, Metallurgy, Properties, and Applications*; Springer: New York, NY, USA, 1979; p. 512.
7. Shen, L.; Li, X.; Lindberg, D.; Taskinen, P. Tungsten extractive metallurgy: A review of processes and their challenges for sustainability. *Miner. Eng.* **2019**, *142*, 105934. [CrossRef]
8. Liu, H.; Liu, H.; Nie, C.; Zhang, J.; Steenari, B.M.; Ekberg, C. Comprehensive treatments of tungsten slags in China: A critical review. *J. Environ.* **2020**, *270*, 110927. [CrossRef] [PubMed]
9. Gomes, J.M.; Uchida, K.; Baker, D.H. *A High-Temperature, Two-Phase Extraction Technique for Tungsten Minerals*; US Department of the Interior and Bureau of Mines: Washington, DC, USA, 1968; Volume 7106, pp. 1–11.
10. Gomes, J.M.; Zadra, J.; Baker, D. *Electrolytic Extraction of Tungsten from Western Scheelite*; US Department of the Interior and Bureau of Mines: Washington, DC, USA, 1964; Volume 6512, pp. 1–12.
11. James, P. Liquid-phase separation in glass-forming systems. *J. Mater. Sci.* **1975**, *10*, 1802–1825. [CrossRef]
12. Yan, B.; Wang, X.; Yang, Z. Experimental study of phase equilibria in the MgO-SiO2-TiOx system. *J. Alloys Compd.* **2017**, *695*, 3476–3483. [CrossRef]
13. Li, J.; Shu, Q.; Chou, K. Phase relations in CaO-SiO$_2$-Al$_2$O$_3$-15 mass pct CaF$_2$ System at 1523 K (1250 °C). *Metall. Mater. Trans. B.* **2014**, *45*, 1593–1599. [CrossRef]
14. Shi, J.; Sun, L.; Qiu, J.; Zhang, B.; Jiang, M. Phase equilibria of CaO-SiO$_2$-5wt% MgO-10wt% Al$_2$O$_3$-TiO$_2$ system at 1300 °C and 1400 °C relevant to Ti-bearing furnace slag. *J. Alloys Compd.* **2017**, *699*, 193–199. [CrossRef]
15. Wang, Z.; Sohn, I. Experimental determination of phase equilibria in the CaO-BaO-SiO$_2$-12 mol pct. Al$_2$O$_3$-13 mol pct. MgO system at 1573 K and 1623 K. *J. Am. Ceram. Soc.* **2019**, *102*, 5632–5644. [CrossRef]
16. Caillet, P. Composition of Alkali Tungstate-Tungstic Anhydride systems. *CRH Acad. Sci.* **1963**, *256*, 1986–1989.
17. Sakka, S. Compound formation in Alkali Tungstate systems. *Bull. Inst. Chem. Res. Kyoto Univ.* **1969**, *46*, 300–308.
18. Chang, L.L.Y.; Sachdev, S. Alkali Tungstates: Stability relations in the systems A$_2$OWO$_3$-WO$_3$. *J. Am. Ceram. Soc.* **1975**, *58*, 267–270. [CrossRef]
19. Štemprok, M. Geological significance of immiscibility in fused silicate systems containing tungsten and molybdenum. *Int. Geol. Rev.* **1975**, *17*, 1306–1316. [CrossRef]
20. Jak, E.; Hayes, P.C. Phase equilibria determination in complex slag systems. *Miner. Process. Extract. Metall. Rev.* **2008**, *117*, 1, 1–17. [CrossRef]
21. Santoso, I.; Taskinen, P. Phase equilibria of the Na$_2$O–SiO$_2$ system between 1173 and 1873 K. *Can. Metall. Q.* **2016**, *55*, 243–250. [CrossRef]
22. Feng, D.; Zhang, J.; Li, M.; Chen, M.; Zhao, B. Phase equilibria of the SiO$_2$–V$_2$O$_5$ system. *Ceram. Int.* **2020**, *46*, 24053–24059. [CrossRef]
23. Wang, D.; Chen, M.; Jiang, Y.; Wang, S.; Zhao, Z.; Evans, T.; Zhao, B. Phase equilibria studies in the CaO–SiO$_2$–Al$_2$O$_3$–MgO system with CaO/SiO$_2$ ratio of 0.9. *J. Am. Ceram. Soc.* **2020**, *103*, 7299–7309. [CrossRef]

24. Ritchie, N.; Newbury, D.; Leigh, S. Breaking the 1% accuracy barrier in EPMA. *Microsc. Microanal.* **2012**, *18*, 1006–1007. [CrossRef]
25. Zaitsev, A.I.; Shelkova, N.E.; Lyakishev, N.P.; Mogutnov, B.M. Thermodynamic properties and phase equilibria in the Na_2O-SiO_2 system. *Phys. Chem. Chem. Phys.* **1999**, *1*, 1899–1907. [CrossRef]
26. Anis, M.; Shirsat, M.D.; Hussaini, S.S.; Joshi, B.; Muley, G.G. Effect of sodium metasilicate on structural, optical, dielectric and mechanical properties of ADP crystal. *J. Mater. Sci Technol.* **2016**, *32*, 62–67. [CrossRef]
27. Morey, G.W.; Bowen, N.L. The Binary System Sodium Metasilicate-silica. *J. Phys. Chem.* **1924**, *29*, 1167–1179. [CrossRef]
28. You, J.; Jiang, G.; Xu, K. High temperature Raman spectra of sodium disilicate crystal, glass and its liquid. *J. Non Cryst Solids* **2001**, *282*, 125–131. [CrossRef]
29. Ringdalen, E. Changes in quartz during heating and the possible effects on Si production. *J. Oper. Manag.* **2015**, *67*, 484–492. [CrossRef]
30. Welche, P.; Heine, V.; Dove, M. Negative thermal expansion in beta-quartz. *Phys. Chem. Miner.* **1998**, *26*, 63–77. [CrossRef]
31. Konnert, J.H.; Appleman, D.E. The crystal structure of low tridymite. *Acta Crystallogr. Sect. B Struct. Sci. Cryst. Eng. Mater.* **1978**, *34*, 391–403. [CrossRef]
32. Dkhilalli, F.; Megdiche, B.S.; Rasheed, M.; Barille, R.; Shihab, S.; Guidara, K.; Megdiche, M. Characterizations and morphology of sodium tungstate particles. *R. Soc. Open Sci.* **2018**, *5*, 172214. [CrossRef] [PubMed]
33. Weaste, R.C.; Suby, S.M.S.; Hodman, C.D. *Handbook of Chemistry and Physics*; CRC Press: Boca Raton, FL, USA, 1979; p. B-129.

 *metals*

Communication

Extraction of Sodium Tungstate from Tungsten Ore by Pyrometallurgical Smelting

Liqiang Xu [1,2] and Baojun Zhao [1,2,*]

[1] Faculty of Materials Metallurgy and Chemistry, Jiangxi University of Science and Technology, Ganzhou 341000, China
[2] International Institute for Innovation, Jiangxi University of Science and Technology, Nanchang 330013, China
* Correspondence: bzhao@jxust.edu.cn

Abstract: Tungsten is one of the strategic metals produced from tungsten ores through sodium tungstate. The hydrometallurgical process is a common technology for extracting sodium tungstate from high-grade tungsten concentrates. The grade of tungsten ore is decreasing, and the mineral processing to produce a high-grade concentrate suitable for the hydrometallurgical process is becoming more difficult. It is desirable to develop a new technology to effectively recover tungsten from the complex low-grade tungsten ores. A fundamental study on the pyrometallurgical processing of wolframite was carried out through thermodynamic calculations and high-temperature experiments. The wolframite was reacted with Na_2CO_3 and SiO_2 at 1050–1200 °C and then leached with water to obtain a sodium tungstate solution as a feed for the traditional process of APT (Ammonium paratungstate). The factors affecting the extraction rate of tungsten from wolframite were investigated in air and neutral atmosphere. The extraction rate of tungsten was found to increase with increasing Na_2O content and decrease with increasing SiO_2 addition and temperature. The extraction rate in argon was higher than that in air for wolframite.

Keywords: tungsten; pyrometallurgical smelting; sodium tungstate; wolframite; FactSage

1. Introduction

Tungsten is regarded as a strategic metal by many countries due to its properties and applications [1,2]. The most common tungsten minerals are wolframite $(Fe,Mn)WO_4$ and scheelite $CaWO_4$. The production of tungsten includes several steps: a mineral process to obtain over 50% WO_3 concentrates from the ores containing 0.1–1% WO_3, leaching from the concentrates to obtain sodium tungstate, conversion and precipitation of intermediate tungsten compound ammonium paratungstate (APT), calcination of APT to obtain WO_3 and reduction of WO_3 at high temperature to obtain tungsten metal powder, which is used for the production of tungsten carbide or other W-containing alloys [3–6]. Tungsten ore is considered as one of the critical minerals due to itseconomic importance and short supply [7]. It is important to use the critical minerals efficiently by maximizing the recovery rate during extraction. The overall recovery rate of tungsten includes the recovery rates during mineral processing and leaching. The recovery rate of leaching tungsten concentrates relies on their WO_3 content. High-grade concentrate can result in a higher recovery rate of leaching. In recent years, the quality of the tungsten ores has been changed significantly with excessive exploitation [8]. In order to obtain high-grade tungsten concentrate, the recovery rate during mineral processing has to be reduced, which will decrease the overall recovery rate of tungsten from the ores [9,10]. On the other hand, different conditions are used for leaching of wolframite or scheelite. Mixed ores are becoming the major source of tungsten, which makes beneficiation and leaching more difficult [11,12]. A number of leaching techniques have been developed to improve the recovery rate for the complex tungsten concentrates [13–15]. However, these hydrometallurgical processes still have to treat high-grade concentrate to achieve a high recovery rate.

Citation: Xu, L.; Zhao, B. Extraction of Sodium Tungstate from Tungsten Ore by Pyrometallurgical Smelting. *Metals* **2023**, *13*, 312. https://doi.org/10.3390/met13020312

Academic Editor: Felix A. Lopez

Received: 28 December 2022
Revised: 27 January 2023
Accepted: 1 February 2023
Published: 3 February 2023

Recent studies on phase equilibria of the Na_2O-WO_3 and Na_2O-SiO_2-WO_3 systems show that several compounds, including Na_2WO_4, $Na_2W_4O_{13}$ and $Na_2W_6O_{19}$, can be formed at high temperature, and they are insoluble with the sodium silicate slag [16,17]. These studies provide a base for pyrometallurgical processing of tungsten ores, as more intensive reaction conditions can be obtained compared to the hydrometallurgical process. The aim of this study is to explore the possibility of recovering tungsten from the ore through a high-temperature process.

2. Experimental Methods and Materials

The materials used in high-temperature extraction of tungsten included wolframite concentrate, analytical sodium carbonate (Na_2CO_3) and silicon dioxide (SiO_2). The chemical composition of the wolframite concentrate used in the present study is shown in Table 1. The concentrate mainly contained WO_3, FeO and MnO, with small amounts of CaO, SiO_2 and sulfur. The concentrate was dried at 120 °C for 24 h and ground for the experiments. An amount of 10 g concentrate was mixed with the required sodium carbonate and silicon dioxide in an agate mortar. Well-mixed samples were pelletized and placed in an alumina crucible. The conditions of the high-temperature experiments are shown in Table 2. Experiments 1–4 investigated the effect of Na_2CO_3 on the extraction rate of wolframite. Note that Na_2O is an effective component at high temperature as a result of Na_2CO_3 decomposition. Experiments 5–7 compared the effect of the reaction time, and experiments 8–10 looked at the effect of SiO_2 on the extraction rate of wolframite. The iron from wolframite can be Fe^{2+} or Fe^{3+} during a high-temperature process. Experiments 11–14 were carried out in argon gas at different Na_2CO_3 additions.

Table 1. The composition of wolframite concentrate analyzed by XRF (wt%).

Wolframite	WO_3	CaO	FeO	MnO	SiO_2	S
wt%	78.04	1.06	10.69	7.88	1.73	0.61

Table 2. Experimental conditions for high-temperature processing of wolframite concentrate.

Exp No	Ore (g)	Temp (°C)	Time (min)	Na_2CO_3 (g)	SiO_2 (g)	Atmosphere
1	10	1200	60	10.5	3	Air
2	10	1200	60	10	3	Air
3	10	1200	60	8.05	3	Air
4	10	1200	60	6.81	3	Air
5	10	1050	60	6.81	4	Air
6	10	1100	60	6.81	4	Air
7	10	1200	60	6.81	4	Air
8	10	1200	60	9.29	3	Air
9	10	1200	60	9.29	4	Air
10	10	1200	60	9.29	5	Air
11	10	1200	60	10.52	3	Ar
12	10	1200	60	9.29	3	Ar
13	10	1200	60	8.05	3	Ar
14	10	1200	60	6.81	3	Ar

The experiments in air were carried out in a muffle furnace. The samples were cooled down in the furnace after the reaction time, as shown in Table 2. The experiments in argon were carried out in a sealed vertical tube furnace similar to the one described in previous studies [16,17]. The crucible with the sample was initially suspended on a Mo wire at the bottom end of the furnace. After the furnace was flashed by 400 mL/min Ar gas flow for 30 min, the sample was raised to the hot zone of the furnace. After the reaction, the sample was lowered to the bottom of the furnace and removed after cooling. The heated sample was ground and leached in water at 50 °C. The water to solid ratio was 5:1, and the mixture was stirred for 120 min before separating the sodium tungstate solution from the residue by filtration. The residue was analyzed by XRF (X-Ray Fluorescence), and the content of WO_3 in the residue was used to evaluate the extraction rate of tungsten from the concentrate.

3. Results and Discussion

3.1. Thermodynamic Considerations of High-Temperature Reactions

FactSage 8.2 [18] is a powerful thermodynamic software and was used to predict high-temperature reactions in the present study. The composition of the wolframite given in Table 1 was used for thermodynamic calculations. The databases of "FactPS" and "FToxid" were used in the "Equilib" module. The solution phases selected in the calculations were "FToxide-SLAGA", "FToxide-SPINC", "FToxide-MeO_A" and "FToxide-OlivA" and "FToxide-Mull".

Figure 1 shows the changes of the phases as a function of sodium oxide addition calculated by FactSage 8.2. SiO_2 addition to 100 g wolframite is 30%, and the temperature is 1200 °C. As can be seen, wolframite decreases and sodium tungstate increases with increasing Na_2O. With approximately 32% Na_2O addition, all wolframite is decomposed and converted to Na_2WO_4. On the other hand, SiO_2 only starts to react with the wolframite when Na_2O addition is greater than 12%, as Na_2O reacts with wolframite first. Na_2O is an essential slag-forming component together with SiO_2. With 16% Na_2O addition, all SiO_2 is dissolved into the liquid slag with the oxides of sodium, iron and manganese. More than 33% Na_2O will only enter the silicate slag. Both Na_2WO_4 and the silicate slag are liquid at 1200 °C, which enables the products to be removed from the furnace easily. During the high-temperature reaction, other components from the concentrate, such as CaO and sulfur, are dissolved in the slag phase.

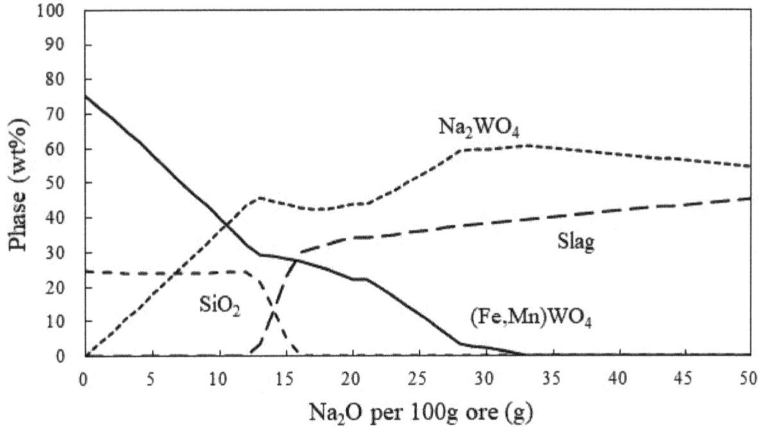

Figure 1. Changes of the phases with the increase in Na_2O content in air at 1200 °C with 30% SiO_2, calculated by FactSage 8.2.

Figure 2 shows the effect of temperature on the decomposition of wolframite at 30 wt% SiO_2 and various Na_2O additions. It can be seen from Figure 2a that decomposition of wolframite increases with increasing temperature at a given Na_2O addition. With 20% and 25% Na_2O addition, the wolframite cannot be completely decomposed even at 1600 °C. With 30% and 35% Na_2O addition, the wolframite can be completely decomposed at 1350 and 1080 °C, respectively. It can be seen from Figure 2b that, at a given Na_2O and SiO_2 addition, a minimum temperature is required to completely decompose the wolframite. For example, 32% Na_2O is required to fully decompose the wolframite at 1200 °C, which is the case shown in Figure 1. In other words, the lowest temperature is required to completely decompose the wolframite at a given Na_2O addition. For example, the temperature must be over 1350 °C to fully decompose the wolframite if 30% Na_2O is added. Na_2O has a much lower melting temperature (1132 °C) than other components. Sufficient Na_2O can form liquid Na_2WO_4 and liquid slag Na_2O-FeO-MnO-SiO_2 to fully decompose the wolframite. If Na_2O is not sufficient, a higher temperature is required to ensure the liquid slag Na_2O-FeO-MnO-SiO_2 is formed.

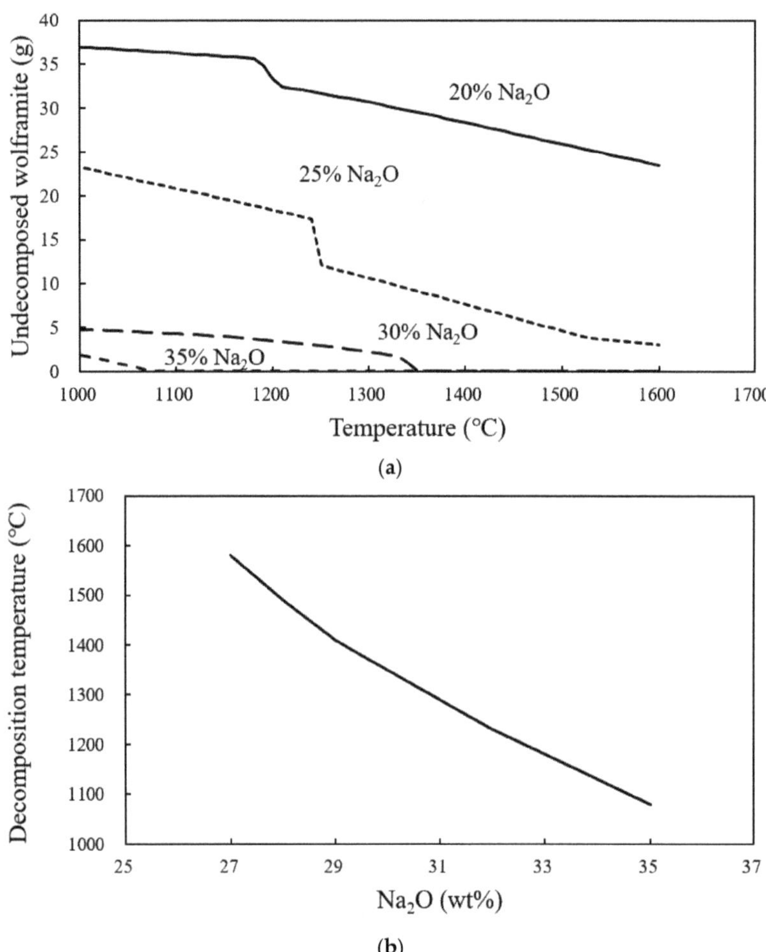

Figure 2. Effects of temperature and Na_2O on undecomposed wolframite with 30 wt% SiO_2, calculated by FactSage 8.2, (**a**) effect of temperature and Na_2O on undecomposed wolframite, (**b**) lowest decomposition temperature corresponding to Na_2O addition.

Wolframite can be decomposed with or without SiO_2:

$$(Fe,Mn)WO_4 + Na_2O \rightarrow Na_2WO_4 + FeO + MnO \tag{1}$$

$$(Fe,Mn)WO_4 + Na_2O + SiO_2 \rightarrow Na_2WO_4 + slag\ (Na_2O\text{-}FeO\text{-}MnO\text{-}SiO_2) \tag{2}$$

If SiO_2 is not present, the FeO and MnO formed from decomposition of the wolframite may cover the surface of the remaining wolframite to stop further reaction between wolframite and Na_2O. In contrast, SiO_2 can react with Na_2O, FeO and MnO to form a liquid slag, which can enhance the decomposition of wolframite. However, it can be seen from Reaction (2) that SiO_2 can also consume Na_2O to form a slag. Figure 3 shows the changes of the phase fractions as a function of SiO_2 addition calculated by FactSage 8.2. At 1200 °C with 30% Na_2O, Na_2WO_4 decreases and slag increases with increasing SiO_2 addition. However, when SiO_2 exceeds 21%, wolframite starts to appear. SiO_2 is a stronger acidic oxide than WO_3. Strong basic oxide Na_2O first reacts with SiO_2. When the Na_2O addition is fixed, more SiO_2 addition consumes Na_2O to form a slag first. The remaining Na_2O is not sufficient to fully decompose the wolframite. A certain amount of SiO_2 is necessary to form a liquid slag with FeO, MnO and Na_2O. Excess SiO_2 will consume more Na_2O and influence the decomposition of wolframite.

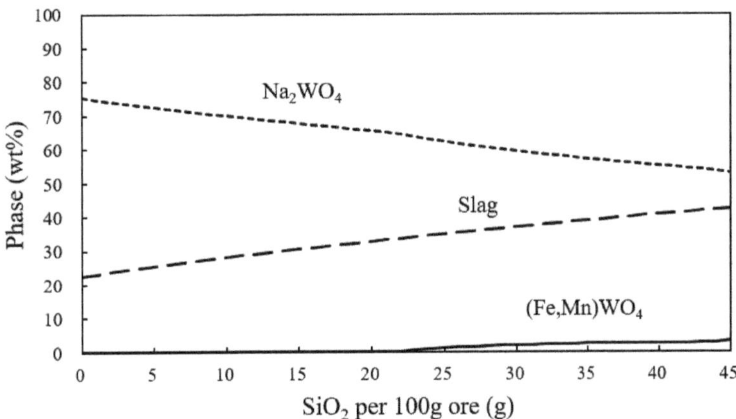

Figure 3. Effect of SiO_2 addition on phase fractions in air at 1200 °C with 30% Na_2O, calculated by FactSage 8.2.

3.2. Experimental Results

High-temperature experiments were carried out to evaluate the effects of Na_2O addition, SiO_2 addition, temperature and oxygen partial pressure on the extraction rate of tungsten from wolframite. The melted samples were leached with water, and the leaching residue was analyzed by XRF. The remaining WO_3 concentration in the leaching residue is used to represent the extraction rate. High WO_3 in the residue means low extraction rate. Figure 4 shows the effect of Na_2O on the WO_3 in the leaching residue in air. An amount of 30% SiO_2 was added, and the temperature was 1200 °C. It can be seen that WO_3 in the leaching residue decreases with the increase in Na_2O addition. When the Na_2O addition is higher than 58.5%, the WO_3 in the leaching residue is 0.36%, which is much lower than that (>1%) in the conventional hydrometallurgical process [19]. The experimental results confirmed the trend predicted by FactSage 8.2, as shown in Figure 1. However, the actual Na_2O required to fully decompose the wolframite is much higher than the predictions. This indicates that the thermodynamic database for a WO_3-containing system needs to be improved.

Figure 5 shows the experimental results on the effect of SiO_2. As can be seen in the figure, the WO_3 in the leaching residue increases with the increase in SiO_2, which confirms the trend predicted by FactSage, as shown in Figure 3. Too much SiO_2 consumed more Na_2O; the remaining Na_2O was not sufficient to fully decompose the wolframite.

Figure 6 shows the effect of temperatures on the remaining WO_3 in the leaching residue with additions of Na_2O and SiO_2 at 39.8% and 40%, respectively. It can be seen in the figure that within the range of 1050–1200 °C, WO_3 in the leaching residue increases with increasing temperature. This can be explained by the solubility of WO_3 in the slag. It seems that the solubility of WO_3 in the silicate slag increases with increasing temperature. The WO_3 dissolved in the silicate slag cannot be leached with water. In the current FactSage database, WO_3 is not included in the slag phase, and the solubility of WO_3 in the slag cannot be predicted.

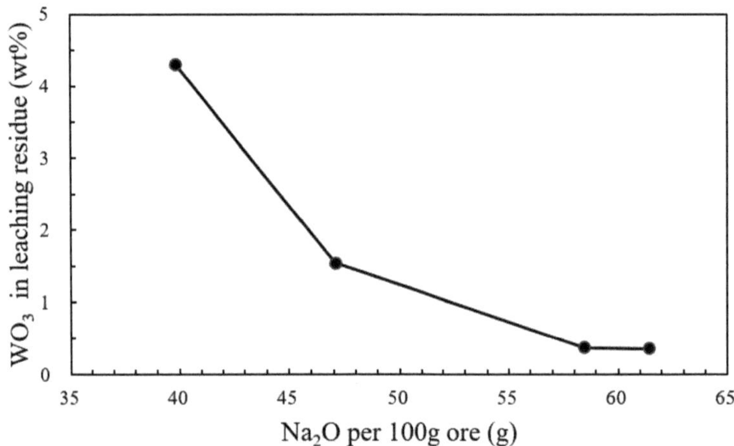

Figure 4. Effect of Na_2O on WO_3 content in leaching residue at 1200 °C in air with 30% SiO_2.

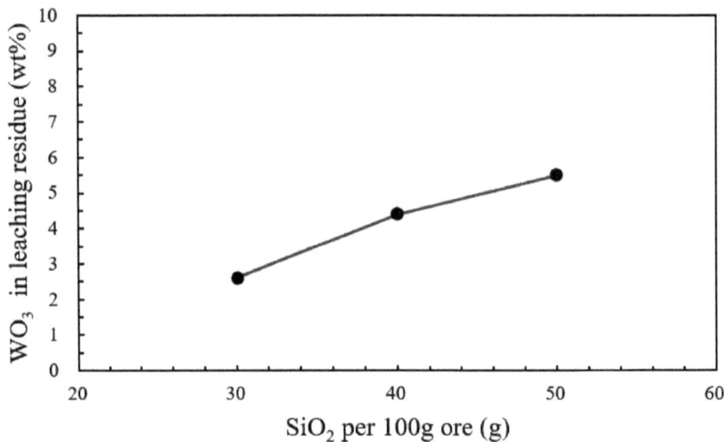

Figure 5. Effect of SiO_2 on WO_3 in leaching residue at 1200 °C in air with 54.3% Na_2O.

Reaction (2) shows the decomposition of wolframite without a variation in the valence state. Iron is initially present in the wolframite as Fe^{2+}, which can be oxidized to Fe^{3+} in air. The above discussions focus on the reactions in air. Figure 7 shows the decomposition of wolframite by Na_2O in argon at 1200 °C, as predicted by FactSage 8.2. The general trend of the reactions in argon is similar to those in air, as shown in Figure 1. However, it can be seen that the Na_2O required to fully decompose wolframite is 30%, which is lower than that in air (30%).

A series of experiments were conducted in argon gas for comparison with those in air. Under the neutral gas flow, iron is present as Fe^{2+}, which is the same as in the wolframite. It can be seen in Figure 8 that WO_3 in the leaching residue is 0.22% with 54% Na_2O addition. If the Na_2O addition is more than 58%, the WO_3 in the leaching residue approaches zero, and the recovery of WO_3 is almost 100%. It is easy to decompose wolframite under reducing conditions.

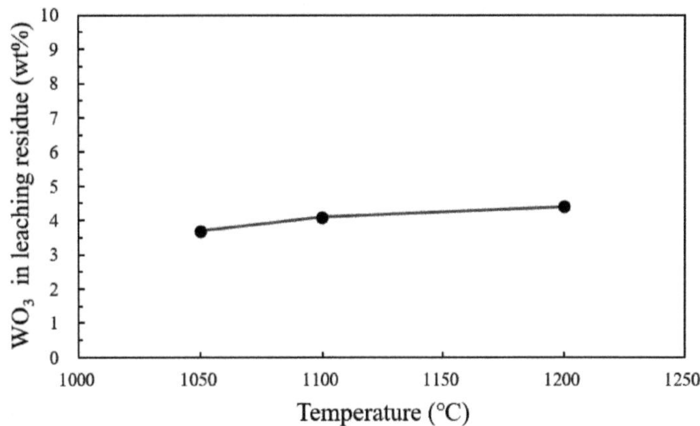

Figure 6. Effect of temperature on WO_3 in leaching residue in air with 39.83% Na_2O and 40% SiO_2.

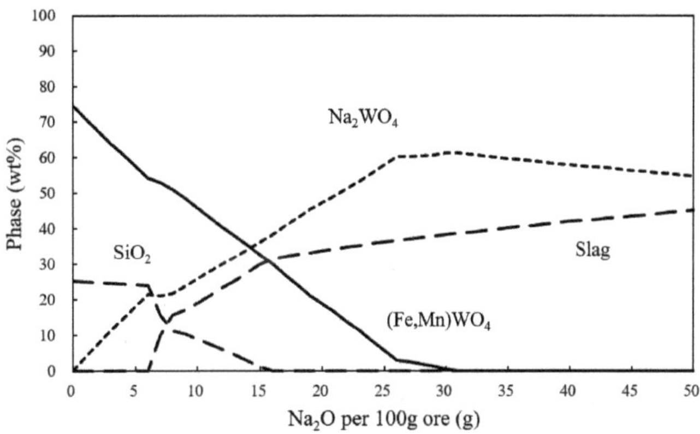

Figure 7. Changes of phases with the increase in Na_2O addition in argon at 1200 °C, calculated by FactSage 8.2.

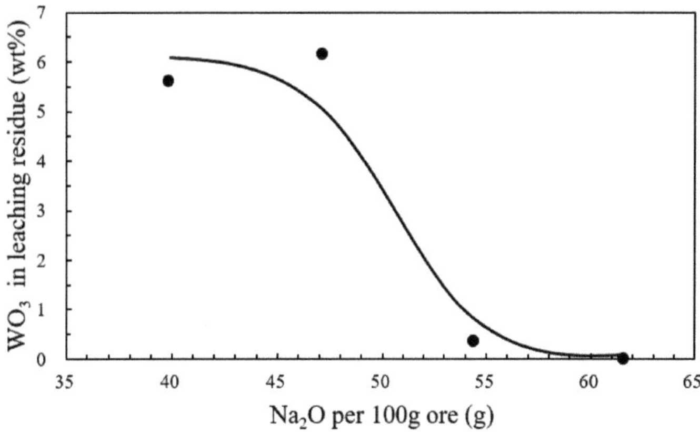

Figure 8. Effect of Na_2O on WO_3 in leaching residue at 1200 °C in argon with 30% SiO_2.

4. Conclusions

Thermodynamic calculations and high-temperature experiments confirmed that WO_3 can be effectively recovered from wolframite by the pyrometallurgical process. The recovery of tungsten from wolframite is much higher than that in the conventional process. SiO_2 can form liquid slag with the oxides of iron, manganese and sodium to enhance the decomposition of wolframite. However, excess SiO_2 consumes Na_2O and reduces the decomposition rate. High Na_2O addition and low temperature are beneficial for maximizing the recovery of tungsten. It is easy to decompose wolframite under reducing conditions where iron is present as Fe^{2+}.

Author Contributions: Conceptualization, B.Z.; methodology, B.Z. and L.X.; Experiments, L.X.; writing—original draft preparation, L.X.; writing—review and editing, B.Z.; supervision, B.Z.; project administration, B.Z. All authors have read and agreed to the published version of the manuscript.

Funding: This research received no external funding.

Data Availability Statement: Not applicable.

Conflicts of Interest: The authors declare no conflict of interest.

References

1. Tkaczyk, A.H.; Bartl, A.; Amato, A.; Lapkovskis, V.; Petranikova, M. Sustainability evaluation of essential critical raw materials: Cobalt, niobium, tungsten and rare earth elements. *J. Phys. D Appl. Phys.* **2018**, *51*, 203001. [CrossRef]
2. Li, X.Y.; Ye, Y.Q.; Zhang, F.L.; Wang, D. Recommended Management Strategies and Resources Status of Chinese Tungsten in the New Period. *Mod. Min.* **2018**, *34*, 25–28.
3. Kvashnin, A.G.; Tantardini, C.; Zakaryan, H.A.; Kvashnina, Y.A.; Oganov, A.R. Computational Search for New W–Mo–B Compounds. *Chem. Mater.* **2020**, *32*, 7028–7035. [CrossRef]
4. Pak, A.Y.; Shanenkov, I.I.; Mamontov, G.Y.; Kokorina, A.I. Vacuumless synthesis of tungsten carbide in a self-shielding atmospheric plasma of DC arc discharge. *Int. J. Refract. Met. Hard Mater.* **2020**, *93*, 105343. [CrossRef]
5. Kvashnin, A.G.; Rybkovskiy, D.V.; Filonenko, V.P.; Bugakov, V.I.; Zibrov, I.P.; Brazhkin, V.V.; Oganov, A.R.; Osiptsov, A.A.; Zakirov, A.Y. WB5−x: Synthesis, Properties, and Crystal Structure—New Insights into the Long-Debated Compound. *Adv. Sci.* **2020**, *7*, 2000775. [CrossRef] [PubMed]
6. Kvashnin, A.G.; Allahyari, Z.; Oganov, A.R. Computational discovery of hard and superhard materials. *J. Appl. Phys.* **2019**, *126*, 040901. [CrossRef]
7. Geoscience Australia. Available online: https://www.ga.gov.au/about/projects/resources/critical-minerals (accessed on 22 December 2022).
8. Gao, Y.D. The characteristics and research advances of mineral processing technologies of Chinese tungsten resources. *China Tungsten Ind.* **2016**, *31*, 35–39.
9. Zhao, Z.W.; Sun, F.L.; Yang, J.H.; Fang, Q.; Jiang, W.W.; Liu, X.H.; Chen, X.Y.; Li, J.T. Status and prospect for tungsten resources, technologies and industrial development in China. *Trans. Nonferrous Met. Soc. China* **2019**, *29*, 1902–1916.
10. Rao, N.K. Beneficiation of tungsten ores in India: A review. *Bull. Mater. Sci.* **1996**, *19*, 201–205.
11. Yao, Z.G. Research on NaF autoclave process for scheelite leaching. *China Tungsten Ind.* **1999**, *14*, 167–170.
12. Zhang, W.J.; Wen, P.C.; Xia, L.; Chen, J.; Che, J.Y.; Wang, C.Y.; Ma, B.Z. Understanding the role of hydrogen peroxide in sulfuric acid system for leaching low-grade scheelite from the perspective of phase transformation and kinetics. *Sep. Purif. Technol.* **2021**, *277*, 119407. [CrossRef]
13. Johansson, Ö.; Pamidi, T.; Shankar, V. Extraction of tungsten from scheelite using hydrodynamic and acoustic cavitation. *Ultrason. Sonochem.* **2020**, *71*, 105408. [CrossRef] [PubMed]
14. Sun, P.M.; Li, H.G.; Li, Y.J.; Zhao, Z.W.; Huo, G.S.; Sun, Z.M.; Liu, M.S. Decomposing scheelite and scheelite-wolframite mixed concentrate by caustic soda digestion. *J. Cent. South Univ. Technol.* **2003**, *10*, 297–300. [CrossRef]
15. Zhao, Z.W.; Li, J.T.; Wang, S.B.; Li, H.G.; Liu, M.S.; Sun, P.M.; Li, Y.J. Extracting tungsten from scheelite concentrate with caustic soda by autoclaving process. *Hydrometallurgy* **2011**, *108*, 152–156. [CrossRef]
16. Zhao, B.J.; Su, K.; Ma, X.D. Phase Equilibria in the Na_2O-WO_3 System. *Ceram. Int.* **2022**, *48*, 15098–15104. [CrossRef]
17. Zhao, B.J.; Su, K.; Ma, X.D. Experimental Determination of Phase Equilibria in the Na_2O-SiO_2-WO_3 System. *Metals* **2021**, *11*, 2014. [CrossRef]

18. Bale, C.W.; Bélisle, E.; Chartrand, P.; Decterov, S.A.; Eriksson, G.; Gheribi, A.E.; Hack, K.; Jung, I.H.; Kang, Y.B.; Melançon, J.; et al. FactSage thermodynamic software and databases 2010–2016. *Calphad* **2016**, *55*, 1–19. [CrossRef]
19. Wang, X.; Ma, X.D.; Liao, C.F.; Zhao, B.J. High Temperature Processing of Tungsten Slag. In *11th International Symposium on High-Temperature Metallurgical Processing*; Peng, Z., Hwang, J.Y., Downey, J.P., Gregurek, D., Zhao, B.J., Yucel, O., Keskinkilic, E., Jiang, T., White, J.F., Mahmoud, M.M., Eds.; Springer International Publishing: Berlin/Heidelberg, Germany, 2020; pp. 289–294; ISBN 978-3030365394.

Article

The Role of Silica in the Chlorination–Volatilization of Cobalt Oxide by Using Calcium Chloride

Peiwei Han [1,2,*] , Zhengchen Li [1,2], Xiang Liu [1,2] , Jingmin Yan [1,2] and Shufeng Ye [1,2,*]

[1] State Key Laboratory of Multiphase Complex Systems, Institute of Process Engineering, Chinese Academy of Sciences, Beijing 100190, China; lizhengchen19@ipe.ac.cn (Z.L.); liuxiang@ipe.ac.cn (X.L.); 18813095233@163.com (J.Y.)
[2] Innovation Academy for Green Manufacture, Chinese Academy of Sciences, Beijing 100190, China
* Correspondence: pwhan@ipe.ac.cn (P.H.); sfye@ipe.ac.cn (S.Y.); Tel.: +86-13811635732; (P.H.); +86-13911828468 (S.Y.)

Abstract: The role of silica in the chlorination–volatilization of cobalt oxide, using calcium chloride, is investigated in this paper. It is found that the Co volatilization percentage of the $CoO-Fe_2O_3-CaCl_2$ system is not larger than 12.1%. Silica plays an important role in the chlorination–volatilization of cobalt oxide by using calcium chloride. In the $CoO-SiO_2-Fe_2O_3-CaCl_2$ system, the Co volatilization percentage is initially positively related to the molar ratio of SiO_2 to $CaCl_2$, and remains almost constant when the molar ratio of SiO_2 to $CaCl_2$ rises from zero to eight. The critical molar ratios of SiO_2 to $CaCl_2$ are 1 and 2 when the molar ratios of $CaCl_2$ to CoO are 8.3 and 16.6, respectively. The Co volatilization percentage remains almost constant with the increase in CaO concentration, and decreases when Al_2O_3 and MgO are added. $Ca_2SiO_3Cl_2$ is determined after roasting at 1073 K and 1173 K, and disappears at temperatures in excess of 1273 K in the calcines from the $CoO-SiO_2-CaCl_2$ system. $CaSiO_3$ always exists in the calcines at temperatures in excess of 973 K.

Keywords: chlorination–volatilization; cobalt oxide; calcium chloride; phases of calcines

Citation: Han, P.; Li, Z.; Liu, X.; Yan, J.; Ye, S. The Role of Silica in the Chlorination–Volatilization of Cobalt Oxide by Using Calcium Chloride. *Metals* **2021**, *11*, 2036. https://doi.org/10.3390/met11122036

Academic Editors: Baojun Zhao, Jianliang Zhang, Xiaodong Ma and Felix A. Lopez

Received: 10 November 2021
Accepted: 10 December 2021
Published: 15 December 2021

Publisher's Note: MDPI stays neutral with regard to jurisdictional claims in published maps and institutional affiliations.

1. Introduction

When roasted at a high temperature, nonferrous metal oxides are converted into corresponding chlorides and volatilize in the form of gaseous chlorides in the presence of chlorinating agents. The chlorination–volatilization method has been used for the recovery of valuable metals from slags or refractory ores, such as Au, Ag, Cu, Pb, and Zn [1–6].

In regards to cobalt recovery, many works have concentrated on chloride roasting as the pretreatment to convert cobalt compounds into soluble chloride, followed by a subsequent hydrometallurgical step [7–14]. The chloridizing agents used for chloride roasting include gaseous Cl_2 and HCl, and solid $MgCl_2 \bullet 6H_2O$, NaCl, $AlCl_3 \bullet 6H_2O$, and NH_4Cl. However, there are few investigations about the chlorination–volatilization of cobalt at present. It was reported that to obtain approximately 50% volatilization percentage, a chlorine consumption of approximately 2.5 times the stoichiometric amount was needed, in the case of producing iron ore pellets from pyrite cinders containing nonferrous metals by using chlorine [15].

Calcium chloride is a popular chloridizing agent because of its high stability and lack of toxicity in practice. The chlorination–volatilization of cobalt, using calcium chloride, was investigated in our previous paper [16]. The effects of different variables on the cobalt volatilization percentage were investigated, including flow rate, oxygen partial pressure and water vapor content of the carrier gas, roasting time, and temperature. The aims of this paper are to investigate the effects of silica, other gangues (Al_2O_3, CaO, and MgO) and $CaCl_2$ dosage on the Co volatilization percentages during the chlorination–volatilization of cobalt oxide, and on the phases of calcines after roasting.

2. Materials and Methods

Cobalt-containing slags were prepared using reagent-grade CoO, SiO_2, Fe_2O_3, Al_2O_3, MgO and CaO powders. Reagent-grade anhydrous $CaCl_2$ was adopted as the chlorinating agent. All the reagents were precisely weighed according to the compositions shown in Table 1 and pressed into briquettes after sufficient mixing. The effects of $CaCl_2$ dosage on the Co volatilization percentage were investigated from No.1 to No.5. The effects of the molar ratio of SiO_2 to $CaCl_2$ on the Co volatilization percentage were also investigated from No.4 to No.14. Finally, the effects of Al_2O_3, MgO and CaO additions on the Co volatilization percentage were investigated from No.14 to No.20.

Table 1. Sample compositions and $CaCl_2$ dosage.

No.	Slag Composition (Mass%)						$CaCl_2$ Dosage (Mass%)	Molar Ratio of $CaCl_2$ to CoO	Molar Ratio of SiO_2 to $CaCl_2$
	CoO	SiO_2	Al_2O_3	MgO	CaO	Fe_2O_3			
1	1	0	0	0	0	99	12.3	8.3	0
2	1	0	0	0	0	99	18.5	12.5	0
3	1	0	0	0	0	99	24.6	16.6	0
4	1	99	0	0	0	0	12.3	8.3	14.9
5	1	99	0	0	0	0	24.6	16.6	7.7
6	1	6.66	0	0	0	92.34	24.6	16.6	0.5
7	1	13.33	0	0	0	85.67	24.6	16.6	1
8	1	26.66	0	0	0	72.34	24.6	16.6	2
9	1	53.31	0	0	0	45.69	24.6	16.6	4
10	1	3.33	0	0	0	95.67	12.3	8.3	0.5
11	1	6.66	0	0	0	92.34	12.3	8.3	1
12	1	13.33	0	0	0	85.67	12.3	8.3	2
13	1	26.66	0	0	0	72.34	12.3	8.3	4
14	1	53.31	0	0	0	45.69	12.3	8.3	8
15	1	53.31	10	0	0	35.69	12.3	8.	8
16	1	53.31	20	0	0	25.69	12.3	8.3	8
17	1	53.31	0	10	0	35.69	12.3	8.3	8
18	1	53.31	0	20	0	25.69	12.3	8.3	8
19	1	53.31	0	0	10	35.69	12.3	8.3	8
20	1	53.31	0	0	20	25.69	12.3	8.3	8

The chlorination–volatilization experiments were carried out in a muffle furnace. Furnace temperature was set to increase at 25 K/min. An alumina boat (length: 60 mm, width: 30 mm) containing approximately 10 g of briquettes was located at the center of the furnace at approximately 873 K. The furnace was turned off after the sample was held at the desired temperatures for 1 h, which was sufficient to reach the maximum cobalt volatilization percentage according to previous work [16]. There was an air inlet under the thermocouple and an air outlet at the top of the furnace hearth to connect with the atmosphere. Hot air in the furnace hearth escaped through the outlet and was exhausted by a negative pressure fan above the furnace into the air after alkaline solution treatment. Samples were cooled inside the furnace to approximately 873 K and then naturally cooled to room temperature.

Samples after roasting were prepared carefully for the chemical and phase analyses. The phases of calcines were detected by X-ray diffraction (XRD, X'PertPro, PANalytical, Almelo, The Netherlands) and identified by comparisons between diffraction peaks of XRD data and no less than three main characteristic peaks of substances. Cobalt concentration was measured by inductively coupled plasma optical emission spectrometer (ICP-OES, Optima 8000, PerkinElmer, Waltham, Massachusetts, MA, USA). The Co volatilization percentage was calculated according to Equation (1).

$$\eta = \frac{c_i m_i - c_f m_f}{c_i m_i} \times 100\%,\tag{1}$$

where c_i and c_f are Co concentrations of samples before and after roasting, respectively. m_i and m_f are the masses of the samples before and after roasting, respectively. The analysis results of cobalt concentrations are within ± 0.015 mass%. Therefore, the errors of volatilization percentages are calculated to be within $\pm 2.5\%$ according to Equation (1).

3. Results and Discussions

3.1. Effect of CaCl₂ Dosage

Figure 1 shows the effect of $CaCl_2$ dosage on Co volatilization percentage in the CoO–Fe_2O_3–$CaCl_2$ and CoO–SiO_2–$CaCl_2$ systems. The Co volatilization percentage obviously increases as the $CaCl_2$ dosage in the CoO–SiO_2–$CaCl_2$ system is strengthened at 1173 K, and slightly increases with increasing $CaCl_2$ dosage at 1273 K and 1373 K. Similar trends are observed in the CoO–Fe_2O_3–$CaCl_2$ system. It should be noted that the Co volatilization percentages in the CoO–SiO_2–$CaCl_2$ system are much larger than those in the CoO–Fe_2O_3–$CaCl_2$ system.

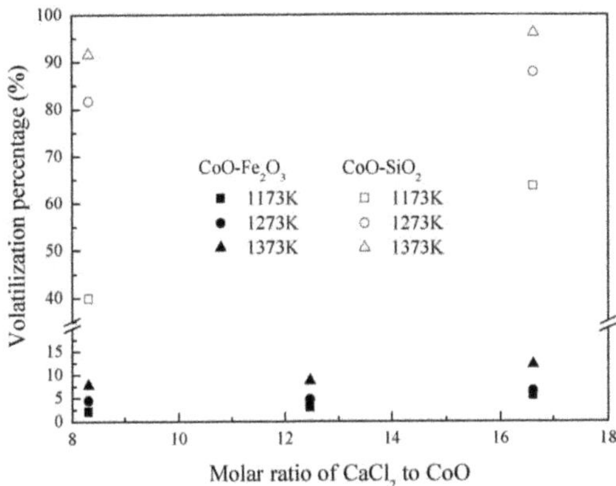

Figure 1. Effect of $CaCl_2$ dosage on the Co volatilization percentage.

Chlorination could be divided into direct chlorination and indirect chlorination using calcium chloride. It reacts directly with cobalt oxide in the former case, which could be represented as follows:

$$CaCl_2 + CoO = CoCl_2 + CaO, \tag{2}$$

$CaCl_2$ firstly reacts with oxygen to release Cl_2 in the latter case, followed by cobalt oxide being chlorinated by Cl_2. It could be expressed as follows:

$$2CaCl_2 + O_2 = 2CaO + 2Cl_2, \tag{3}$$

$$2CoO + 2Cl_2 = 2CoCl_2 + O_2, \tag{4}$$

Figure 1 shows that the Co volatilization percentage of the CoO–Fe_2O_3–$CaCl_2$ system is not larger than 12.1%, even with a higher $CaCl_2$ dosage, and it becomes much larger in the CoO–SiO_2–$CaCl_2$ system. This indicates that SiO_2 is more beneficial to the chlorination-volatilization of cobalt oxide compared with Fe_2O_3 in both the cases of direct and indirect chlorination. The direct chlorination of cobalt oxide by calcium chloride, in the presence of SiO_2, could be expressed as follows:

$$CaCl_2 + CoO + SiO_2 = CoCl_2 + CaSiO_3, \tag{5}$$

The generation of Cl_2 by the decomposition of $CaCl_2$, in the presence of SiO_2, could be expressed as follows:

$$2CaCl_2 + O_2 + 2SiO_2 = 2CaSiO_3 + 2Cl_2, \tag{6}$$

The standard Gibbs free energies of Equations (2)–(6) are calculated according to the data from [17], and are shown in Figure 2. Equations (7)–(9) in Section 3.4 are also listed in Figure 2. $CoCl_2$ is considered to be in the gaseous state. The standard Gibbs free energies of Equations (5) and (6) are much lower than those of Equations (2) and (3), respectively, which means that the chlorination of cobalt oxide is promoted in the presence of SiO_2. A larger $CaCl_2$ dosage is helpful to improve the Co volatilization percentage in the presence of SiO_2 in both the cases of direct and indirect chlorination.

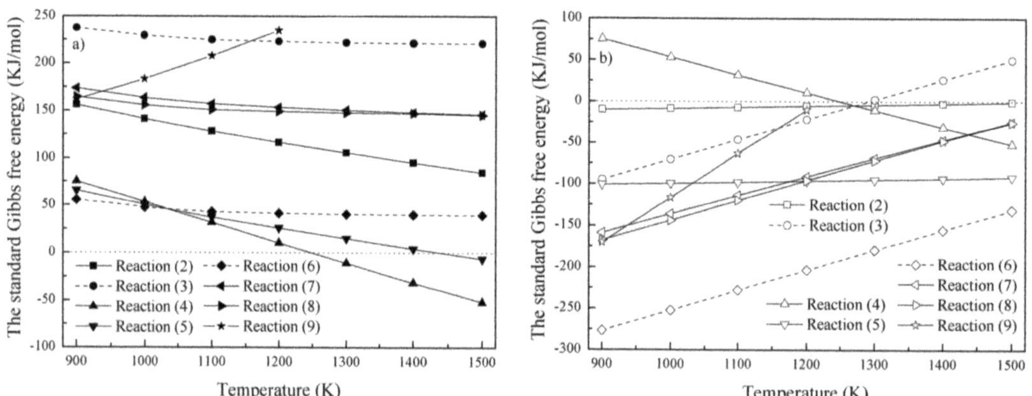

Figure 2. The standard Gibbs free energy of reactions: (**a**) solid or liquid $CaCl_2$; (**b**)gaseous $CaCl_2$.

3.2. Effect of the Molar Ratio of SiO_2 to $CaCl_2$

Figure 3 shows the effect of the molar ratio of SiO_2 to $CaCl_2$ on the Co volatilization percentage in the CoO–SiO_2–Fe_2O_3–$CaCl_2$ system. When the molar ratio of $CaCl_2$ to CoO is 8.3, the Co volatilization percentage increases significantly, as the molar ratio of SiO_2 to $CaCl_2$ rises from 0 to 1, and remains almost constant when the molar ratio of SiO_2 to $CaCl_2$ increases from 1 to 8 at all the desired temperatures. Similar trends are observed when the molar ratio of $CaCl_2$ to CoO is 16.6 at 1273 K and 1373 K. The Co volatilization percentage increases significantly and remains almost constant when the critical molar ratio of SiO_2 to $CaCl_2$ is two. However, it increases significantly as the molar ratio of SiO_2 to $CaCl_2$ is amplified at 1173 K.

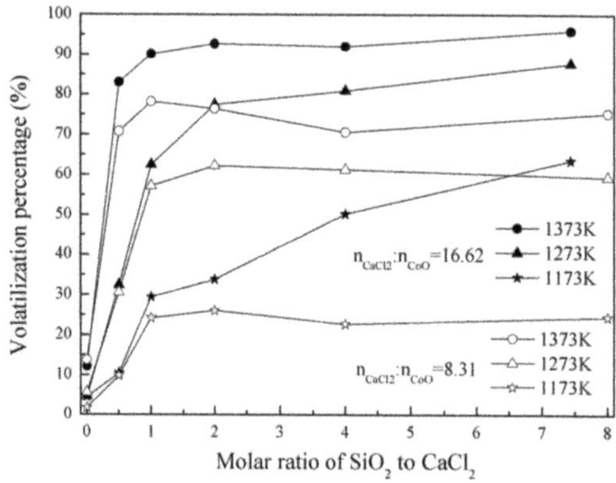

Figure 3. Effect of the molar ratio of SiO_2 to $CaCl_2$ on the Co volatilization percentage.

Equation (5) is initially promoted with an increase in the molar ratio of SiO_2 to $CaCl_2$ from zero. The enhancement becomes weaker when the molar ratio of SiO_2 to $CaCl_2$ is large enough. As shown in Equation (6), SiO_2 plays an important role in the generation of

Cl$_2$, by the decomposition of CaCl$_2$. This result is consistent with the works of Liu et al. [18], Zhu et al. [19], and Ding [20]. The Co volatilization percentage is very low when the molar ratio of SiO$_2$ to CaCl$_2$ is zero. Most of the CaCl$_2$ is lost in the form of volatilization in this case. This also means that the role of Fe$_2$O$_3$ in the generation of Cl$_2$, by the decomposition of CaCl$_2$, is much weaker than that of SiO$_2$. More Cl$_2$ is released when the molar ratio of SiO$_2$ to CaCl$_2$ is increased, which leads to an increase in the Co volatilization percentage. There is sufficient Cl$_2$ at a high molar ratio of SiO$_2$, resulting in a constant Co volatilization percentage.

3.3. Effect of Al$_2$O$_3$, MgO and CaO

Figure 4 shows the effects of Al$_2$O$_3$, MgO, and CaO on the Co volatilization percentage in the CoO–SiO$_2$–(Al$_2$O$_3$/MgO/CaO)–Fe$_2$O$_3$–CaCl$_2$ systems. The Co volatilization percentage remains almost constant with increasing CaO concentration, with a molar ratio of CaCl$_2$ to CoO of 8.3 and a molar ratio of SiO$_2$ to CaCl$_2$ of eight. This could be attributed to the sufficient amount of SiO$_2$ existing in the system. The molar ratio of SiO$_2$ to CaCl$_2$ is still larger than two, taking into consideration that some SiO$_2$ could react with CaO. However, it decreases with increasing Al$_2$O$_3$ and MgO concentrations under the same conditions. The reasons for this need further study. Table 2 shows the comparisons of the effects of Al$_2$O$_3$, MgO, CaO, and SiO$_2$ on the volatilization percentages of CoO, ZnO [19], Zn•Fe$_2$O$_3$ [19], Cu$_2$O [21], and CuO [21]. It should be noted that the molar ratios of CaCl$_2$ to ZnO, CaCl$_2$ to Zn•Fe$_2$O$_3$, CaCl$_2$ to Cu$_2$O, and CaCl$_2$ to CuO are much lower than the molar ratio of CaCl$_2$ to CoO in this study.

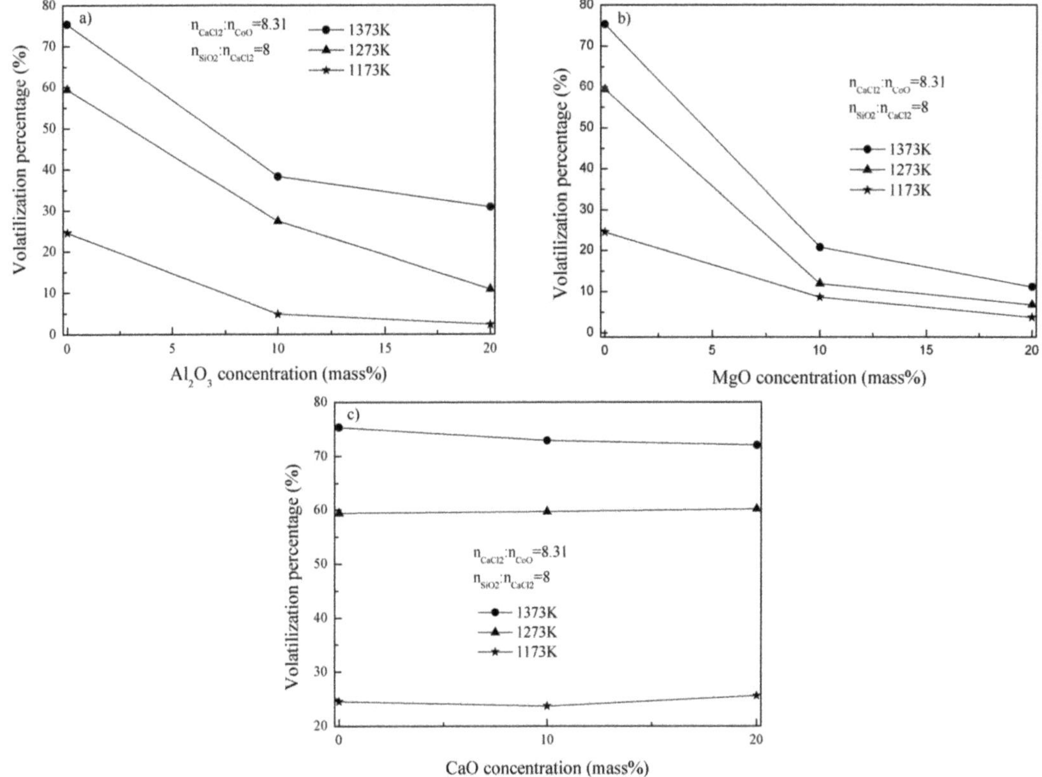

Figure 4. Effect of (**a**) Al$_2$O$_3$, (**b**) MgO and (**c**) CaO on Co volatilization percentage.

Table 2. The effects of Al_2O_3, MgO, CaO and SiO_2 on the volatilization percentage.

Phases	System	Al_2O_3	MgO	CaO	SiO_2	Reference
CoO	$CoO–SiO_2–Fe_2O_3–Al_2O_3–CaCl_2$ $CoO–SiO_2–Fe_2O_3–MgO–CaCl_2$ $CoO–SiO_2–Fe_2O_3–CaO–CaCl_2$ $CoO–SiO_2–Fe_2O_3–CaCl_2$	Negatively	Negatively	None	Positively	**This study**
ZnO	$ZnO–Fe_2O_3–Al_2O_3–CaCl_2$ $ZnO–Fe_2O_3–MgO–CaCl_2$ $ZnO–Fe_2O_3–CaO–CaCl_2$ $ZnO–Fe_2O_3–SiO_2–CaCl_2$	Negatively	Negatively	Positively	Positively	[19]
$Zn•Fe_2O_3$	$Zn•Fe_2O_3–Fe_2O_3–Al_2O_3–CaCl_2$ $Zn•Fe_2O_3–Fe_2O_3–MgO–CaCl_2$ $Zn•Fe_2O_3–Fe_2O_3–CaO–CaCl_2$ $Zn•Fe_2O_3–Fe_2O_3–SiO_2–CaCl_2$	Negatively	Negatively	Negatively	Negatively	[19]
Cu_2O	$Cu_2O–Fe_2O_3–Al_2O_3–CaCl_2$ $Cu_2O–Fe_2O_3–MgO–CaCl_2$ $Cu_2O–Fe_2O_3–CaO–CaCl_2$ $Cu_2O–Fe_2O_3–SiO_2–CaCl_2$	Negatively	Negatively	Negatively	Negatively	[21]
CuO	$CuO–Fe_2O_3–Al_2O_3–CaCl_2$ $CuO–Fe_2O_3–MgO–CaCl_2$ $CuO–Fe_2O_3–CaO–CaCl_2$ $CuO–Fe_2O_3–SiO_2–CaCl_2$	None	Negatively	Positively	Positively	[21]

3.4. Phases of Calcines

Figure 5 shows the XRD results of the calcines from the $CoO–Fe_2O_3–CaCl_2$ system after roasting with a molar ratio of $CaCl_2$ to CoO of 16.6. Fe_2O_3, $CaFe_2O_4$, $Ca_2Fe_2O_5$, and $CaCO_3$ are identified in the calcines at 973 K, which is a temperature lower than the melting point of $CaCl_2$ (1055 K). The diffraction peaks of $CaCO_3$ disappear at temperatures higher than 1173 K, which could be attributed to the decomposition of $CaCO_3$. The generation of Cl_2, by the decomposition of $CaCl_2$, in the $CoO–Fe_2O_3–CaCl_2$ system could be expressed as follows:

$$2CaCl_2 + O_2 + Fe_2O_3 = Ca_2Fe_2O_5 + 2Cl_2, \tag{7}$$

$$2CaCl_2 + O_2 + 2Fe_2O_3 = 2CaFe_2O_4 + 2Cl_2, \tag{8}$$

$$2CaCl_2 + O_2 + 2CO_2 = 2CaCO_3 + 2Cl_2, \tag{9}$$

where Equation (9) occurs at temperatures no more than 1173 K. The equilibrated chlorine partial pressure of Equation (9) is calculated according to Equation (10), and is shown in Figure 6, where the O_2 and CO_2 partial pressures are 0.209 and 3.1×10^{-4}, respectively. The chlorine partial pressure is approximately 10^{-9} when the activity of solid or liquid $CaCl_2$ is set to one. This means that Equation (9) could theoretically take place when the chlorine partial pressure is less than 10^{-9}. The equilibrated chlorine partial pressure becomes much larger when $CaCl_2$ changes from the solid or liquid state to the gaseous state.

$$\frac{\Delta G_9^0}{-RT} = \ln \frac{(P_{Cl_2}/P^0)^2}{a_{CaCl_2}(P_{CO_2}/P^0)^2(P_{O_2}/P^0)^{0.5}} \ or = \ln \frac{(P_{Cl_2}/P^0)^2}{(P_{CaCl_2}/P^0)^2(P_{CO_2}/P^0)^2(P_{O_2}/P^0)^{0.5}} \tag{10}$$

Figure 7 shows the XRD results of the calcines from the $CoO–SiO_2–CaCl_2$ system after roasting with a molar ratio of $CaCl_2$ to CoO of 16.6. There is newly generated $CaSiO_3$, and unreacted SiO_2 and $CaCl_2$, at 973 K. Newly generated $Ca_2SiO_3Cl_2$ exists alongside $CaSiO_3$ at 1073 K and 1173 K. The formation reaction of $Ca_2SiO_3Cl_2$ could be expressed as follows:

$$4CaCl_2 + O_2 + 2SiO_2 = 2Ca_2SiO_3Cl_2 + 2Cl_2, \tag{11}$$

The diffraction peaks of $Ca_2SiO_3Cl_2$ disappear at temperatures in excess of 1273 K, and only newly generated $CaSiO_3$ and unreacted SiO_2 remain in the calcines. This could be attributed to the instability of $Ca_2SiO_3Cl_2$ at higher temperatures. The reaction could be expressed as follows:

$$2Ca_2SiO_3Cl_2 + O_2 + 2SiO_2 = 4CaSiO_3 + 2Cl_2, \tag{12}$$

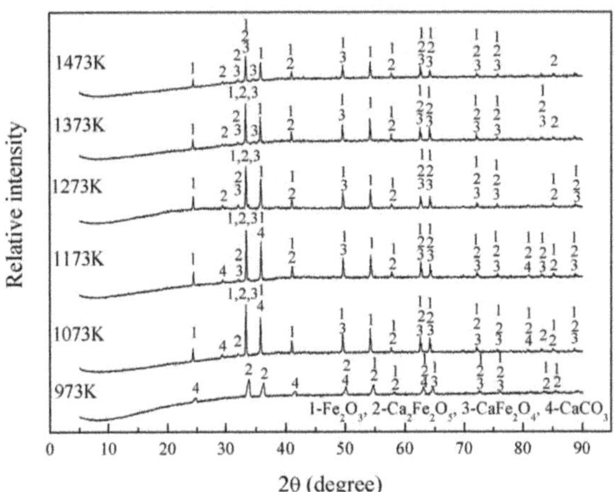

Figure 5. XRD results of calcines from the CoO–Fe$_2$O$_3$–CaCl$_2$ system.

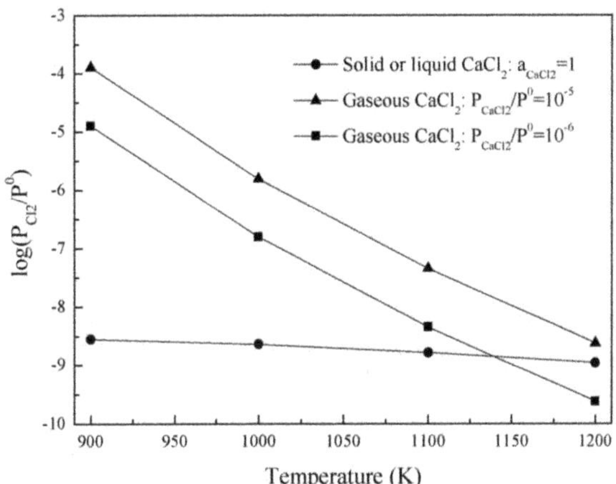

Figure 6. The equilibrated chlorine partial pressure of the following reaction: $2CaCl_2 + O_2 + 2CO_2 = 2CaCO_3 + 2Cl_2$.

Ca$_2$SiO$_3$Cl$_2$ was formed when the molar ratios of SiO$_2$ to CaCl$_2$ were two and one at 1023 K and 1073 K, respectively [22]. In this study, Ca$_2$SiO$_3$Cl$_2$ is formed when the molar ratio of SiO$_2$ to CaCl$_2$ is approximately 7.4 at 1073 K and 1173 K. Considering the results presented by Zhang [22], along with those from this study, it is suggested that the intermediate product Ca$_2$SiO$_3$Cl$_2$ is formed when the molar ratio of SiO$_2$ to CaCl$_2$ is larger than one at temperatures between 1023 K and 1173 K.

CaSiO$_3$, CaFe$_2$O$_4$, Ca$_2$Fe$_2$O$_5$, and Ca$_3$Fe$_2$(SiO$_4$)$_3$ are generated in the CoO–SiO$_2$–Fe$_2$O$_3$–CaCl$_2$ system at 1273 K, with a molar ratio of CaCl$_2$ to CoO of 8.3 and a molar ratio of SiO$_2$ to CaCl$_2$ of eight [16]. CaAl$_2$Si$_2$O$_8$ and CaMg(SiO$_3$)$_2$ are generated after the addition of Al$_2$O$_3$ and MgO, respectively, as shown in Figure 8.

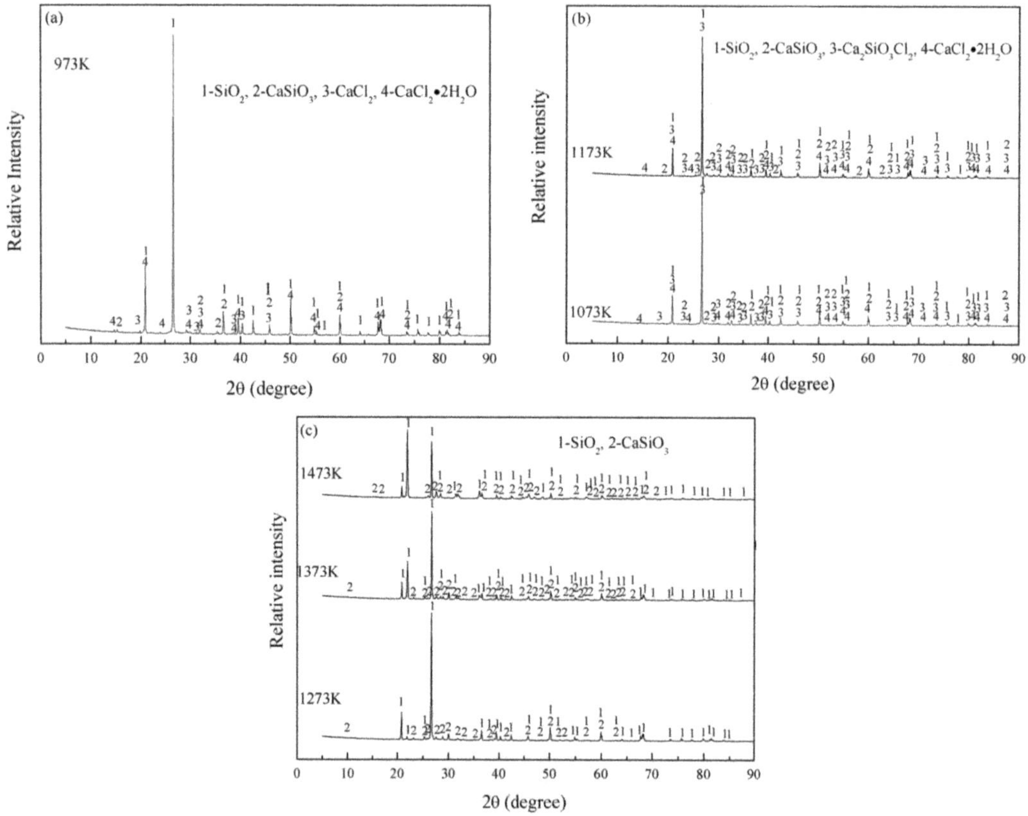

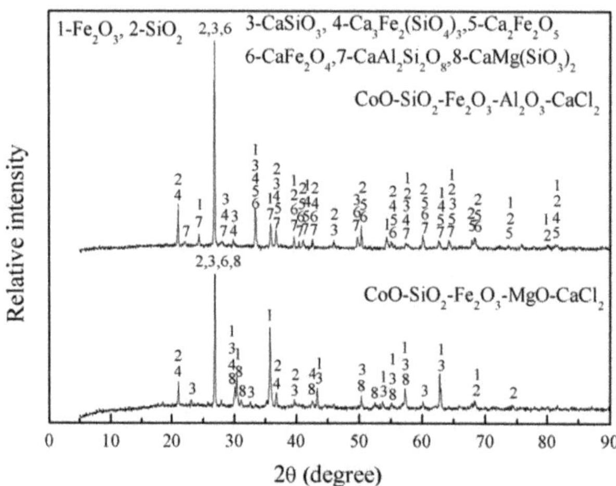

Figure 7. XRD results of calcines from the $CoO-SiO_2-CaCl_2$ system (**a**) at 973 K; (**b**) at 1073 K and 1173 K; (**c**) between 1273 K and 1473 K.

Figure 8. XRD results of calcines from the $CoO-SiO_2-Fe_2O_3-(Al_2O_3/MgO)-CaCl_2$ system at 1273 K.

4. Conclusions

The following conclusions could be drawn from this study:

1. The Co volatilization percentage of the CoO–Fe_2O_3–$CaCl_2$ system is not larger than 12.1%.
2. SiO_2 plays an important role in the chlorination–volatilization of cobalt oxide, using calcium chloride. In the CoO–SiO_2–Fe_2O_3–$CaCl_2$ system, the Co volatilization percentage is initially positively related to the molar ratio of SiO_2 to $CaCl_2$, and remains almost constant when the molar ratio of SiO_2 to $CaCl_2$ rises from zero to eight. The critical molar ratios of SiO_2 to $CaCl_2$ are 1 and 2 when the molar ratios of $CaCl_2$ to CoO are 8.3 and 16.6, respectively.
3. The Co volatilization percentage remains almost constant with increasing CaO concentration, and decreases when Al_2O_3 and MgO are added, with a molar ratio of $CaCl_2$ to CoO of 8.3 and a molar ratio of SiO_2 to $CaCl_2$ of eight.
4. Fe_2O_3, $CaFe_2O_4$, $Ca_2Fe_2O_5$, and $CaCO_3$ are identified in the calcines from the CoO–Fe_2O_3–$CaCl_2$ system at 973 K, and $CaCO_3$ disappears at temperatures higher than 1173 K.
5. $Ca_2SiO_3Cl_2$ exists in the calcines from the CoO–SiO_2–$CaCl_2$ system at 1073 K and 1173 K, and disappears at temperatures in excess of 1273 K. $CaSiO_3$ always exists in the calcines at temperatures in excess of 973 K.

Author Contributions: Methodology, P.H. and S.Y.; formal analysis, P.H.; investigation, P.H., Z.L., X.L. and J.Y.; data curation, P.H.; writing—original draft preparation, P.H.; writing—review and editing, S.Y.; supervision, S.Y.; funding acquisition, S.Y. All authors have read and agreed to the published version of the manuscript.

Funding: The authors gratefully acknowledge the research funding from the National Natural Science Foundation of China (No.51804292) and Innovation Academy for Green Manufacture, Chinese Academy of Sciences (No. IAGM-2019-A05).

Institutional Review Board Statement: Not applicable.

Informed Consent Statement: Not applicable.

Data Availability Statement: The datasets analyzed or generated during the study are available from the corresponding author on reasonable request.

Conflicts of Interest: The authors declare no conflict of interest.

References

1. Jena, P.K.; Brocchi, E.A. Metal extraction through chlorine metallurgy. *Miner. Process. Extr. Metall. Rev.* **1997**, *16*, 211–237. [CrossRef]
2. Yazawa, A.; Kameda, M. The removal of impurities from pyrite cinders by a chloride volatilization process. *Can. Metall. Q.* **1967**, *6*, 263–280. [CrossRef]
3. Panias, D.; Neou-Syngouna, P. Gold extraction from pyrite cinders by high temperature chlorination. *Erzmetall* **1990**, *43*, 41–44.
4. Ojeda, M.W.; Perino, E.; Ruiz, M.C. Gold extraction by chlorination using a pyrometallurgical process. *Miner. Eng.* **2009**, *22*, 409–411. [CrossRef]
5. Guo, T.; Hu, X.; Matsuura, H.; Tsukihashi, F.; Chou, K. Kinetics of Zn removal from ZnO-Fe$_2$O$_3$-CaCl$_2$ system. *ISIJ Int.* **2010**, *50*, 1084–1088. [CrossRef]
6. Ding, J.; Han, P.; Lv, C.; Qian, P.; Ye, S.; Chen, Y. Utilization of gold-bearing and iron-rich pyrite cinder via a chlorination-volatilization process. *Int. J. Miner. Metall. Mater.* **2017**, *24*, 1241–1250. [CrossRef]
7. Gaballah, I.; Djona, M. Processing of spent hydrorefining catalysts by selective chlorination. *Metall. Mater. Trans. B* **1994**, *25*, 481–490. [CrossRef]
8. Gaballah, I.; Djona, M. Recovery of Co, Ni, Mo, and V from unroasted spent catalysts by selective chlorination. *Metall. Mater. Trans. B* **1995**, *26*, 41–50. [CrossRef]
9. Fan, C.; Zhai, X.; Fu, Y.; Chang, Y.; Li, B.; Zhang, T. Extraction of nickel and cobalt from reduced limonitic laterite using a selective chlorination—Water leaching process. *Hydrometallurgy* **2010**, *105*, 191–194. [CrossRef]
10. Fan, C.; Zhai, X.; Fu, Y.; Chang, Y.; Li, B.; Zhang, T. Kinetics of selective chlorination of pre-reduced limonitic nickel laterite using hydrogen chloride. *Miner. Eng.* **2011**, *24*, 1016–1021. [CrossRef]
11. Zhang, M.; Zhu, G.; Zhao, Y.; Feng, X. A study of recovery of copper and cobalt from copper–cobalt oxide ores by ammonium salt roasting. *Hydrometallurgy* **2012**, *129–130*, 140–144. [CrossRef]

12. Li, J.; Li, Y.; Gao, Y.; Zhang, Y.; Chen, Z. Chlorination roasting of laterite using salt chloride. *Int. J. Miner. Process.* **2016**, *148*, 23–31. [CrossRef]
13. Xu, C.; Cheng, H.; Li, G.; Lu, C.; Lu, X.; Zou, Z.; Xu, Q. Extraction of metals from complex sulfide nickel concentrates by low-temperature chlorination roasting and water leaching. *Int. J. Miner. Metall. Mater.* **2017**, *24*, 377–385. [CrossRef]
14. Cui, F.; Mu, W.; Wang, S.; Xin, H.; Shen, H.; Xu, Q.; Zhai, Y.; Luo, S. Synchronous extractions of nickel, copper, and cobalt by selective chlorinating roasting and water leaching to low-grade nickel-copper matte. *Sep. Purif. Technol.* **2018**, *195*, 149–162. [CrossRef]
15. Lippert, K.K.; Pietsch, H.B.; Roeder, A.; Walden, H.W. Recovery of non-ferrous metal impurities from iron ore pellets by chlorination (CV or LDK process). *Trans. Inst. Min. Metall. Sect. C* **1969**, *78*, 98–107.
16. Han, P.; Xiao, L.; Wang, Y.; Lu, Y.; Ye, S. Cobalt recovery by the chlorination-volatilization method. *Metall. Mater. Trans. B* **2019**, *50*, 1128–1133. [CrossRef]
17. Barin, I. *Thermochemical Data of Pure Substance*, 3rd ed.; WILEY-VCH Verlag Gmbh: Weinheim, Germany, 1995.
18. Liu, J.; Wen, S.; Chen, Y.; Liu, D.; Bai, S.; Wu, D. Process optimization and reaction mechanism of removing copper from an Fe-rich pyrite cinder using chlorination roasting. *J. Iron Steel Res. Int.* **2013**, *20*, 20–26. [CrossRef]
19. Zhu, D.; Chen, D.; Pan, J.; Zheng, G. Chlorination behaviors of zinc phases by calcium chloride in high temperature oxidizing-chloridizing roasting. *ISIJ Int.* **2011**, *51*, 1773–1777. [CrossRef]
20. Ding, J. Research on the Extraction of Gold from Gold-Bearing Pyrite Cinder by High-Temperature Chlorination Method. Ph.D. Thesis, University of Chinese Academy of Sciences, Beijing, China, 2018.
21. Zhu, D.; Chen, D.; Pan, J.; Chun, T.; Zheng, G.; Zhou, X. Chlorination behaviors of copper phases by calcium chloride in high temperature oxidizing-chloridizing roasting. In Proceedings of the 3rd International Symposium on High-Temperature Metallurgical Process, Orlando, FL, USA, 11–15 March 2012.
22. Zhang, L. Research on the Chloridizing Segregation Process of Nickel Laterites. Master's Thesis, Central South University, Changsha, China, 2011.

metals [MDPI]

Article

Effects of CaO, Al$_2$O$_3$ and MgO on Kinetics of Lead-Rich Slag Reduction

Sui Xie [1], Chunfa Liao [1] and Baojun Zhao [1,2,*]

[1] Faculty of Materials Metallurgy and Chemistry, Jiangxi University of Science and Technology, Ganzhou 341000, China; 7120210006@mail.jxust.edu.cn (S.X.); liaochfa@163.com (C.L.)
[2] Sustainable Minerals Institute, University of Queensland, Brisbane 4072, Australia
* Correspondence: bzhao@jxust.edu.cn

Abstract: Lead-rich slag as a green feedstock can be used in a blast furnace or smelting reduction furnace to produce lead metal. It is desirable to understand the reduction mechanisms of lead-rich slag to optimize the reduction operations. The volume of CO/CO$_2$ gases was continuously measured in the experiments to determine the reduction degree of lead-rich slag by carbon. The effects of CaO, Al$_2$O$_3$ and MgO on reaction kinetics of lead-rich slag with carbon were investigated in the temperature range 1073 to 1473 K. The activation energies of the reduction were determined experimentally in the chemically controlled and diffusion-controlled stages, and the reduction mechanism is analyzed using experimental results and thermodynamic calculations. It was found that the activation energies at the chemically controlled and diffusion-controlled stages decrease with increasing optical basicity. CaO and MgO have a similar behavior to accelerate the reduction of the lead-rich slag by carbon. In contrast, Al$_2$O$_3$ can increase the activation energies at both chemically controlled and diffusion-controlled stages resulting in a slow reaction.

Keywords: lead-rich slag; kinetics; fluxes; FactSage; reduction mechanisms

Citation: Xie, S.; Liao, C.; Zhao, B. Effects of CaO, Al$_2$O$_3$ and MgO on Kinetics of Lead-Rich Slag Reduction. *Metals* **2022**, *12*, 1235. https://doi.org/10.3390/met12081235

Academic Editors: Mark E. Schlesinger and Lauri Holappa

Received: 10 May 2022
Accepted: 19 July 2022
Published: 22 July 2022

Publisher's Note: MDPI stays neutral with regard to jurisdictional claims in published maps and institutional affiliations.

1. Introduction

Sinter plant-blast furnace is a traditional route to produce primary lead from the concentrates [1]. Fine lead sulphide concentrates are sintered in a Dwight Lloyd machine to remove the sulphur and form agglomerated oxide sinter lump for the blast furnace. In the sintering process the concentration of SO$_2$ in the gas is low, creating a difficulty for producing acid. The capture of SO$_2$ and lead-bearing dust during the sintering process is not efficient as the moving sintering machine results in significant leakage. The productivity of the lump sinter suitable for the blast furnace is low in the sinter plant due to the low physical strength of the sinter, and approximately 30% of sinter needs to return to the sintering process. The sinter lumps fed into the blast furnace are not uniform in the composition and microstructure, and to overcome these disadvantages several lead-smelting technologies have been developed in recent years [1–6]. These smelting technologies, including top submerged lance technology (TSL), bottom-blowing smelting technology (SKS) and side-blowing smelting technology, have a greatly simplified operation with more efficient capture and utilization of SO$_2$ and lead-bearing dust. The lead-rich slag produced from the smelting process is low in sulphur and high in uniformity. Lead metal is also produced simultaneously in these processes making the smelting furnaces more productive.

The molten lead-rich slag can be casted and sent to a blast furnace or a smelting reduction furnace to produce lead metal. The lead-rich slag has the similar composition and reactions during the reduction process; however, the dense slag has different macrostructure from the sinter due to the different technologies used for their production and it is necessary to study the kinetics and mechanism of the lead-rich slag reduction by carbon. Kinaev et al., [7] studied the reduction kinetics of lead oxide by graphite and coke at high temperatures. The synthetic slags $PbO-FeO-Fe_2O_3-CaO-SiO_2$ were used for the study and the volume of the product gases was measured to monitor the reaction progress. Due to the limitation of the technique used, only initial 300 s reaction was studied. It was found that the rate-limiting step was the chemical reaction at the gas/slag interface. The composition of the product gas (CO/CO_2 ratio) was not determined in this study [7]. Upadhya [8] studied reaction kinetics of the $PbO-CaO-Al_2O_3-SiO_2$ slags by carbon in iron at 1693 K. The reaction rate was determined by the PbO concentration in the reduced slag collected at different times. It was found that the reaction rate for the initial stage reduction was controlled by the chemical reaction and the second stage reduction was best explained by the mass transport control [8]. The study was limited to the slags with up to 6.6 wt% PbO. The composition of the product gases was not determined in this study. However, Upadhya adopted the results of Sommerville and Bullard that the product gas was almost CO because CO_2 was consumed by the excess carbon [8]. All of these studies [7,8] were not directly related to the lead-rich slags that only appeared in recent years. The study on the reduction of synthetic lead-rich slags and industrial lead sinters by CO gas indicated that liquid phase played an important role in the reduction process [9]. Both synthetic lead-rich slags and industrial lead sinters showed similar behaviors during the reduction if they have the same composition. The reduction kinetics of the synthetic lead-rich slag and industrial lead sinter by graphite were compared by Hou et al. [10]. It was found that rapid reduction was associated with the formation of the liquid slag that was a function of composition and temperature. The effect of PbO concentration on the reduction kinetics of lead-rich slags by graphite was studied recently by Xie et al. [11]. They found that a high PbO concentration in the slag is favorable to the reduction. During the production and reduction of lead-rich slag, CaO is usually added as a flux to decrease the liquidus temperature and viscosity of the slag. Al_2O_3 and MgO from the concentrates, carbon-containing materials and refractory are always present in the slag. The effects of CaO, Al_2O_3 and MgO on the reduction kinetics of lead-rich slags remain unclear. Additionally, optical basicity has been associated with the properties of the metallurgical slags with different compositions [12–15]. The concept of optical basicity has been adopted to describe the effects of CaO, Al_2O_3 and MgO on the reduction mechanisms and kinetics of lead-rich slags by graphite. The reduction mechanisms are analyzed by experimental data and FactSage 8.1 software [16]. The remodified shrinking core model is also used to analyze the reduction kinetics of the lead-rich slag by graphite [17–20].

2. Research Methodology

The lead-rich slags used for the reduction experiments were first prepared and characterized. The ground slag samples were then reacted with graphite in a properly sealed system and the volume of the product gas was accurately measured to determine the progress of the reduction reactions. The FactSage software was used to interpret the experimental results from thermodynamic point of view.

3. Preparation of Lead-Rich Slags

Pure PbO, ZnO, Fe_2O_3, $CaCO_3$, SiO_2, Al_2O_3 and MgO chemicals were used to prepare the lead-rich slags Y1, Y2 and Y3 in an induction furnace in air. A 1000 g pelletized mixture prepared from pure powders was placed in a mullite crucible. The slag and crucible were melted between 1100 to 1150 °C for 60 min by a Heraeus 25 KW induction generator through a graphite susceptor. The temperature of the slag was monitored by an R-type thermocouple (Pt/Pt-13%Rh) that was placed in an alumina sheath. Argon gas with a 5000 mL·min^{-1} flow rate was used to stir the slag throughout the experiment, and the molten slag was cooled naturally after the induction furnace was turned off. The slow-cooled lead-rich slags were mounted, polished and examined using optical and scanning electron microscopy. The compositions of the phases present in the samples were measured by a JEOL 8800 L Electron Probe Microanalyser (EPMA) with Wavelength-Dispersive

Spectrometer (WDS). An accelerating voltage of 15 kV and a probe current of 15 nA were applied. The average accuracy of the EPMA measurements is within 1 wt%.

4. Reduction of Lead-Rich Slag by Graphite

The schematic diagram of the experimental system including a vertical tube furnace, gas collection and data record system is shown in Figure 1. The experiments were carried out in a 19 mm ID corundum reaction tube. A platform consisting of a Pt/Pt-13%Rh thermocouple was inserted from the bottom to support a graphite crucible (18 mm OD, 14 mm ID, and 40 mm high) located in the hot zone of the furnace. A water-containing pressure device was developed to capture the gas generated from the reaction and it displaced the same volume of water into a container sitting on a balance. The weight of the water was recorded by a computer and then converted to the gas volume.

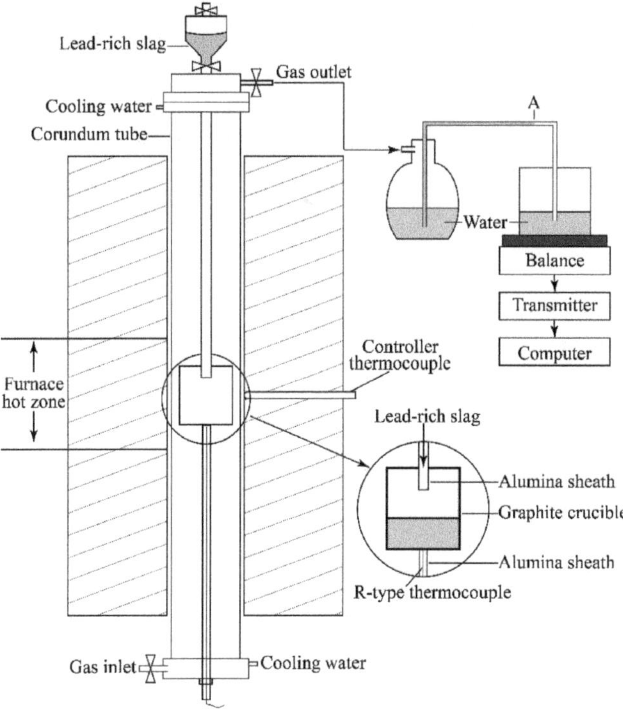

Figure 1. Schematic diagram of the experimental system used in this study.

The graphite crucible was raised from the bottom to the hot zone of the reaction tube after 30 min flushing by ultrahigh purity argon. The gas flushing was stopped, and the reaction tube was sealed properly. The argon gas was then turned on carefully to enable the water to stay at point A indicated in Figure 1 and then the argon gas was stopped. The balance was adjusted to zero and the computer was started to record the weight changes. A 5.00 g lead-rich slag (~1 mm diameter size) was added into the graphite crucible from the airtight funnel that was placed on the top of the furnace. The reduction reaction started when the sample touched the graphite crucible at the required temperature. The generated CO/CO_2 gases were captured by the water-containing pressure device and the weight of water was continuously recorded by the computer in 5 s interval. When the experiment finished, up to 400 mL CO/CO_2 gases were generated and displaced approximately 400 g H_2O into a container sitting on a balance. The reduction was stopped when the graphite crucible was lowered rapidly to the bottom-end of the furnace.

5. Thermodynamic Calculations

FactSage 8.1 software was used to predict the compositions of liquid slags, and activities of PbO and ZnO in liquid slags at high temperature [16]. The databases selected were "FactPs" and "Ftoxide". The solution phases selected included "FToxid-SLAGA", "FToxid-SPINA", "FToxid-MeO", "FToxid-cPyrA", "FToxid-PyrA", "FToxid-WOLLA", "FToxid-Bred", "FToxid-bC2SA", "FToxid-aC2SA", "FToxid-Mel", "FToxid-Oliv", "FToxid-Mull", "FToxid-CORU", "FToxid-ZNIT", "FToxid-WILL", "FToxid-PbO" and "FToxid-PCSi".

The optical basicity, the electron-donating power of the oxygen in an oxidic glass, was defined by Duffy [12,13]. The Equation (1) below is usually used to calculate the optical basicity of an oxide slag [14]:

$$\Lambda = \frac{\sum x_1 n_1 \Lambda_1 + x_2 n_2 \Lambda_2 + \cdots}{\sum x_1 n_1 + x_2 n_2 + \cdots} \tag{1}$$

where Λ is the optical basicity of the slag, $\Lambda_{1,2\ldots}$ is the optical basicity of an individual oxide as shown in Table 1, x is the mole fraction of the individual oxide and $n_{1,2\ldots}$ is the number of oxygen atoms associated with the oxide.

Table 1. The optical basicity of individual oxide [13,15].

Oxide	PbO	ZnO	Fe_2O_3	CaO	SiO_2	Al_2O_3	MgO
Λ	0.95	0.95	0.69	1	0.47	0.68	0.92

6. Results and Discussion

6.1. Characterization of Slags

The target and actual compositions of the lead-rich slags are shown in Table 2. The $ZnO/Fe_2O_3/CaO/SiO_2$ ratios are approximately the same in three samples.

Table 2. The target and actual compositions of lead-rich slags for reduction experiments.

Slag	PbO	ZnO	Fe_2O_3	CaO	SiO_2	Al_2O_3	MgO
Y1-planned	52.4	11.5	17.2	6.8	12.1	0.0	0.0
Y1-actual	51.8	11.4	17.0	6.7	11.9	1.3	0.0
Y2-planned	50.5	11.0	16.5	6.5	11.5	4.0	0.0
Y2-actual	49.8	10.9	16.3	6.4	11.4	5.1	0.0
Y3-planned	51.5	11.2	16.8	6.7	11.8	0.0	2.0
Y3-actual	50.8	11.1	16.6	6.6	11.6	1.3	2.0

The typical microstructures of Y2 and Y3 are shown in Figure 2 as the typical microstructures of Y1 have been reported in a previous study [10]. It appears that 5.1 wt% Al_2O_3 in Y2 and 2.0 wt% MgO in Y3 did not make a significant difference in the microstructure. The major phases presented in Y2 and Y3 are spinel, melilite and lead silicate $6PbO \cdot FeO_{1.5} \cdot SiO_2$ (PFS) [21]. In addition, Ca_2SiO_4 (C2S) and larsenite $PbZnSiO_4$ (PZS) [22,23] are also present in both samples. The spinel crystals are present on the surface of the melilite indicating that spinel is the primary phase precipitated first. These lead-rich slags are therefore located in the spinel primary phase field. Spinel and melilite were precipitated at relatively higher temperatures and the other phases were precipitated at lower temperatures. The compositions of the phases present in these slags were measured by EPMA and are shown in Table 3.

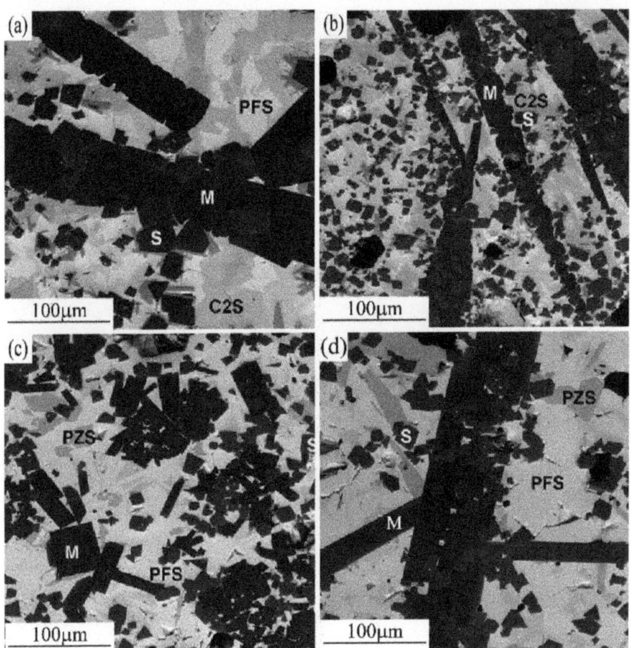

Figure 2. Typical microstructures of slow-cooled lead-rich slags: Y2 (**a,b**) and Y3 (**c,d**). S = spinel, M = melilite, PFS = 6PbO·FeO$_{1.5}$·SiO$_2$, C2S = Ca$_2$SiO$_4$, PZS = PbZnSiO$_4$.

Table 3. Compositions of the phases present in the melted slags measured by EPMA.

Slag	Phase	Composition (wt%)						
		Fe$_2$O$_3$	ZnO	Al$_2$O$_3$	MgO	CaO	SiO$_2$	PbO
Y2	Spinel	49.4	34.7	15.4	0.2	0.1	0.0	0.2
	Melilite	2.7	17.2	6.1	0.1	33.7	33.6	6.6
	PFS	5.3	0.6	0.8	0.0	0.0	5.4	87.9
	Ca$_2$SiO$_4$	0.2	0.4	0.0	0.0	17.2	18.4	63.7
Y3	Spinel	64.1	24.0	5.1	6.6	0.1	0.0	0.1
	Melilite	0.7	14.8	1.1	4.5	34.4	37.5	6.9
	PFS	3.8	0.9	0.4	0.0	0.0	5.2	89.7
	PbZnSiO$_4$	0.2	21.5	0.2	1.3	1.5	18.3	57.0

6.2. Reduction

Thermodynamically both CO and CO$_2$ can be formed during the reduction of lead-rich slag by carbon. However, excess carbon can rapidly consume CO$_2$ to form CO which has been confirmed by the experimental measurements [8]. CO is therefore considered to be the only exit gas from the graphite crucible in the present study. The main reactions between carbon and liquid slag containing Fe$_2$O$_3$, PbO and ZnO include:

$$PbO(l) + C(s) = Pb(l) + CO(g) \qquad (2)$$

$$Fe_2O_3(l) + C(s) = 2FeO(s) + CO(g) \qquad (3)$$

$$ZnO(l) + C(s) = Zn(g) + CO(g) \qquad (4)$$

The reduction degree is calculated according to Equation (5):

$$\alpha = V_m \times 100\% / V_T \tag{5}$$

where V_m is the experimentally measured gas volume (mL), V_T is the total volume of the theoretical gas volume generated from Equations (2)–(4) (mL) by 5.00 g slag, and α is the reduction degree (%).

Isothermal reduction experiments have been carried out in the temperature range of 1073 to 1473 K. The reduction degrees of Y1, Y2 and Y3 as a function of reaction time are shown in Figures 3–5, respectively. It can be seen from these figures that the reduction degree increases rapidly at the beginning of the reduction and then increases slowly with time. The reduction process includes two stages: the chemically controlled stage and the diffusion-controlled stage that has been reported in previous studies [8,10,11]. Additionally, at temperatures below 1173 K the reaction of both Y2 and Y3 with carbon is limited. Once the temperature is above 1173 K, the reduction degree has a significant increase at the same reaction time.

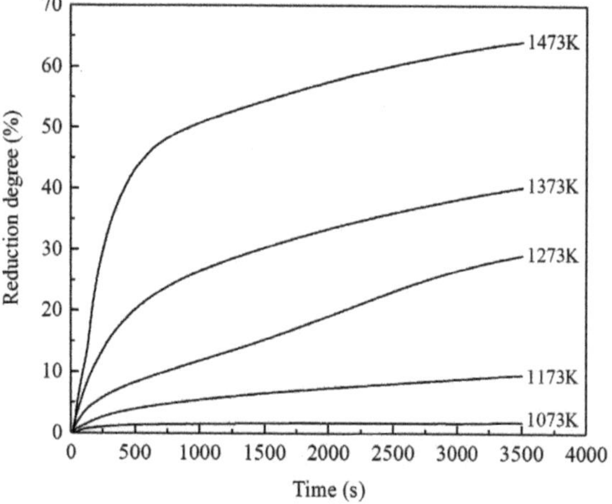

Figure 3. Reduction degree of Y1 as a function of reaction time [10].

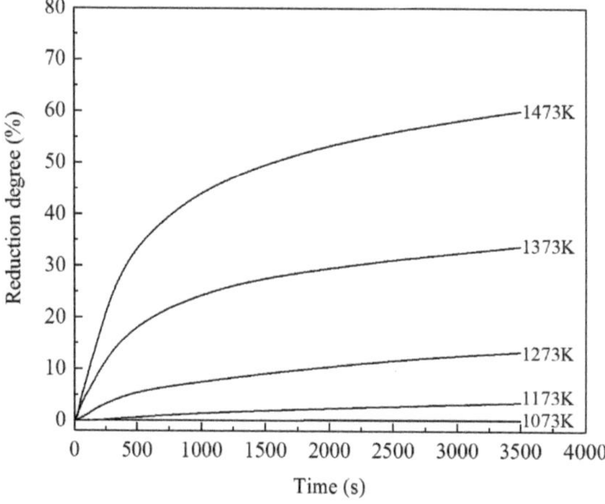

Figure 4. Reduction degree of Y2 (Al$_2$O$_3$ addition) as a function of reaction time.

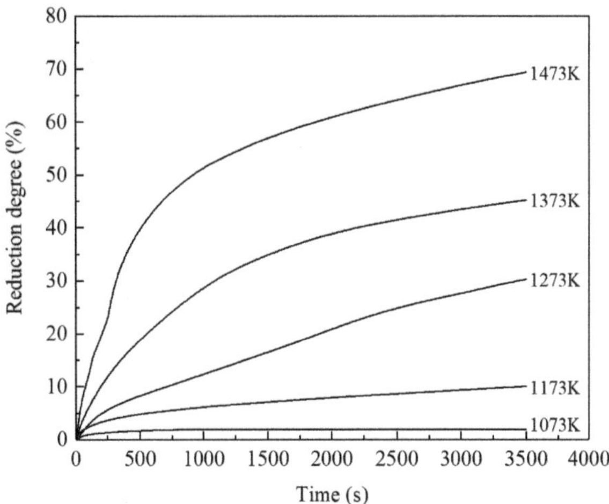

Figure 5. Reduction degree of Y3 (MgO addition) as a function of reaction time.

Figures 6 and 7 illustrate the relationship between the reduction degree and temperature at the chemically controlled stage (50 s) and diffusion-controlled stage (3500 s). They show that the reduction degree of three slags increases with increasing temperature in both the chemically controlled stage and diffusion-controlled stage. The increment is more significant at higher temperatures. The addition of MgO (Y3) seems to increase the reduction degree slightly. The presence of Al_2O_3 in the lead-rich slag significantly decreased the reduction degree and the reaction rates at temperatures below 1173 K are very low. It has been reported [24] that the presence of 4 wt% Al_2O_3 in the lead-rich slag can increase the liquidus temperature in the spinel primary phase field. As a result, the proportion of the liquid phase is lowered with the addition of Al_2O_3 which causes a low reaction rate between the slag and carbon.

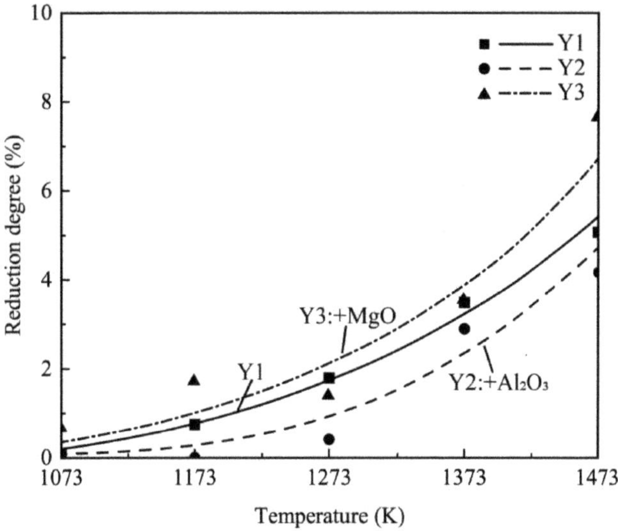

Figure 6. Comparison of reduction degree of three slags after 50 s reaction; the symbols are experimental results and the lines are data fitting.

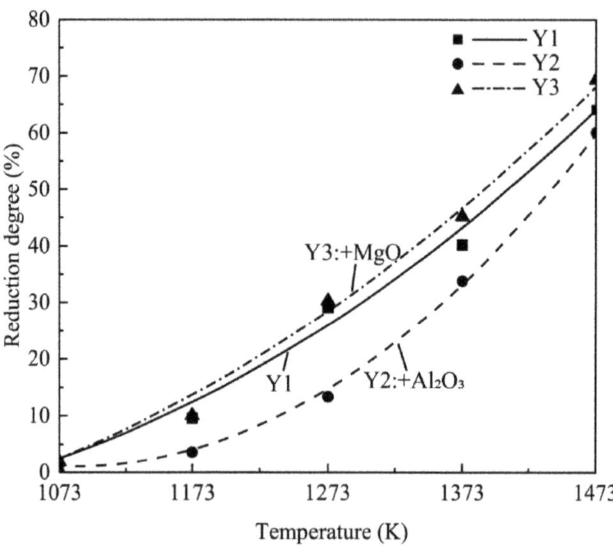

Figure 7. Comparison of reduction degree of three slags after 3500 s reaction; the symbols are experimental results and the lines are data fitting.

The typical microstructures of Y2 and Y3 after the reaction at 1173 K and 1373 K for 3500 s are shown in Figure 8. The compositions of the phases present in the samples were measured by EPMA and are given in Table 4. It can be seen from Figure 8a that a large proportion of glass phase is present in the high-Al_2O_3 slag after the reduction at 1173 K. The presence of lead metal indicates that significant amount of lead oxide had been reduced. However, the extent of the PbO reduction was limited, resulting in a large proportion of PbO-containing liquid retained at 1173 K. The microstructure is shown in Figure 8a and also indicates that this part of the sample was cooled quickly and low-melting point phases did not precipitate. It can be seen from Figure 8b that after the reduction at 1373 K for 3500 s, only a small proportion of the PbO-containing glass is present and more lead metal was formed. Figure 8c shows the reduced MgO-containing slag at 1173 K. It can be seen that the proportion of the glass in this sample is lower than that shown in Figure 8a indicating that more PbO was reduced when MgO is present. After the reduction at 1373 K, it can be seen from Figure 8d that the glass phase is not present in the sample. Two new phases, olivine $(Fe,Ca)_2SiO_4$ and willemite $(Zn,Fe)_2SiO_4$ appeared. It is clear from the microstructures that the extent of the PbO reduction in the MgO-containing sample (Y3) is higher than other samples because less PbO-containing phases are present after the reduction. This is confirmed from the phase compositions shown in Table 4. All phases present in the reduced Y3 from 1373 K do not contain significant PbO. In contrast, ZnO is present in all phases indicating that ZnO was not easy to be reduced.

Figure 8. Microstructure of slow-cooled slag of Y2 and Y3 after reaction at 1173 K and 1373 K for 3500 s: (**a**) Y2 1173 K; (**b**) Y2 1373 K; (**c**) Y3 1173 K; (**d**) Y3 1373 K. G = glass, S = spinel, M = melilite, Pb = lead metal, PZS = PbZnSiO$_4$, PFS = 6PbO·FeO$_{1.5}$·SiO$_2$, Wil = willemite, O = olivine.

Table 4. Compositions of the phases present in the slags after reduction measured by EPMA.

Slag	Temperature (K)	Phase	Composition (wt%)						
			FeO	ZnO	Al$_2$O$_3$	MgO	CaO	SiO$_2$	PbO
Y2	1173	Glass	4.2	1.2	2.7	0.0	0.2	12.2	79.6
		Spinel	42.9	36.8	19.5	0.3	0.2	0.0	0.3
		Melilite	2.2	18.0	5.2	0.1	33.1	34.5	6.8
	1373	Glass	22.2	14.1	4.5	0.1	13.6	33.9	11.6
		Spinel	45.9	33.7	19.9	0.2	0.1	0.1	0.0
		Melilite	3.2	19.1	3.1	0.1	33.2	36.3	5.1
Y3	1173	Spinel	61.5	25.4	6.2	6.5	0.2	0.1	0.2
		Melilite	0.6	16.9	1.0	3.3	32.8	37.7	7.7
		PFS	3.6	1.3	0.3	0.0	0.0	5.8	89.0
		PbZnSiO$_4$	1.0	19.3	0.1	2.0	1.3	18.4	57.9
	1373	Spinel	65.5	21.7	9.1	3.0	0.2	0.5	0.1
		Melilite	3.3	16.7	0.7	2.6	34.8	38.5	3.3
		Willemite	10.8	48.6	0.2	9.5	0.2	30.7	0.0
		Olivine	48.0	10.0	0.1	5.7	4.7	31.6	0.0

6.3. Kinetics

As shown in Figures 3–5, the reduction of the lead-rich slag by carbon includes chemically controlled stage and the diffusion-controlled stage. The reaction between the lead-rich slags and graphite is mainly liquid-solid reaction. The experimental results of the lead-rich slag reduction are analyzed in two steps.

(1) The reduction degrees can be calculated as a function of time at a fixed temperature. At the chemically controlled stage, the reaction rate constant is commonly calculated using Equation (6) [17–20]:

$$1 - (1 - \alpha)^{1/3} = kt \tag{6}$$

where α is the reduction degree, t is reaction time, and k is the reaction rate constant.

Several models have been developed to describe the diffusion-controlled reactions. The equation suggested by Dickinson [17] was found to be the best fitting one for the experimental data and is used to simulate the experimental results in the diffusion-controlled stage.

$$[(1 - \alpha)^{-1/3} - 1]^2 = kt \tag{7}$$

The definition of α, t and k in the diffusion-controlled stage is the same as the Equation (6). Kinetic curves are calculated from the experimental data using Equations (6) and (7). The reaction rate constant is determined by the slope of the tangent line of the kinetic curve. The experimentally determined reaction rate constants in the chemically controlled stage and the diffusion-controlled stage are given in Table 5.

Table 5. Experimentally determined reaction rate constants.

Slag	Chemically Controlled Stage					Diffusion Controlled Stage				
	1073 K	1173 K	1273 K	1373 K	1473 K	1073 K	1173 K	1273 K	1373 K	1473 K
Y1	3.8×10^{-5}	3.9×10^{-5}	1.2×10^{-4}	3.1×10^{-4}	5.3×10^{-4}	3.6×10^{-9}	4.3×10^{-7}	5.1×10^{-6}	9.8×10^{-6}	4.0×10^{-5}
Y2	-	7.2×10^{-6}	4.8×10^{-5}	1.8×10^{-4}	3.2×10^{-4}	2.3×10^{-10}	5.8×10^{-8}	5.8×10^{-7}	4.2×10^{-6}	2.9×10^{-5}
Y3	4.6×10^{-5}	9.8×10^{-5}	8.9×10^{-5}	2.1×10^{-4}	3.4×10^{-4}	-	3.9×10^{-7}	6.9×10^{-6}	1.1×10^{-5}	7.2×10^{-5}

(2) The activation energy in the chemically controlled stage or the diffusion-controlled stage can be calculated using the Arrhenius equation, where the reaction rate constant as a function of temperature can be expressed as:

$$k = Ae^{-E/RT} \tag{8}$$

where A is the pre-exponential factor (s^{-1}), E is the activation energy ($J \cdot moL^{-1}$), R is the gas constant ($8.314 \, J \cdot moL^{-1} \cdot K^{-1}$) and T is the absolute temperature (K).

The activation energy can be obtained from a linear relationship between $ln(k)$ and $1/T$. Equation (8) can be rewritten as:

$$ln(k) = -\frac{E}{RT} + A \tag{9}$$

Figures 9 and 10 illustrate the relationships between $ln(k)$ and $1/T$ in the chemically controlled stage and diffusion-controlled stage, respectively. In the chemically controlled stage shown in Figure 9, the activation energies of Y1, Y2 and Y3 are determined to be 95, 185 and 62 kJ/moL, respectively. The activation energies of Y1 and Y3 are significantly lower than those of Y2. It seems that both CaO and MgO can decrease the activation energy in the chemically controlled stage to accelerate the reaction. On the other hand, the presence of Al_2O_3 in the slag suppresses the reaction rate of the reduction.

It can be seen from Figure 10 that in the diffusion-controlled stage, the activation energies of Y1, Y2 and Y3 are determined to be 294, 371 and 233 kJ/moL, respectively. It is clear that the activation energies in the diffusion-controlled stage are much higher than those in the chemically controlled stage. Similarly, CaO and MgO can accelerate the reaction and Al_2O_3 suppresses the reduction in the diffusion-controlled stage. Hou et al., [10] reported that the activation energy was 83.8 and 102.9 kJ/moL at chemically controlled stage for lead sinter and synthetic lead-rich slag, respectively. At the diffusion-controlled stage, the reaction activation energy was 224.9 and 259.4 kJ/moL for lead sinter and synthetic lead-rich slag, respectively. The activation energies obtained in the present study are in the same range with those of industrial lead sinter [10].

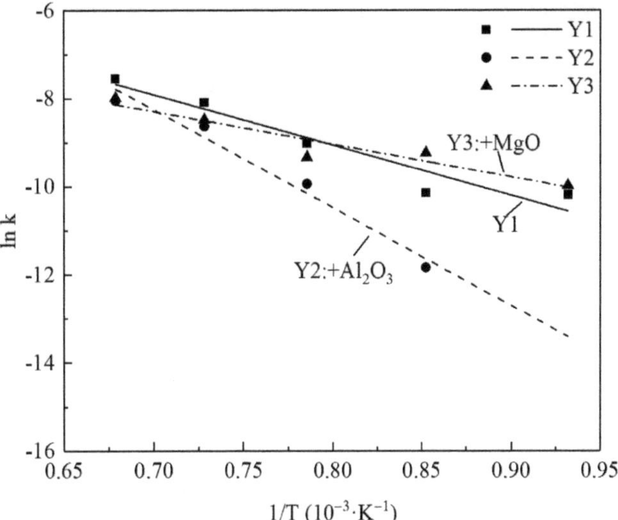

Figure 9. The relationship between ln(k) and 1/T for the reduction reaction in the chemically controlled stage; the symbols are experimental results and the lines are data fitting.

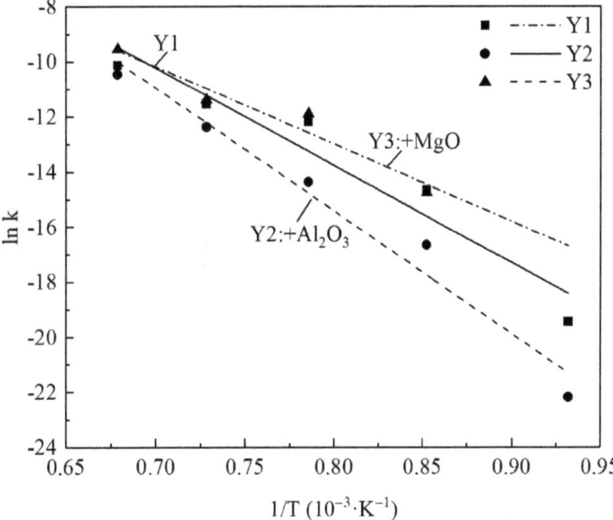

Figure 10. The relationship between ln(k) and 1/T for the reduction reaction in the diffusion-controlled stage; the symbols are experimental results and the lines are data fitting.

6.4. Thermodynamic Analysis

The optical basicity is a useful parameter to characterize the properties of the silicate slag. Based on the data given in Table 1 and the bulk compositions of the lead-rich slags in Table 2, the apparent optical basicity of Y1, Y2 and Y3 is calculated to be 0.727, 0.723 and 0.735, respectively. The relationships between the activation energies and slag apparent optical basicity are shown in Table 6 and Figure 11. It can be seen from Figure 11 that the activation energies at the chemically and diffusion-controlled stages decrease with increasing the slag apparent optical basicity. It suggests that increasing the apparent optical basicity can stimulate the progress of the reduction.

Table 6. The apparent optical basicity and activation energies of the lead-rich slags.

Slag	Apparent Optical Basicity	Chemically Controlled Stage (KJ/moL)	Diffusion-Controlled Stage (KJ/moL)
Y1	0.727	95	294
Y2	0.723	185	371
Y3	0.735	62	233

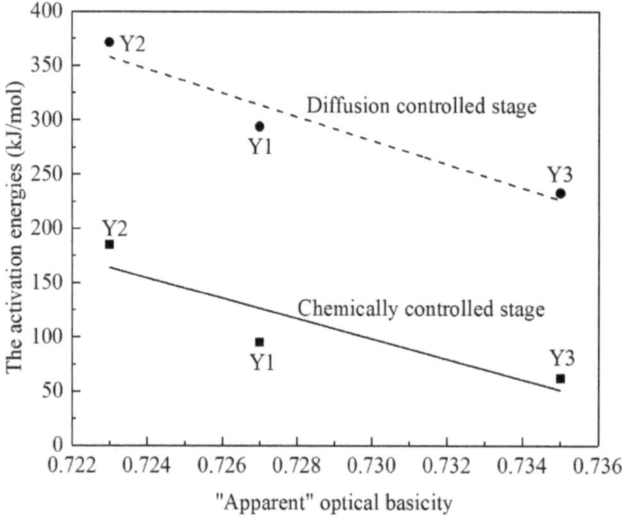

Figure 11. The relationships between the activation energies and slag apparent optical basicity at the chemically and diffusion-controlled stages.

It is clear from Figure 11 that the activation energies at the chemically controlled stage are much lower than those at diffusion-controlled stage. This difference can be explained by the concentrations and activities of PbO and ZnO in the liquid slag. Previous studies [7,8,10,11] confirmed that the reduction reactions mainly occurred between the liquid slag and reductants. The compositions of the liquid slag and the activities of PbO and ZnO in the liquid can be calculated by the thermodynamic model FactSage 8.1 [16] for a given bulk slag composition and temperature. The optical basicity of a liquid slag can be calculated from its composition. Table 7 shows the compositions of the liquid slag for Y1, Y2 and Y3 at 1473 K and their liquid optical basicity. The activities of PbO and ZnO in the liquid slag are also calculated from the FactSage. It can be seen from Figure 12 that the activities of PbO are much higher than those of ZnO and they all increase with increasing the optical basicity of the liquid slag. High activity of a component in the liquid slag indicates that the component can be relatively easier to be reduced. In the initial stage of the reduction the main reaction is the reduction of PbO to Pb in the liquid slag by carbon. High concentration and activity of PbO in the liquid (Table 7) results in a fast reaction. After PbO in the liquid is rapidly reduced, ZnO starts to be reduced. As a result of PbO reduction, the proportion of the liquid decreases and the viscosity of the liquid increases. Therefore, the reduction of ZnO is a diffusion-controlled reaction and this reaction is much slower than the reduction of PbO. On the other hand, it can be seen from Figure 12 that the activities of PbO and ZnO increase with increasing the optical basicity of the liquid slag. This explains the trends shown in Figure 11 where the activation energies at the chemically controlled and diffusion-controlled stages decrease with increasing apparent optical basicity.

Table 7. The compositions and activities of the liquid slags at 1473 K are calculated by FactSage; the liquid optical basicity is calculated from liquid compositions and the data in Table 1.

Slag	Compositions (wt%)							Liquid Optical Basicity	Activity	
	PbO	ZnO	Fe_2O_3	CaO	SiO_2	Al_2O_3	MgO		PbO	ZnO
Y1	57.0	9.2	11.7	7.4	13.1	1.3	0.0	0.723	0.21	0.094
Y2	57.7	7.4	8.9	7.4	13.2	5.1	0.0	0.717	0.19	0.075
Y3	59.0	9.1	8.0	7.7	13.5	1.3	1.4	0.731	0.27	0.1

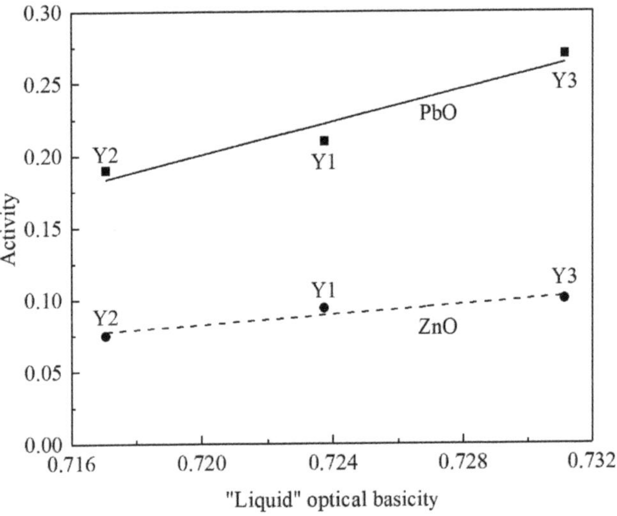

Figure 12. The relationship between optical basicity of liquid slags and the activity of PbO and ZnO at 1473 K.

7. Conclusions

The isothermal reduction of lead-rich slag by carbon has been investigated in the temperature range of 1073–1473 K by accurate measurements of the gas volume generated from the reactions. Experimental results and thermodynamic calculations are combined to explain the effects of CaO, MgO and Al_2O_3 on the reduction kinetics and mechanisms. It was found that the activation energy was 62–185 kJ/moL at the chemically controlled stage and 280–392 kJ/moL at the diffusion-controlled stage for the lead-rich slags investigated. All activation energies increase with increasing optical basicity of the slag. Both CaO and MgO can decrease the activation energy and accelerate the reduction of the lead-rich slag by carbon. In contrast, Al_2O_3 increases the activation energies at both chemically controlled and diffusion-controlled stages resulting in a lower reaction rate. As a result, MgO-containing lead concentrate and refractories are encouraged, and in contrast, use of Al_2O_3-containing lead concentrate and refractories should be carefully considered.

Author Contributions: Conceptualization, B.Z.; methodology, S.X. and B.Z.; software, B.Z.; validation, S.X., B.Z. and C.L.; formal analysis, S.X.; investigation, S.X. and B.Z.; resources, B.Z.; data curation, S.X. and B.Z.; writing—original draft preparation, S.X.; writing—review and editing, S.X., B.Z. and C.L. All authors have read and agreed to the published version of the manuscript.

Funding: This research received no external funding.

Conflicts of Interest: The authors declare that they have no conflict of interest.

References

1. Liu, A.; Reed, M.E.; Close, R.; Thompson, L. Lead Plant Transformations. In Proceedings of the Pb-Zn 2020: 9th International Symposium on Lead and Zinc Processing, San Diego, CA, USA, 23–27 February 2020; pp. 165–172.
2. Gao, W.; Wang, C.; Yin, F.; Chen, Y.; Yang, W. Situation and Technology Progress of Lead Smelting in China. *Adv. Mater. Res.* **2012**, *581–582*, 904–911. [CrossRef]
3. Li, Y. Application of oxygen side blow lead smelting technology. *Nonferr. Met. (Metall.)* **2012**, *11*, 13–15. [CrossRef]
4. Wu, W.; Xin, P.; Wang, J. The Latest Development of oxygen bottom blowing lead smelting technology. In Proceedings of the Pb-Zn 2020: 9th International Symposium on Lead and Zinc Processing, San Diego, CA, USA, 23–27 February 2020; pp. 327–336.
5. Li, W.; Zhan, J.; Fan, Y.; Wei, C.; Zhang, C.; Hwang, J. Research and Industrial Application of a Process for Direct Reduction of Molten High-Lead Smelting Slag. *JOM* **2017**, *69*, 784–789. [CrossRef]
6. Chen, L.; Hao, Z.; Yang, T.; Liu, W.; Zhang, D.; Zhang, L.; Bin, S.; Bin, W. A Comparison Study of the Oxygen-Rich Side Blow Furnace and the Oxygen-Rich Bottom Blow Furnace for Liquid High Lead Slag Reduction. *JOM* **2015**, *67*, 1123–1129. [CrossRef]
7. Kinaev, N.; Jak, E.; Hayes, P. Kinetics of reduction of lead smelting slags with solid carbon. *Scand. J. Metall.* **2005**, *34*, 150–157. [CrossRef]
8. Upadhya, K. Kinetics of Reduction of Lead Oxide in Liquid Slag by Carbon in Iron. *Metall. Mater. Trans. B* **2005**, *17*, 271–279. [CrossRef]
9. Zhao, B.; Errington, B.; Jak, E.; Hayes, P. Gaseous Reduction of Synthetic Lead Slags and Industrial Lead Sinters. *Can. Metall. Q.* **2010**, *49*, 241–248. [CrossRef]
10. Hou, X.; Chou, K.; Zhao, B. Reduction kinetics of lead-rich slag with carbon in the temperature range of 1073 to 1473K. *J. Min. Metall. Sect. B Metall.* **2013**, *49*, 201–206. [CrossRef]
11. Xie, S.; Liao, C.; Zhao, B. Experimental Studies on Reduction Mechanisms of Lead-Rich Slag with Different PbO Concentrations. In Proceedings of the 12th International Symposium on High-Temperature Metallurgical Processing, Anaheim, CA, USA, 27 February–3 March 2022. [CrossRef]
12. Duffy, J. A review of optical basicity and its applications to oxidic systems. *Geochim. Cosmochim. Acta* **1993**, *57*, 3961–3970. [CrossRef]
13. Lebouteiller, A.; Courtine, P. Improvement of a Bulk Optical Basicity Table for Oxidic Systems. *J. Solid State Chem.* **1998**, *137*, 94–103. [CrossRef]
14. Ghosh, D.; Krishnamurthy, V.; Sankaranarayanan, S. Application of optical basicity to viscosity of high alumina blast furnace slags. *J. Min. Metall. Sect. B-Metall.* **2010**, *46*, 41–49. [CrossRef]
15. Sommerville, I.; Sosinsky, D. Solubility, capacity and stability of species in metallurgical slags and glasses. In *Pyrometallurgy for Complex Materials and Wastes*; TMS Publication: Pittsburgh, PA, USA, 1994; pp. 73–91.
16. Bale, C.; Bélisle, E.; Chartrand, P.; Decterov, S.; Eriksson, G.; Gheribi, A.; Hack, K.; Jung, I.; Kang, Y.; Melançon, J.; et al. FactSage thermochemical software and databases, 2010–2016. *Calphad* **2016**, *54*, 35–53. [CrossRef]
17. Dickinson, C.; Heal, G. Solid–liquid diffusion-controlled rate equations. *Thermochim. Acta* **1999**, *340*, 89–103. [CrossRef]
18. Bidari, E.; Aghazadeh, V. Investigation of Copper Ammonia Leaching from Smelter Slags: Characterization, Leaching and Kinetics. *Metall. Mater. Trans. B* **2015**, *46*, 2305–2314. [CrossRef]
19. Vegliò, F.; Trifoni, M.; Pagnanelli, F.; Toro, L. Shrinking core model with variable activation energy: A kinetic model of manganiferous ore leaching with sulphuric acid and lactose. *Hydrometallurgy* **2001**, *60*, 167–179. [CrossRef]
20. Pritzker, D. Model for parallel surface and pore diffusion of an adsorbate in a spherical adsorbent particle. *Chem. Eng. Sci.* **2003**, *58*, 473–478. [CrossRef]
21. Shevchenko, M.; Jak, E. Experimental liquidus studies of the Pb-Fe-Si-O system in equilibrum with metallic Pb. *Metall. Mater. Trans. B* **2017**, *49*, 159–180. [CrossRef]
22. Jak, E.; Zhao, B.; Liu, N.; Hayes, P.C. Experimental study of phase equilibria in the system PbO-ZnO-SiO$_2$. *Metall. Mater. Trans. B* **1999**, *30*, 21–27. [CrossRef]
23. Shevchenko, M.; Jak, E. Experimental liquidus study of the binary PbO-ZnO and ternary PbO-ZnO-SiO$_2$ systems. *Ceram. Int.* **2019**, *45*, 6795–6803. [CrossRef]
24. Liao, J.; Zhao, B. Phase Equilibria Study in the system "Fe$_2$O$_3$"-ZnO-Al$_2$O$_3$-(PbO+CaO+SiO$_2$) in air. *Calphad* **2021**, *74*, 102282. [CrossRef]

Review

Development of Bottom-Blowing Copper Smelting Technology: A Review

Baojun Zhao * and Jinfa Liao 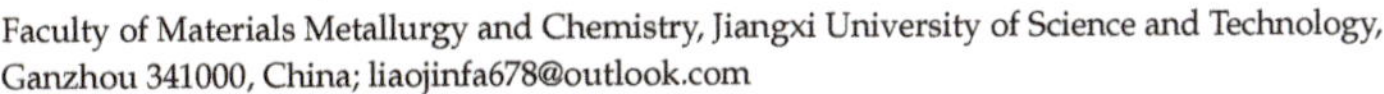

Faculty of Materials Metallurgy and Chemistry, Jiangxi University of Science and Technology, Ganzhou 341000, China; liaojinfa678@outlook.com
* Correspondence: bzhao@jxust.edu.cn

Abstract: Bottom-blowing copper smelting technology was initiated and developed in China in the 1990s. Injection of oxygen-enriched high-pressure gas strongly stirs the molten bath consisting of matte and slag. Rapid reaction at relatively lower temperatures and good adaptability of the feed materials are the main advantages of this technology. Development and optimisation of bottom-blowing copper smelting technology were supported by extensive studies on the thermodynamics of the slag and the fluid dynamic of the molten bath. The history of technological development and fundamental studies related to this technology are reviewed in this paper.

Keywords: copper smelting; bottom-blowing; SKS; slag; fluid dynamic

Citation: Zhao, B.; Liao, J. Development of Bottom-Blowing Copper Smelting Technology: A Review. *Metals* **2022**, *12*, 190. https://doi.org/10.3390/met12020190

Academic Editor: Mark E. Schlesinger

Received: 30 December 2021
Accepted: 19 January 2022
Published: 20 January 2022

Publisher's Note: MDPI stays neutral with regard to jurisdictional claims in published maps and institutional affiliations.

1. Introduction

Smelting is one of the important processes in the pyrometallurgical production of blister copper in which copper–iron sulphides are oxidised to form molten matte and slag. Flash smelting and bath smelting are the major smelting technologies in copper production. Copper concentrates react with oxygen directly in the flash smelting process which has the advantages of high capacity and automatic control. However, fine and dry feeds are required for the flash smelting furnace to enable fast reactions. As a result, considerable feed preparation is required, and the dust rate is relatively higher. It has limited ability to treat scrap and other copper-containing materials with large sizes. Bath smelting is an alternative technology to flash smelting which involves the reactions of copper concentrate with oxygen in the molten bath. Several technologies have been developed based on the bath smelting principles that include IsaSmelt/Ausmelt, Noranda/El Teniente, Vanyukov, Mitsubishi and recently developed bottom-blowing smelting (BBS) process [1–4]. Due to its unique technological features, such as good adaptability to raw materials, high oxygen utilisation and thermal efficiency, and flexible capacity, BBS technology has attracted strong interest from the copper industry [5–14]. In 2016, 13 BBS furnaces were constructed or were under construction with the capacity of 1600 kt/a copper production. The furnace size ranged from Ø3.8 m × 11.5 m to Ø5.8 m × 30 m [5,14]. The fundamental studies including thermodynamics of the slag and fluid dynamic of the molten bath have been extensively conducted in recent years to understand and support the new technology. This review summarises the development of the copper BBS including the history, features and relevant fundamental studies.

2. History of Technology Development

From November 1991 to June 1992, a plant trial was conducted at Shuikoushan (SKS) Smelter for a period of 217 days to smelt copper concentrate in a bottom-blowing furnace [15,16]. The name of SKS Smelting Technology was therefore used initially to represent the bottom-blowing copper smelting technology. Figure 1 below shows the flowsheet of the SKS trial [16]. It can be seen from the figure that carbon fuel was used during the trial, and the feeds were palletised during the trial. The smelting slag was treated by flotation process to recover copper and the concentrate from the flotation was sent back to the bottom-blowing furnace (BBF). Only the coarse flue dust from the BBF and PS converter was sent back to the BBF. The detailed operating parameters are given in Table 1. It can be seen from Table 1 that the BBF treated 50 t feeds per day during the trial and Cu in the concentrate was approximately 20%. Matte with 50% Cu was produced and Cu in the slag was in the range of 1–3%. Oxygen utilization at 100% was claimed and a lance life of 5000 h was estimated.

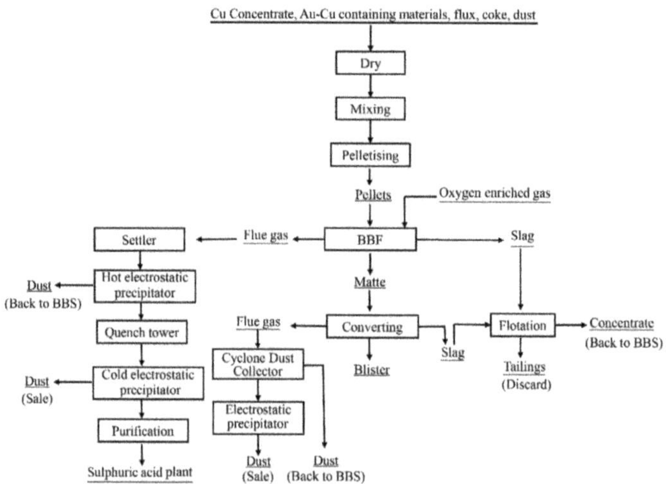

Figure 1. Flowsheet of SKS copper smelting technology, adapted from [16].

Table 1. Detailed operating parameters during SKS trial. Data derived from [16].

Parameter	Unit	SKS
Capacity	t/d	50
Cu in Concentrate	%	~20
Fe in Concentrate	%	~26
S in Concentrate	%	25~30
Matte grade	%	~50
Cu in slag	%	1~3
Slag clean	-	flotation
Cu in tailing	%	0.34
Fe/SiO$_2$ in slag	-	1.5~1.7
O$_2$ in gas	vol%	60~70
Oxygen pressure	Mpa	0.5~0.7
O$_2$ utilisation	%	100
Lance life	hour	5000
Production rate	%	81.4
SO$_2$ in gas	vol%	>20
Cu direct recovery	%	93
Cu total recovery	%	98
Furnace campaign life	day	>330

In 2005 the first industrial scope BBF was built in Sin Quyen smelter in Vietnam. The size of the furnace was Ø3.8 m × 11.5 m and the capacity was designed to be 10,000 t Cu per year [6,17]. However, the BBF in Sin Quyen smelter seemed not operating properly as no operating details were reported. In 2008 the first real commercialised BBF started in Dongying Fangyuan Nonferrous Metals (Fangyuan) [7]. The main equipment was a horizontal cylindrical furnace shown in Figure 2. The size of the furnace was Φ 4.4 m × 16.5 m and it was lined with 380 mm thick chrome-magnesia bricks. The BBF had nine gas lances arranged in two rows on the bottom, five lances at 7 degrees and four lances at 22 degrees towards the outside. Each lance consisted of an inside tube and an external shroud. The inside tube injected pure oxygen and the shroud injected airflow as a coolant to protect the lance. The furnace operated with a rotation mechanism so the lances could be rolled above the molten bath during maintenance, repair, power failure or another emergency. Mixed feed materials with 7–10 wt% moisture were continuously transported by a belt conveyor into the high-temperature melt in the furnace through a feed mouth located above the reaction zone. High-pressure oxygen and air were blown constantly by the lances into the copper matte, in which iron and sulphur were rapidly oxidised. The by-product SO_2 was sent to the acid plant and the slag was tapped regularly followed by the flotation process to recover copper. Matte was tapped regularly and sent to the PS converters by a ladle.

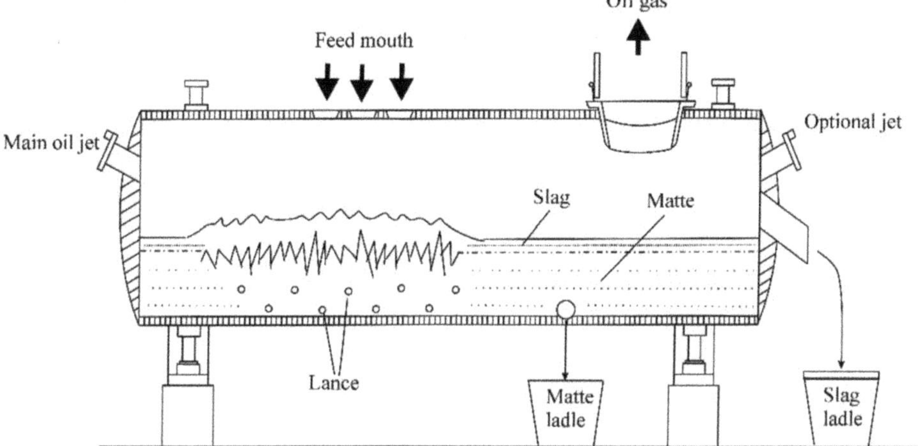

Figure 2. The bottom-blowing furnace at Dongying Fangyuan Nonferrous Metals. Reprinted with permission from ref. [7]. Copyright 2021 John Wiley and Sons.

Table 2 shows key process parameters in the Fangyuan smelting plant in January 2012 compared with the initial design. The effort made by the management and engineers enabled unexpected performance to be achieved in the first commercialised BBF. Differently from the trials in Shuikoushan, the following advantages were demonstrated in the Fangyuan BBF:

(1).   No carbon fuel was used—no CO_2 was generated.
(2).   Feed preparation was not required—up to 100 mm size could be fed into the BBF directly.
(3).   High-grade matte with 70% Cu was produced—more concentrate could be treated.
(4).   Concentrate feed rate increased from 32 to 75 dry t/h—the furnace capacity was more than doubled.

Table 2. Typical process parameters for Fangyuan smelting plant in January 2012. Data derived from [7].

Parameter	Unit	Design	January 2012
Maximum concentrate feed rate	dry t/h	32	75
Average concentrate feed rate	dry t/h	32	70
Average Cu content in concentrate	%	25	22
Average moisture of the feed	%	8	7
Average silica flux feed rate	dry t/h	-	8
Average coal feed rate	dry t/h	2.46	0–0.8
Total average feed to the furnace	wet t/h	-	90
Average copper matte grade	%	55	70
Average Fe/SiO_2 in slag	-	1.7	1.8
Average Cu in smelting slag	%	4	2.6
Average Cu in flotation tailing slag	%	0.42	0.3
Average oxygen + air flow rate	Nm^3/s	-	4.2
Average oxygen enrichment	%	70	72
Bath temperature range	°C	1180–1200	1150–1170

Compared with other bath smelting technologies, BBF shows several advantages:

(1).　Less preparation of feeds

Top blowing technology (Ausmelt and Isasmelt) requires pelletised feeds and side blowing technology (Teniente Converter) needs to inject dry and fine particle feeds. BBF can treat a wide range of size feeds up to 100 mm diameter with moisture up to 10%.

(2).　High gas pressure and long lance life

Much higher gas pressure in BBF has many advantages confirmed by Kapusta et al. [18–23]. As one of the significant developments at Fangyuan, a series of trials were undertaken to identify the optimum pressures of the oxygen and air in the lances. It was found that optimum gas pressures were related to the feeding rate, the viscosity of the slag, thicknesses of the slag and matte layers. The following requirements need to be met: (1) no backflow of matte into the lances; (2) enough stirring for rapid reactions and generation of the surface wave; (3) no strong splashing of the slag and matte to block the feed mouth and (4) a protection accretion is formed at the tip of the lance. Figure 3 shows the "mushroom" formed at the tip of the lances which enabled the lance life to be over 10 months [7]. These mushroom tips not only protect the lances but also distribute the injected gases to small bubbles.

(3).　Autogenous Smelting

Autogenous operation was achieved in the BBF at Fangyuan for smelting of normal copper concentrates. This operation reduced energy consumption and CO_2 emission. The autogenous smelting was obtained by (1) low-temperature operation; (2) low slag rate resulting in a high Fe/SiO_2 ratio; (3) low off-gas volume resulting in high oxygen concentration; (4) high thermal efficiency as most of the oxidisation reactions occur in the lower part of the bath, the heat generated from these reactions can be efficiently absorbed by the molten matte and slag.

(4).　Low-Temperature Smelting

The slag temperatures measured at the taphole of the Fangyuan BBF were usually in the range of 1150 to 1170 °C which was much lower than the liquidus temperature of the slag [7,24]. Approximately 20% solid spinel was present in the slag which increased the slag viscosity significantly. High-pressure gases injected from the bottom of the molten bath can generate surface waves that push the viscous slag out of the furnace through the taphole. The advantages of low-temperature smelting are (1) no extra fuels are required to maintain smelting temperature (although oil jets were available on both ends of the BBF, they were rarely used); (2) low consumption of the refractory. There was no significant erosion of the refractory on most parts of the furnace after one and a half years operation since spinel accretion was formed to protect the refractory [24–27]; (3) relatively higher viscosities of the matte and slag allowed higher pressures of the injected gases without strong splashing; (4) a higher-grade matte can be produced at a lower temperature at a given oxygen partial pressure.

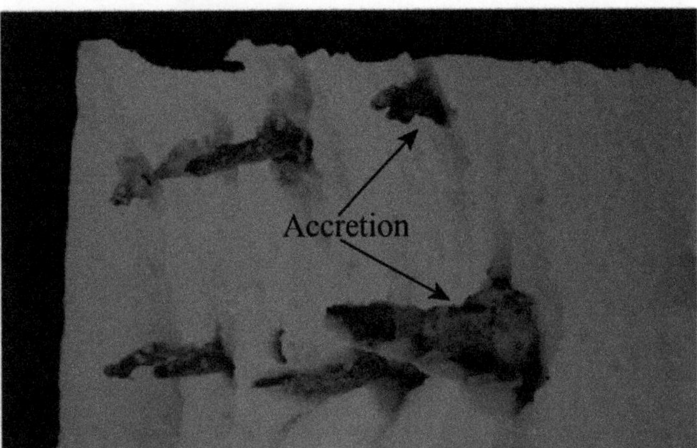

Figure 3. "Mushroom" tips formed on the lances of the Fangyuan BBF. Reprinted with permission from ref. [7]. Copyright 2021 John Wiley and Sons.

3. Fundamental Studies to Support Development of BBS Technology

3.1. Slag Chemistry

In the conventional theory of pyrometallurgy, the slag temperature must be higher than its liquidus temperature to avoid the formation of the solid phase. In Fangyuan BBF operation, high matte grade and high Fe/SiO_2 in the slag resulted in a high liquidus temperature of the smelting slag. In contrast, the BBF slag temperature was lower than other copper smelting technologies [7,28–30]. It was doubted widely that either slag composition or temperature was not reliable. Researchers from the University of Queensland measured the slag tapping temperatures carefully using a K-type thermocouple and collected quenched slag samples. In combination with electron probe X-ray microanalysis (EPMA) and thermodynamic analyses [31], it was confirmed that it is possible to operate BBF with a viscous slag containing solid phase.

Figure 4 shows a typical microstructure of the quenched BBF slag compared with the FSF slag [7]. It can be seen from Figure 4a that liquid, spinel and matte phases were present in the slag. The shape and size of the spinel phase indicate that the spinel was the primary phase presented at high temperatures. The slag tapping temperature of the BBF was therefore lower than its liquidus temperature. The matte droplets in different sizes were also present in the slag. In contrast, it can be seen from Figure 4b that no solid phase was found in the quenched flash smelting slag. The compositions of the phases present in a quenched BBF slag were measured by EPMA and are shown in Table 3 together with the bulk composition measured by XRF. Both EPMA and XRF can only measure elemental compositions. The compositions shown in the table were calculated by assuming that the elements were present in certain forms of oxides or elements in the slag. It is estimated from the compositions that the proportion of the spinel is approximately 16.8 wt%.

Reheating experiments were carried out using the slag shown in Table 3 under ultra-high purity Ar flow. The experiments were undertaken at 1150, 1200, 1250 and 1300 °C respectively using Pt foil. In these experiments, the oxygen partial pressure was not controlled, it was assumed that there was no oxygen exchange between the slag and Ar gas. FactSage was used to simulate the reheating experiments. The experimentally determined proportion of liquid phase from mass balance was compared with FactSage predictions. It can be seen from Figure 5 that both experimental data and FactSage predictions showed the proportion of liquid phase decreased slowly with decreasing temperature between

1150 and 1300 °C [7]. This indicates that equilibrium was obtained during the reheating experiments and the FactSage predictions were relatively reliable under the condition used. Reheating experiments were not undertaken below 1150 °C. It can be seen from the figure that, according to the FactSage predictions, the proportion of liquid phase decreases sharply if the temperature is lower than 1150 °C due to the formation of other solid phases such as olivine and silica.

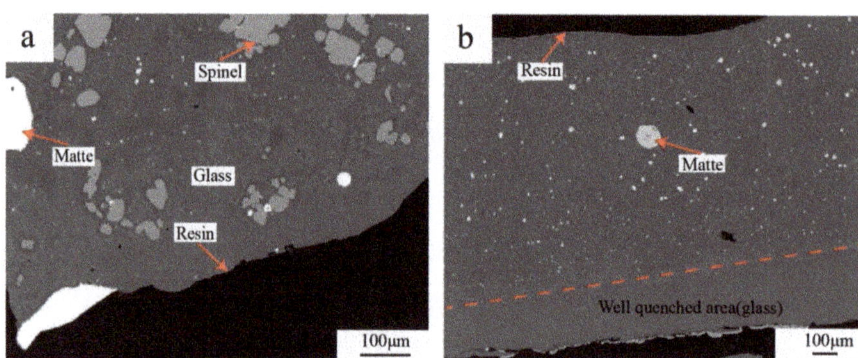

Figure 4. Typical microstructures of quenched smelting slag from Fangyuan BBF (**a**) (Reprinted with permission from ref. [7]. Copyright 2021 John Wiley and Sons) and a flash furnace (**b**).

Table 3. Compositions of the phases present in a quenched BBF slag measured by EPMA and bulk composition measured by XRF (wt%). Data derived from [7].

Phases	FeO	Cu_2O	CaO	SiO_2	Al_2O_3	As_2O_3	MgO	S	PbO	ZnO	MoO_3
bulk-XRF	62.2	3.20	1.00	24.2	3.10	0.100	0.600	1.70	0.500	3.10	0.200
glass	58.4	0.800	1.20	30.5	3.20	0.100	0.700	1.10	0.500	3.30	0.200
spinel	93.7	0.100	0.000	0.600	3.40	0.000	0.300	0.00	0.100	1.70	0.100
matte	10.1	68.9	0.000	0.000	0.00	0.100	0.000	20.3	0.100	0.200	0.300

3.2. Phase Equilibria

It can be seen from Table 3 that significant amounts of Al_2O_3, CaO, MgO and ZnO are present in the BBF slag which can influence the liquidus temperature of the slag. Systematic studies were conducted under controlled oxygen partial pressures to evaluate the effects of these components on liquidus temperature of the copper smelting slag [32–39]. Figure 6 shows liquidus temperature as a function of Fe/SiO_2 weight ratio at $P_{O_2} = 10^{-8}$ atm [32]. The lines in the figure were calculated by FactSage 6.2 [31] and the symbols represent experimental data. It can be seen that spinel and tridymite are the primary phases in the composition range investigated for "FeO"–SiO_2, "FeO"–SiO_2–CaO and "FeO"–SiO_2–CaO–Al_2O_3 slags. Olivine and wustite primary phase fields replaced spinel primary phase fields when 4 wt% CaO and 4 wt% MgO are present in the slag. Additions of CaO and Al_2O_3 move the low-liquidus region towards a low Fe/SiO_2 direction and increase the liquidus temperatures in the spinel primary phase field. FactSage predictions showed similar trends on the liquidus temperature as the experimental results. However, there are significant differences in liquidus temperatures between FactSage predictions and experimental results in some areas indicating that the FactSage database needs to be improved.

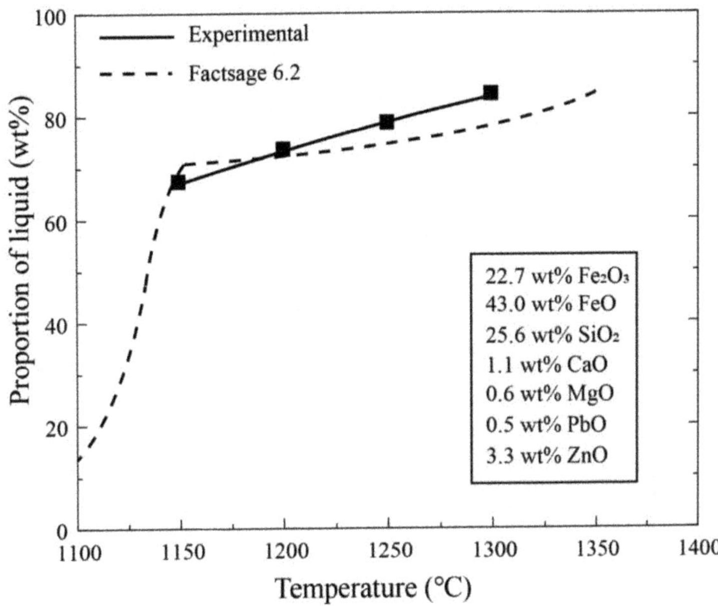

Figure 5. Proportion of liquid as a function of temperature for a typical BBF slag, in comparison between experimental results and FactSage 6.2 predictions. Reprinted with permission from ref. [7]. Copyright 2021 John Wiley and Sons.

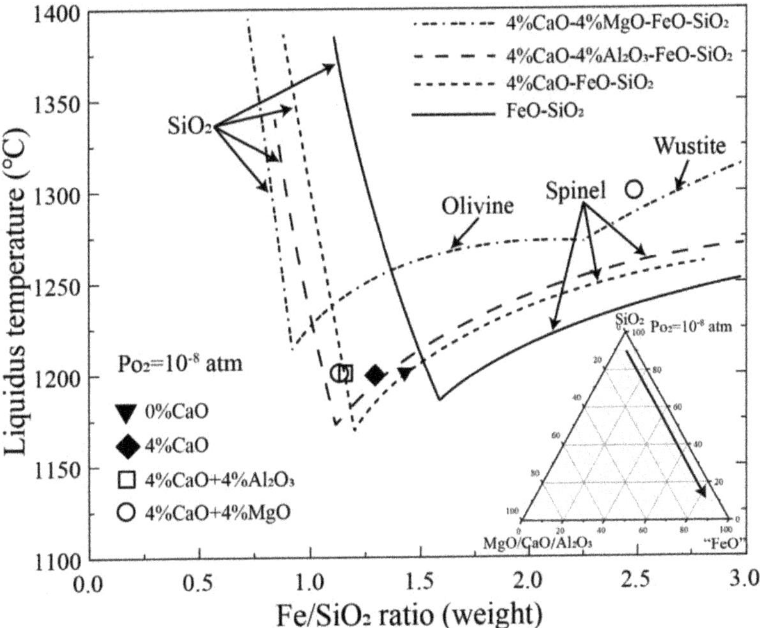

Figure 6. Liquidus temperature as a function of Fe/SiO$_2$ weight ratio, P$_{O_2}$ = 10^{-8} atm, lines were calculated by FactSage 6.2, symbols are experimental results in spinel (1200 °C) and wustite (1300 °C) primary phase fields [32].

The phase equilibria in the system ZnO–"FeO"–SiO_2–Al_2O_3–CaO–MgO were systematically studied under controlled oxygen partial pressure by Liu et al. [33–36]. Figure 7 shows the effects of ZnO, Al_2O_3, CaO and MgO on liquidus temperatures at fixed FeO/SiO_2 = 1.9 and P_{O_2} 10^{-8} atm [33]. Spinel is the only primary phase at the given conditions without MgO. Olivine is the primary phase when 4 wt% MgO and less than 2 wt% ZnO are present. It can be seen from the figure that, the liquidus temperatures of the copper smelting slag always increase with increasing ZnO concentration in the slag. The introduction of 4 wt% CaO, MgO or Al_2O_3 into the ZnO–"FeO"–SiO_2 slag significantly increases the liquidus temperature in the spinel primary phase field. It seems that the liquidus temperatures are more sensitive to the CaO and MgO concentrations than that of Al_2O_3.

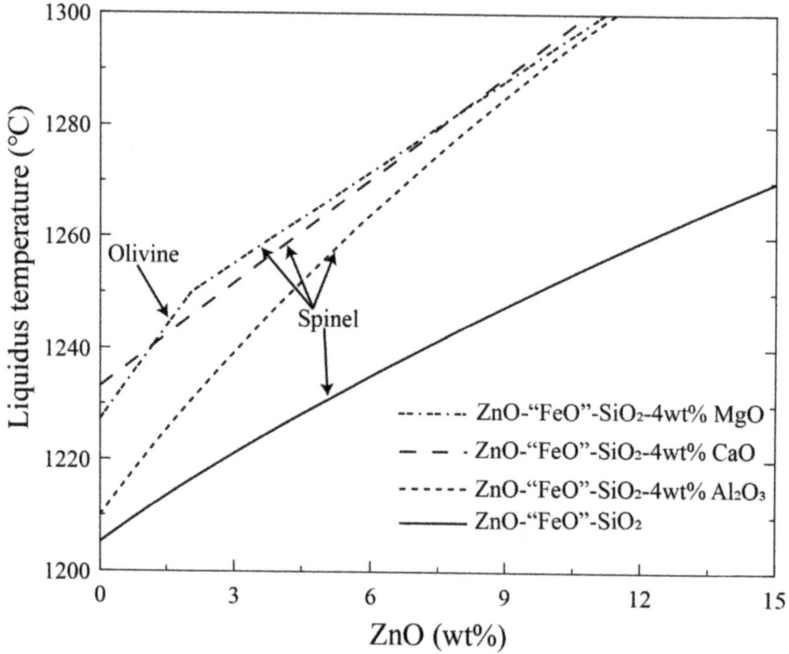

Figure 7. Effects of 4 wt% Al_2O_3, CaO or MgO on liquidus temperatures of the ZnO–"FeO"–SiO_2 slag at fixed FeO/SiO_2 = 1.9 (in mass) and P_{O_2} 10^{-8} atm. Reprinted with permission from ref. [33]. Copyright 2021 Elsevier.

An experimental technique was further developed to determine phase equilibria of the copper smelting slags under conditions closer to the smelting process. The phase equilibria in the systems "FeO"–SiO_2, "FeO"–SiO_2–CaO and "FeO"–SiO_2–CaO–MgO were experimentally investigated at fixed $P(SO_2)$ = 0.3 and 0.6 atm and a fixed matte grade 72 wt% Cu [37–39]. Figure 8 shows the effects of 2 wt% CaO or MgO on the liquidus temperatures in the spinel primary phase field at fixed $P(SO_2)$ = 0.3 atm with a fixed matte grade of 72 wt% Cu [38]. The predictions by FactSage 7.3 were also shown in the figure for comparison. It can be seen from the figure that, both the predictions and experimental results indicate that the liquidus temperatures decrease with increasing SiO_2 concentration in the spinel primary phase field. The presence of 2 wt% CaO or MgO significantly increases the liquidus temperatures. To keep a constant liquidus temperature, more SiO_2 flux is required which will increase the slag volume. FactSage 7.3 predicted much lower liquidus or less SiO_2 flux than the experimental results as no experimental data were reported in such complex conditions for FactSage optimisation.

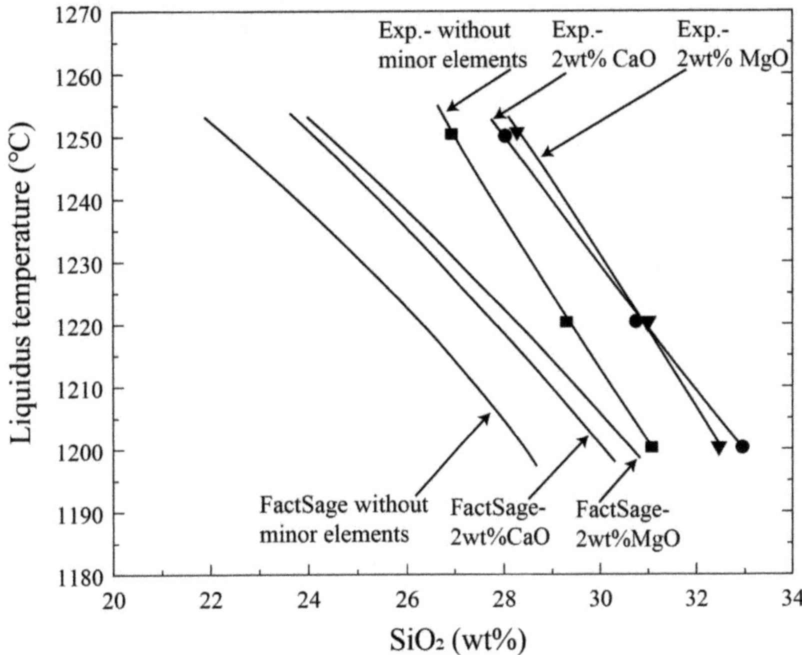

Figure 8. Effects of 2 wt% CaO or MgO on the liquidus temperatures in the spinel primary phase field at a fixed P(SO$_2$) = 0.3 atm and matte grade 72 wt% Cu, in comparison between experimental (symbols) and predicted (lines) results by FactSage 7.3. Reprinted with permission from ref. [38]. Copyright 2021 Elsevier.

With increased impurities in copper concentrates, control of the minor elements is an important issue in the copper smelting process. Based on the principle of Gibbs energy minimisation and the technological features of the BBS process, a multiphase equilibrium model SKS simulation software (SKSSIM) was developed [40,41]. Distributions of As, Bi, Sb, Pb and Zn as a function of matte grade can be calculated among slag, matte and gas. The calculated distributions in the BBS process were different from that in the Isasmelt and the flash smelting processes [42–44].

3.3. Fluid Dynamic Studies

The major difference between BBS and other bath smelting technologies is the direction of gas injection and gas pressure. High-pressure gas injected from different directions significantly influences stirring energy, surface waves, plume eye and splashing. Fluid dynamics of the molten bath is an important issue to understand the advanced performance of the BBS. Extensive studies were conducted on the fluid dynamic behaviours of the BBF using a water model and CFD (Computational Fluid Dynamic) simulation [45–68]. Mixing behaviour [45–51], surface wave [52,53], plume eye [54], bubble behaviour [55–58], fluctuating behaviour [59] were studied by water model to simulate the molten bath in BBF. Lance arrangement [60–62], gas–liquid multi-phase flows [63–65], flow and mixing behaviour [66] and multiphase interface behaviours [67,68] in the BBF were studied by the CFD method. A summary of the research on the fluid dynamic of the molten bath in BBF is presented in Table 4. Some examples are given in the following sections.

Table 4. Summary of the research on the fluid dynamic studies for the molten bath of BBF.

Reference	Research Method	Objects	Main Findings
Wang et al. [46,47]	Water model	Mixing behaviour	Effects of bath height, installation angles of nozzle and gas flow rate on velocity distribution
Shui et al. [48]	Water model	Mixing behaviour	Effects of gas flow rate and bath height on mixing time
Shui et al. [49]	Water model	Mixing behaviour	Effects of industry-adjustable variables on bath mixing time
Jiang et al. [50,51]	Water model	Mixing behaviour	Effect of horizontal distance between tuyeres, gas flow rate and bath height on mixing time
Shui et al. [52]	Water model	Surface wave	Effect of blowing angle, gas flow rate and bath height on mixing time on 1st asymmetric standing wave
Shui et al. [53]	Water model	Surface wave	Amplitude and frequency of surface longitudinal waves
Jiang et al. [54]	Water model	Plume eye	Effects of different operating parameters on the sizes of the plume eyes
Wang et al. [55,56]	Water model	Average diameter of bubbles	An empirical formula developed for average bubble diameter
Cheng et al. [57,58]	Water model	Copper matte attachment behaviour	Effects of bubble on the copper losses to the smelting slags
Luo et al. [59]	Water model	Fluctuation behaviour	Effects of diameter, inclination angle, gas flow rate and liquid surface on the violent level of liquid level.
Yan et al. [60]	CFD	Lance arrangement	An optimised lance arrangement (diameter, spacing, inclination)
Zhang et al. [61]	CFD	Tuyere structure parameters	An optimised tuyere arrangement (spacing, size, angle)
Guo et al. [62]	CFD	oxygen lances	An optimised oxygen lance layout
Zhang et al. [63]	CFD, Water model	Gas–liquid multi-phase flows	The effects of bubble parameters, gas holdup distribution, inlet pressure variations and the fluid level fluctuation on oxygen–copper matte flow
Tang et al. [64]	CFD	Gas–liquid phase interaction	Gas residence time and liquid copper matte splashing phenomena under varying gas flow rates
Li et al. [65]	CFD	Gas–liquid multiphase flow	The characteristics of each flow region in the furnace are obtained
Shao and Jiang [66]	CFD, Water model	Flow and mixing behaviour	The effect of nozzle arrangement and gas flow rate on mixing time
Zhang et al. [21]	CFD	Bubble behaviour	The relation between mixing efficiency and bubble characteristics
Guo et al. [67]	CFD	Mechanism and multiphase interface behaviour	The capacity of BBS can be raised by reasonably controlling the potential value in different layers and regions
Guo et al. [68]	CFD	Optimisation of smelting process	Matte grade and slag type have deep effect on copper in slag

In a molten bath, the first step of the reactions is the melting of the concentrate particles and decomposition of the sulphides, such as FeS_2 and $CuFeS_2$. These sulphides are then oxidised to form FeS–Cu_2S matte and iron oxides (FeO, Fe_2O_3). The iron oxides generated by the reactions continuously react with SiO_2 to form slag which is insoluble to matte and floats on the top of the matte layer. On the other hand, the injected oxygen reacts with iron, copper and sulphur on the bottom of the bath to generate all heat to maintain the temperature of the molten bath. The mass and thermal transfer inside the molten bath significantly affect the reaction efficiency of the BBF. Cai et al. [69] did experiments in a physical model of a bottom-blowing lead smelting furnace to find an optimal lance arrangement and configuration. An empiric calculation formula was concluded by normal hydraulic model test:

$$S/W = 26.224 \cdot (W/D_0)^{-0.629} \cdot (\mathrm{Fr}')^{0.122} \cdot (H/D)^{0.523} \tag{1}$$

where S is the effective stirring diameter of the oxygen lance, W is the space between oxygen lances, D_0 is the inner diameter of the export oxygen lance, Fr' is the modified Froude number, H is the depth of bath pool, D is the inner diameter of the furnace. In a late study, Shui et al. [48] defined effective stirring range D_e quantitatively and developed the 1st empiric equation to characterise the mixing behaviour of the copper BBF.

Figure 9 shows the mixing time as a function of distance from the lance at different depths of the bath [48]. In the water model experiments, 4 M KCl solution as a tracer of electric conductivity was added directly on the top of the gas injection lance. An electrode connected to a potentiostat was put in the required position to measure the electric conductivity of that location. The time to reach a stable electric conductivity was defined as "mixing time". It can be seen that within 220 mm the mixing time on the surface decreases slightly and then increases sharply with increasing the distance between the lance and electrode. However, D_e is different at different depths of the bath and increases with increasing the depth. When the horizontal distance from the lance is greater than D_e, the mixing time increases rapidly with increasing the distance. The maximum distance between two neighbour lances is the D_e on the surface to ensure the whole bath in the reaction zone can be efficiently stirred.

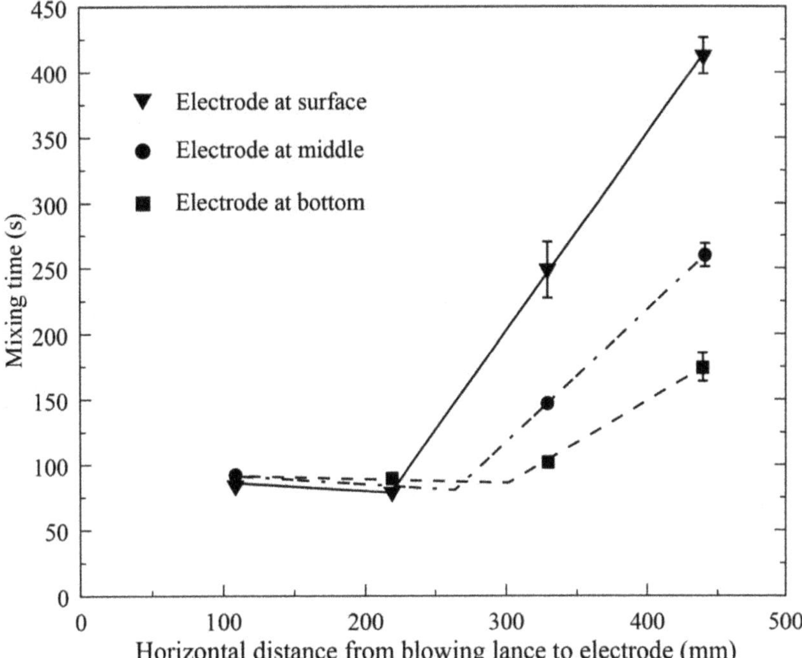

Figure 9. Mixing time vs. horizontal distance, at gas flow rate 150 mL/s, bath height 10 cm. Reprinted with permission from ref. [48]. Copyright 2021 Springer Nature.

The empiric equation to characterise the mixing behaviour of the copper BBF was obtained from the experimental measurements of the water model [48]

$$\tau = 37.5Q^{-0.39}h^{-1.08} \tag{2}$$

where τ is mixing time, Q is gas flow rate and h is bath height. Several important findings from this study can be directly applied to BBS:

1. The effective stirring range can be accurately determined for a single lance which is one of the important parameters to design the number of lances and distance between the lances. The optimum number and distance of the lances enable the bath to be efficiently mixed with minimum energy and gas consumption.
2. Within the effective stirring range, the mixing time is not sensitive to the vertical locations. However, the effective stirring range is smaller on the surface and mixing energy decreases much faster at the surface beyond the range. In copper BBF, many reactions occur on the surface of the bath as all solid materials are fed from the top. It is therefore important to ensure the surface area in the reaction zone is fully covered within the effective stirring area by proper arrangement of the lances.
3. Equation (2) is the first quantitative expression of the mixing time for copper BBF. Required injection gas flow rate and bath height can be calculated from this equation according to productivity.

In copper BBF operation, molten matte is covered by a thick slag layer. The fluid dynamic studies for a single layer bath are not enough to accurately describe the molten bath of the copper BBF. Jiang et al. [50] and Shui et al. [49] studied mixing behaviours of two-layer baths by using silicon oil and water to simulate slag and matte, respectively. It can be seen from Figure 10 that the changes in mixing time are not significant when the water thickness increases from 0.07 m to 0.09 m, but there is a sharp decrease when the water thickness exceeds 0.09 m. This indicates that a minimum matte height is required to generate enough stirring energy to mix the bath efficiently. In addition, it can be seen that the trends are similar at different gas flow rates.

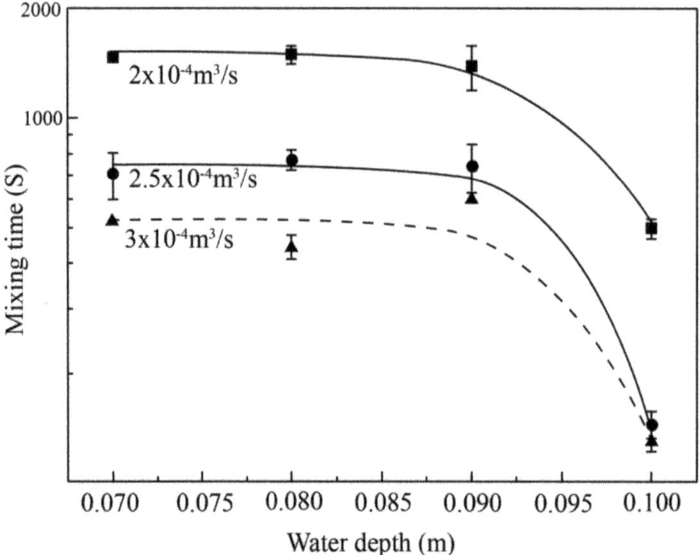

Figure 10. Effect of water height on mixing time at different gas flow rates, 0.03 m thick oil with viscosity 5×10^{-5} m^2/s on the top, symbols are experimental points. Reprinted with permission from ref. [49]. Copyright 2021 Springer Nature.

An empiric equation was also generated for a two-layer bath in copper BBF [49]. Four variables H, Q, h, v_s were converted into SI units and correlated with the experimental mixing time in an overall multiple regression:

$$\tau = 7.31 \times 10^{-5} H^{-3.10} Q^{-2.29} h^{2.32} v_s^{0.27} \tag{3}$$

where τ is mixing time, Q is gas flow rate, H is water height, h is oil height and v_s is oil viscosity.

The waves formed on the bath surface play important role in the BBF operations. Tapping of the viscous slag, corrosion of the refractory around the surface and settlement of the matte droplets in the slag are all associated with the bath waves. Simulation experiments were carried out to investigate the features of the waves formed on the bath surface of the BBF [52,53]. It was found that the ripples, the 1st asymmetric standing wave and the 1st symmetric standing wave can occur in the BBF bath. Empirical occurrence boundaries were determined from water model experiments. The amplitude of the 1st asymmetric standing wave was found to be much greater than the 1st symmetric standing wave and the ripples. The amplitudes increase with increasing bath height and flow rate but decrease with blowing angle. The frequency of the 1st asymmetric standing wave was found to increase with increasing bath height but be independent of gas flowrate and blowing angle. The dimensionless number We can be expressed as

$$We = \frac{\rho_L Q_g^2}{\sigma D^3} \tag{4}$$

where ρ_L is liquid density (kg/m^3), σ is the surface tension of the liquid (dyn/cm), D is container inner diameter (mm), Q_g is gas flowrate (mL/s).

Figure 11 shows the conditions for the occurrence of the 1st asymmetric standing waves [53]. The solid curves show the correlated boundaries while the points show the experimental results. It can be seen that, in general, the boundaries of different blowing angles are similar in shape. On the right side of the boundary, the 1st asymmetric standing wave is present, while on the left side only minor ripples occur.

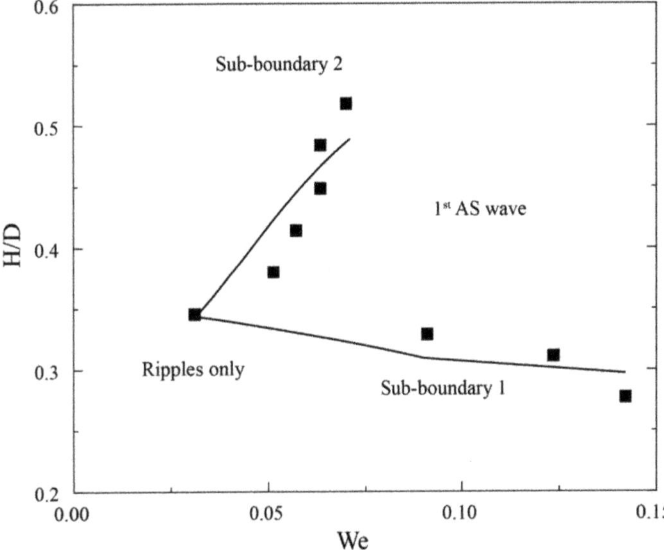

Figure 11. Occurrence condition of the 1st asymmetric standing wave at lance angle 7°, symbols are experimental points. Reprinted with permission from ref. [53]. Copyright 2021 Springer Nature.

During the Fangyuan BBF operation, it was observed from the feeding mouth that a plume eye was formed on the surface of the bath. The plume eye was initially observed and reported in the steelmaking process to describe the exposure of the lower liquid as the upper-level liquid was pushed away by the bottom injected gas. Through the water model experiments, it was found that the sizes of the plume eyes increase with increasing gas flow rate and lower liquid thickness, decrease with increasing upper liquid thickness [54]. Figure 12 shows the effects of gas flow rate and water height on the plume eye area at a fixed oil thickness.

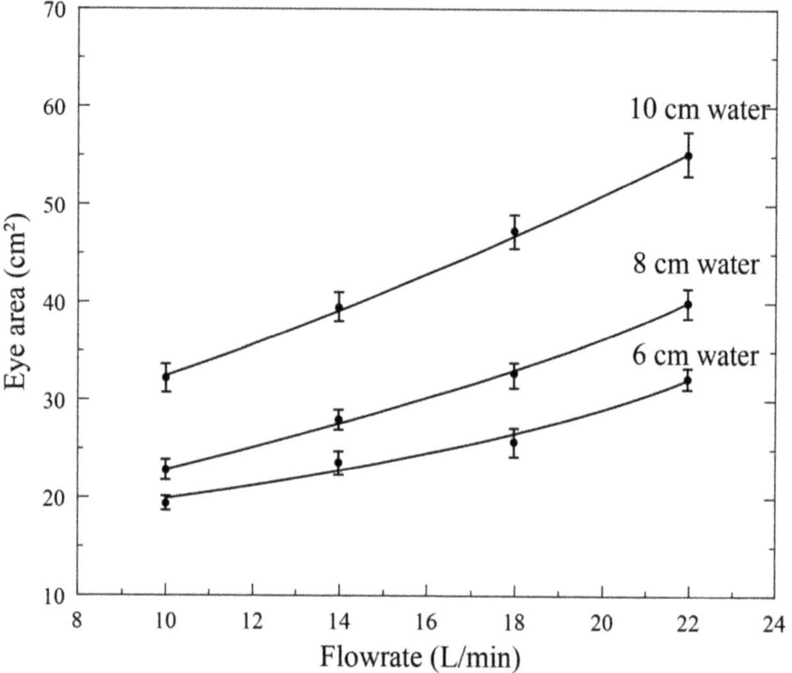

Figure 12. Effects of gas flow rate and water height on the plume eye area at a fixed oil thickness 4 cm. Reprinted with permission from ref. [54]. Copyright 2021 Springer Nature.

A modified model was developed in this study from the experimental data to predict the size of the plume eye [54]. Formation of the plume eye is important to understand the rapid reaction of the BBF. The presence of a plume eye below the feeding mouth means that the dropped copper concentrate (considered as low-grade matte) can enter and melt rapidly inside the matte layer where the oxygen partial pressure is relatively high. The formed iron and sulphur oxides in the matte raise to the slag layer.

4. Conclusions

Commercialised bottom-blowing copper smelting technology started in 2008. In less than 15 years this new technology has been developed rapidly to be the second-largest copper smelting technology. Development history and features of the new technology were reviewed from extensive publications. It is demonstrated that fundamental studies on slag chemistry and bath fluid dynamics have played an important role in supporting the development of this new technology. High-pressure gas injected from the bottom of the molten bath showed several advantages: (1.) generating a plume eye to allow the copper concentrate to react with matte directly; (2.) the surface waves assist in the tapping of the viscus slag so that low-temperature operation to produce high-grade matte with high Fe/SiO$_2$ slag is possible; (3.) strong stirring energy to enable fast reactions in the molten bath resulting in high capacity and low-copper in the slag; (4.) high efficiency of oxygen utilization and heat absorption. The flexibility of the BBS on feeding materials and productivity is expected to not only increase its application in the copper industry to treat more complicated concentrates but also to enable the exploration of more applications of this technique in other high-temperature processes.

Author Contributions: Conceptualization, B.Z.; resources, B.Z.; data curation, B.Z. and J.L.; writing—original draft preparation, B.Z.; writing—review and editing, B.Z.; All authors have read and agreed to the published version of the manuscript.

Funding: This research received no external funding.

Data Availability Statement: Not applicable.

Conflicts of Interest: The authors declare no conflict of interest.

References

1. Schlesinger, M.; Sole, K.; Davenport, W.; Alvear, G. Chapter 5—Theory to practice: Pyrometallurgical industrial processes. In *Extractive Metallurgy of Copper*, 6th ed.; Elsevier: Oxford, UK, 2021; pp. 95–117. ISBN 9780128218754.
2. Gonzales, T.W.; Walters, G.; White, M. Comparison of Smelting Technologies. In Proceedings of the 58th Annual Conference of Metallurgists (COM) Hosting the 10th International Copper Conference 2019, Vancouver, BC, Canada, 18–21 August 2019; p. 602421.
3. Wang, J. Copper smelting: 2019 world copper smelting data. In Proceedings of the 58th Annual Conference of Metallurgists (COM) Hosting the 10th International Copper Conference 2019, Vancouver, BC, Canada, 18–21 August 2019; p. 592732.
4. Watt, J.; Kapusta, J.P.T. The 2019 copper smelting survey. In Proceedings of the 58th Annual Conference of Metallurgists (COM) Hosting the 10th International Copper Conference 2019, Vancouver, BC, Canada, 18–21 August 2019; p. 595947.
5. Cheng, L.; Wang, Z.; Jiang, J.; Huang, Q. The SKS copper smelting process in China. In *Volume VI-Smelting, Technology Development, Process Modelling and Fundamentals, Proceedings of Copper 99-Cobre International Conference, Phoenix, AZ, USA, 10–13 October 1999*; Diaz, C., Landolt, C., Utigard, T., Eds.; The Minerals, Metals, and Materials Society: Warrendale, PA, USA; pp. 83–91. ISBN 0873394356.
6. Liang, S. A review of oxygen bottom blowing process for copper smelting and converting. In Proceedings of the 9th International Copper Conference (Copper 2016), Kobe, Japan, 13–16 November 2016; pp. 1008–1014.
7. Zhao, B.; Cui, Z.; Wang, Z. A new copper smelting technology—Bottom blown oxygen furnace developed at Dongying Fangyuan Nonferrous Metals. In Proceedings of the 4th International Symposium on High-Temperature Metallurgical Processing, San Antonio, TX, USA, 3–7 March 2013; pp. 3–10.
8. Cui, Z.; Shen, B.; Wang, Z.; Li, W.; Bian, R. New process of copper smelting with oxygen enriched bottom blowing technology. *Nonferr. Met. Extr. Metall.* **2010**, *3*, 17–20. (In Chinese)
9. Chen, Z. The application of oxygen bottom-blown bath smelting of copper. *China Nonferr. Metall.* **2009**, *5*, 16–22. (In Chinese)
10. Qu, S.; Li, T.; Dong, Z.; Luan, H. Plant practice of and design discussion on oxygen enriched bottom blowing smelting. *Nonferr. Met. Extr. Metall.* **2012**, *3*, 10–13. (In Chinese) [CrossRef]
11. Du, X.; Zhao, G.; Wang, H. Industrial application of oxygen bottom-blowing copper smelting technology. *China Nonferr. Metall.* **2018**, *4*, 4–6. (In Chinese) [CrossRef]
12. Wang, H.; Yuan, J.; He, R.; Wang, X. Plant practice of copper oxygen enrichment bottom blowing smelting. *Nonferr. Met. Extr. Metall.* **2013**, *10*, 15–19. (In Chinese)
13. Su, G.L. Discussion on practice of "blowing oxygen at the end of the catch matte" smelting. *China Nonferr. Metall.* **2010**, *4*, 5–8. (In Chinese)
14. Yan, J. Recent operation of the oxygen bottom-blowing copper smelting and continuous copper converting technologies. In Proceedings of the 9th International Copper Conference (Copper 2016), Kobe, Japan, 13–16 November 2016; pp. 13–16.
15. He, S.; Li, D. Discussion on issues related to Shuikoushan Copper Smelting process. *China Nonferr. Metall.* **2005**, *34*, 19–22. (In Chinese)
16. He, S.; Li, D. SKS copper smelting process. *China Nonferr. Metall.* **2006**, *35*, 6–9. (In Chinese)
17. Liang, S.; Chen, Z. Application and development of oxygen enriched bottom-blowing copper smelting technology furnace. *Energy Sav. Nonferr. Metall.* **2013**, *4*, 16–19. (In Chinese)
18. Kapusta, J.P.T. Implementation of air liquid shrouded injector (ALSI) technology at the Thai copper industries smelter. In *2007 International Copper Conference: The Carlos Diaz Symposium On Pyrometallurgy, Toronto, ON, Canada, 25–30 August 2007*; Warner, A.E.M., Newman, C.J., Vahed, A., George, D.B., Mackey, P.J., Warczok, A., Eds.; Canadian Institute of Mining, Metallurgy and Petroleum: Montreal, QC, Canada, 2007; Volume 3, pp. 483–500. ISBN 1-894475-73-9.

19. Kapusta, J.P.T.; Larouche, F.; Palumbo, E. Adoption of high oxygen bottom blowing in copper matte smelting: Why is it taking so long? In *Proceedings of the Torstein Utigard Memorial Symposium, Toronto, Canada, 23–26 August 2015*; The Materials and Metallurgical Society of CIM: Montreal, QC, Canada, 2015; pp. 1–25.

20. Kapusta, J.P.T. Submerged gas jet penetration: A study of bubbling versus jetting and side versus bottom blowing in copper bath smelting. *JOM* **2017**, *69*, 970–979. [CrossRef]

21. Coursol, P.; Mackey, P.J.; Kapusta, J.P.T.; Cardona Valencia, N. Energy consumption in copper smelting: A new Asian horse in the race. *JOM* **2015**, *67*, 1066–1074. [CrossRef]

22. Kapusta, J.P.T. Gas jet penetration, smelting intensity, and oxygen efficiency in side blowing versus bottom blowing. In Proceedings of the 9th International Copper Conference (Copper 2016), Kobe, Japan, 13–16 November 2016; pp. 1282–1305.

23. Kapusta, J.P.T.; Lee, R.G.H. The Savard-Lee shrouded injector: A review of its adoption and adaptation from ferrous to non-ferrous pyrometallurgy. In Proceedings of the 6th International Copper Conference (Copper 2013), Santiago, Chile, 1–4 October 2013; pp. 1115–1151.

24. Chen, M.; Cui, Z.; Wei, C.; Zhao, B. Degradation mechanisms of refractories in a bottom blown copper smelting furnace. In Proceedings of the 9th International Symposium on High-Temperature Metallurgical Processing, Phoenix, AZ, USA, 11–15 March 2018; pp. 149–157.

25. Chen, M.; Jiang, Y.; Cui, Z.; Wei, C.; Zhao, B. Chemical degradation mechanisms of magnesia–chromite refractories in the copper smelting furnace. *JOM* **2018**, *70*, 2443–2448. [CrossRef]

26. Chen, M.; Zhao, B. Investigation of the accretions in the bottom blown copper smelting furnace. In Proceedings of the 9th International Copper Conference (Copper 2016), Kobe, Japan, 13–16 November 2016; pp. 1184–1194.

27. Xu, L.; Chen, M.; Wang, N.; Gao, S. Chemical wear mechanism of magnesia-chromite refractory for an oxygen bottom-blown copper-smelting furnace: A post-mortem analysis. *Ceram. Int.* **2021**, *47*, 2908–2915. [CrossRef]

28. Chen, M.; Cui, Z.; Zhao, B. Slag Chemistry of Bottom Blown Copper Smelting Furnace at Dongying Fangyuan. In Proceedings of the 6th International Symposium on High-Temperature Metallurgical Processing, Orlando, FL, USA, 15–19 March 2015; pp. 257–264.

29. Cui, Z.; Wang, Z.; Zhao, B. Features of the bottom blown oxygen copper smelting technology. In Proceedings of the 4th International Symposium on High-Temperature Metallurgical Processing, San Antonio, TX, USA, 3–7 March 2013; pp. 351–360.

30. Chen, M.; Zhao, B. Comparison of slag chemistry between Teniente converter and flash smelting furnace. In Proceedings of the 9th International Copper Conference (Copper 2016), Kobe, Japan, 13–16 November 2016; pp. 967–975.

31. Bale, C.W.; Bélisle, E.; Chartrand, P.; Decterov, S.A.; Eriksson, G.; Gheribi, A.E.; Hack, K.; Jung, I.H.; Kang, Y.B.; Melançon, J.; et al. Reprint of: FactSage thermochemical software and databases, 2010–2016. *Calphad* **2016**, *55*, 1–19. [CrossRef]

32. Zhao, B.; Hayes, P.C.; Jak, E. Effects of CaO, Al$_2$O$_3$ and MgO on liquidus temperatures of copper smelting and converting slags under controlled oxygen partial pressures. *J. Min. Metall. Sect. B Metall.* **2013**, *49*, 153–159. [CrossRef]

33. Liu, H.; Cui, Z.; Chen, M.; Zhao, B. Phase equilibria in the ZnO-"FeO"-SiO$_2$-CaO system at P$_{O_2}$ 10^{-8} atm. *Calphad* **2018**, *61*, 211–218. [CrossRef]

34. Liu, H.; Cui, Z.; Chen, M.; Zhao, B. Phase equilibria study of the ZnO-"FeO"-SiO$_2$-Al$_2$O$_3$ system at P$_{O_2}$ 10^{-8} atm. *Metall. Mater. Trans. B* **2016**, *47*, 1113–1123. [CrossRef]

35. Liu, H.; Cui, Z.; Chen, M.; Zhao, B. Phase equilibria study of the ZnO-'FeO'-SiO$_2$-MgO system at P$_{O_2}$ 10^{-8} atm. *Miner. Process. Extr. Metall.* **2018**, *127*, 242–249. [CrossRef]

36. Liu, H.; Cui, Z.; Chen, M.; Zhao, B. Phase equilibrium study of ZnO-"FeO"-SiO$_2$ system at fixed P$_{O_2}$ 10^{-8} atm. *Metall. Mater. Trans. B* **2015**, *47*, 164–173. [CrossRef]

37. Chen, M.; Sun, Y.; Balladares, E.; Pizarro, C.; Zhao, B. Experimental studies of liquid/spinel/matte/gas equilibria in the Si-Fe-O-Cu-S system at controlled P(SO$_2$) 0.3 and 0.6 atm. *Calphad* **2019**, *66*, 101642. [CrossRef]

38. Sun, Y.; Chen, M.; Balladares, E.; Pizarro, C.; Contreras, L.; Zhao, B. Effect of MgO on the liquid/spinel/matte/gas equilibria in the Si–Fe–Mg–O–Cu–S system at controlled P(SO$_2$) 0.3 and 0.6 atm. *Calphad* **2020**, *70*, 101803. [CrossRef]

39. Sun, Y.; Chen, M.; Balladares, E.; Pizarro, C.; Contreras, L.; Zhao, B. Effect of CaO on the liquid/spinel/matte/gas equilibria in the Si–Fe–O–Cu–S system at controlled P(SO$_2$) 0.3 and 0.6 atm. *Calphad* **2020**, *69*, 101751. [CrossRef]

40. Guo, X.; Wang, S.; Wang, Q.; Tian, Q. Development and application of oxygen bottom blowing copper smelting simulation software SKSSIM. *Nonferr. Met. Sci. Eng.* **2017**, *27*, 946–953. [CrossRef]

41. Wang, Q.; Guo, X.; Tian, Q.; Jiang, T.; Chen, M.; Zhao, B. Development and application of SKSSIM simulation software for the oxygen bottom blown copper smelting process. *Metals* **2017**, *7*, 431. [CrossRef]

42. Wang, Q.; Guo, X.; Tian, Q.; Chen, M.; Zhao, B. Reaction mechanism and distribution behavior of arsenic in the bottom blown copper smelting process. *Metals* **2017**, *7*, 302. [CrossRef]

43. Wang, Q.; Guo, X.; Tian, Q.; Jiang, T.; Chen, M.; Zhao, B.J. Effects of matte grade on the distribution of minor elements (Pb, Zn, As, Sb, and Bi) in the bottom blown copper smelting process. *Metals* **2017**, *7*, 502. [CrossRef]

44. Wang, Q.; Wang, Q.; Tian, Q.; Guo, X. Simulation study and industrial application of enhanced arsenic removal by regulating the proportion of concentrates in the SKS copper smelting process. *Processes* **2020**, *8*, 385. [CrossRef]

45. Song, K.; Jokilaakso, A. Transport phenomena in copper bath smelting and converting processes–A review of experimental and modeling studies. *Miner. Process. Extr. Metall. Rev.* **2020**, *42*, 107–121. [CrossRef]

46. Wang, D.; Liu, Y.; Zhang, Z.; Shao, P.; Zhang, T. Experimental study of bottom blown oxygen copper smelting process for water model. *AIP Conf. Proc.* **2013**, *1542*, 1304–1307. [CrossRef]
47. Wang, D.; Liu, Y.; Zhang, Z.; Zhang, T.; Li, X. PIV measurements on physical models of bottom blown oxygen copper smelting furnace. *Can. Metall. Q.* **2017**, *56*, 221–231. [CrossRef]
48. Shui, L.; Cui, Z.; Ma, X.; Rhamdhani, M.; Nguyen, A.; Zhao, B. Mixing phenomena in a bottom blown copper smelter: A water model study. *Metall. Mater. Trans. B* **2015**, *46*, 1218–1225. [CrossRef]
49. Shui, L.; Cui, Z.; Ma, X.; Jiang, X.; Chen, M.; Xiang, Y.; Zhao, B. A Water model study on mixing behavior of the two-layered bath in bottom blown copper smelting furnace. *JOM* **2018**, *70*, 2065–2070. [CrossRef]
50. Jiang, X.; Cui, Z.; Chen, M.; Zhao, B. Mixing behaviors in the horizontal bath smelting furnaces. *Metall. Mater. Trans. B* **2018**, *50*, 173–180. [CrossRef]
51. Ma, X.; Cui, Z.; Contreras, L.; Jiang, X.; Chen, M.; Zhao, B. Fluid dynamics studies of bottom-blown and side-blown copper smelting furnaces. In Proceedings of the 58th Annual Conference of Metallurgists (COM) Hosting the 10th International Copper Conference 2019, Vancouver, BC, Canada, 18–21 August 2019; pp. 1–8.
52. Shui, L.; Ma, X.; Cui, Z.; Zhao, B. An investigation of the behavior of the surficial longitudinal wave in a bottom-blown copper smelting furnace. *JOM* **2018**, *70*, 2119–2127. [CrossRef]
53. Shui, L.; Cui, Z.; Ma, X.; Rhamdhani, M.A.; Nguyen, A.V.; Zhao, B. Understanding of bath surface wave in bottom blown copper smelting furnace. *Metall. Mater. Trans. B* **2015**, *47*, 135–144. [CrossRef]
54. Jiang, X.; Cui, Z.; Chen, M.; Zhao, B. Study of Plume eye in the copper bottom blown smelting furnace. *Metall. Mater. Trans. B* **2019**, *50*, 782–789. [CrossRef]
55. Wang, D.; Liu, T.; Liu, Y.; Zhu, F. Water model study of bubble behavior in matter smelting process with oxygen bottom blowing. *J. Northeast. Univ. Nat. Sci.* **2013**, *34*, 1755–1758.
56. Wang, D.; Liu, Y.; Zhang, Z.; Shao, P.; Zhang, T. Dimensional analysis of average diameter of bubbles for bottom blown oxygen copper furnace. *Math. Probl. Eng.* **2016**, *2016*, 4170371. [CrossRef]
57. Cheng, X.; Cui, Z.; Contreras, L.; Chen, M.; Nguyen, A.; Zhao, B. Introduction of matte droplets in copper smelting slag. In Proceedings of the 8th International Symposium on High-Temperature Metallurgical Processing, San Diego, CA, USA, 26 February–2 March 2017; pp. 385–394.
58. Cheng, X.; Cui, Z.; Contreras, L.; Chen, M.; Nguyen, A.; Zhao, B. Matte entrainment by SO_2 bubbles in copper smelting Slag. *JOM* **2019**, *71*, 1897–1903. [CrossRef]
59. Luo, Q.; Yan, H.; Jin, J.; Huang, Z.; Gong, H.; Liu, L. Gold model experiment on fluctuation behavior of liquid level in setting zone of bottom-blowing furnace. *Nonferr. Met. Eng.* **2020**, *10*, 46–53. [CrossRef]
60. Yan, H.; Liu, F.; Zhang, Z.; Gao, Q.; Liu, L.; Cui, Z.; Shen, D. Influence of lance arrangement on bottom-blowing bath smelting process. *Chin. J. Nonferr. Met.* **2012**, *22*, 2393–2400. [CrossRef]
61. Zhang, Z.; Yan, H.; Liu, F.; Wang, J. Optimization analysis of lance structure parameters in oxygen enriched bottom-blown furnace. *Chin. J. Nonferr. Met.* **2013**, *23*, 1471–1478. [CrossRef]
62. Guo, X.; Yan, S.; Wang, Q.; Tian, Q. Layout optimization of oxygen lances of oxygen bottom blown furnace. *Chin. J. Nonferr. Met.* **2018**, *28*, 2450–2550. [CrossRef]
63. Zhang, Z.; Chen, Z.; Yan, H.; Liu, F.; Liu, L.; Cui, Z.; Shen, D. Numerical simulation of gas-liquid multi-phase flows in oxygen enriched bottom-blown furnace. *Chin. J. Nonferr. Met.* **2012**, *22*, 1826–1834. [CrossRef]
64. Tang, G.; Silaen, A.K.; Yan, H.; Cui, Z.; Wang, Z.; Wang, H.; Tang, K.; Zhou, P.; Zhou, C. CFD study of gas-liquid phase interaction inside a submerged lance smelting furnace for copper smelting. In Proceedings of the 8th International Symposium on High-Temperature Metallurgical Processing, San Diego, CA, USA, 26 February–2 March 2017; pp. 101–111.
65. Li, D.; Dong, Z.D.; Xin, Y.; Cheng, L.; Guo, T.; Li, B.; Li, P. Research of gas–liquid multiphase flow in oxygen-enriched bottom blowing copper smelting furnace. In Proceedings of the 11th International Symposium on High-Temperature Metallurgical Processing, San Diego, CA, USA, 23–27 February 2020; pp. 975–986.
66. Shao, P.; Jiang, L. Flow and mixing behavior in a new bottom blown copper smelting furnace. *Int. J. Mol. Sci.* **2019**, *20*, 5757. [CrossRef]
67. Guo, X.; Wang, Q.; Tian, Q.; Zhang, Y. Mechanism and multiphase interface behavior of copper sulfide concentrate smelting in oxygen-enriched bottom blowing furnace. *Nonferr. Met. Sci. Eng.* **2014**, *5*, 28–34. [CrossRef]
68. Guo, X.; Wang, Q.; Tian, Q.; Zhao, B. Performance analysis and optimization of oxygen bottom blowing copper smelting process. *Chin. J. Nonferr. Met.* **2016**, *26*, 689–698. [CrossRef]
69. Cai, C.; Liang, Y.; Qian, Z. A model study on the continuous leadmaking process with bottom oxygen injection. *Chin. J. Process. Eng.* **1985**, *4*, 113–121. (In Chinese)

Article

Enhanced Productivity of Bottom-Blowing Copper-Smelting Process Using Plume Eye

Jinfa Liao [1], Keqin Tan [2] and Baojun Zhao [1,3,*]

1   Faculty of Materials Metallurgy and Chemistry, Jiangxi University of Science and Technology, Ganzhou 341000, China
2   Dongying Fangyuan Nonferrous Metals, Dongying 257000, China
3   Sustainable Minerals Institute, University of Queensland, Brisbane 4072, Australia
*   Correspondence: bzhao@jxust.edu.cn

Abstract: Bottom-blowing copper smelting is a bath smelting technology recently developed in China. It has the advantages of good adaptability of raw materials, high oxygen utilization and thermal efficiency, and flexible production capacity. Plume eye is a unique phenomenon observed in the bottom-blowing copper-smelting furnace where the slag on the surface of the bath is pushed away by the high-pressure gas injected from the bottom. The existence of plume eye was first confirmed by analyzing the quenched industrial samples collected above the gas injection area and then investigated by laboratory water model experiments. Combining the plant operating data and the smelting mechanism of the copper concentrate, the role of the plume eye in bottom-blowing-enhanced smelting is analyzed. It reveals that the direct dissolution of copper concentrate as a low-grade matte into the molten matte can significantly accelerate the reactions between the concentrate and oxygen. The productivity of the bottom-blowing furnace is therefore increased as a result. The effects of the gas flow rate and thickness of the matte and of the slag layer on the diameter of the plume eye were studied using water-model experiments. It was found that increasing the gas flow and the thickness of the matte and reducing the thickness of the slag can increase the diameter of the plume eye. This work is of great significance for further understanding the copper bottom-blowing smelting technology and optimizing industrial operations.

Keywords: copper metallurgy; plume eye; bottom-blowing copper smelting; water model

Citation: Liao, J.; Tan, K.; Zhao, B. Enhanced Productivity of Bottom-Blowing Copper-Smelting Process Using Plume Eye. *Metals* **2023**, *13*, 217. https://doi.org/10.3390/met13020217

Academic Editor: Antoni Roca

Received: 20 December 2022
Revised: 13 January 2023
Accepted: 19 January 2023
Published: 23 January 2023

1. Introduction

The concept of the plume eye was first proposed in steelmaking [1–9]. Plume eye was a description of the exposure of lower-layer liquid due to upper-layer liquid being pushed away because of high-pressure gas injection [2]. In a steel ladle, a large plume eye is not desirable because it can result in oxygen and nitrogen pick up by the steel, which can affect the steel quality [3]. Similarly, a large plume eye in steel tundish can lead to the formation of inclusion and slag entrainment because of the metal-liquid re-oxidation reactions [4]. To investigate the nature of the plume eye in steel ladle, the researchers used water to simulate the metal and oil to simulate slag [2,4–7]. The control of the plume eye size was the primary interest in the steel ladle [3,4,8,9].

Bottom blowing is a copper-smelting technology that was proposed in China in the 1990s [10]. Its unique technical characteristics, including good raw material adaptability, high oxygen utilization rate and thermal efficiency, and flexible production capacity [11–15], have attracted strong interest in the copper industry [16–24]. From November 1991 to June 1992, a plant trial was conducted at Shuikoushan (SKS) Smelter for 217 days to smelt copper concentrate in a bottom-blowing furnace [16,17]. In 2005, the first industrial scope BBF (Bottom-blowing furnace) was built in the Sin Quyen smelter in Vietnam. The size of the furnace was Ø3.8 m × 11.5 m and the capacity was designed to be 10,000 t Cu per year [18]. However, the BBF in the Sin Quyen smelter seemed to be operating incorrectly, as no operating details were reported. In 2008, the first real commercialized BBF started in Dongying Fangyuan Nonferrous Metals (Fangyuan, Shangdong) [19,20]. The oxygen-enriched bottom-blown matte smelting of Shandong Humon Smelting Co., Ltd. (Yantai, China) was successfully put into production in April 2010. After nearly five months of trial production, the unique advantages of this process in energy conservation, environmental protection, safety, and other aspects have been fully demonstrated [21]. In 2011, the bottom-blown furnace (Ø3.8 m × 13.5 m) of Baotou Huading Copper Industry Development Co., Ltd. (Baotou, China) was officially put into operation, with an average feeding capacity of 28 t/h and a matte grade of about 50% [22,23]. Thirteen BBS furnaces were constructed or were under construction with the capacity of 1600 kt/a copper production until 2016. The furnace size ranged from Ø3.8 m × 11.5 m to Ø5.8 m × 30 m [24]. During the operation of oxygen bottom-blowing bath smelting, high-pressure oxygen-enriched air is injected into liquid matte from the bottom to provide oxygen for copper-concentrate smelting reactions and to strongly stir the bath. It has been observed by the authors directly from the feeding mouth that the plume eye exists in the oxygen bottom-blowing furnace (BBF). The high-pressure oxygen-enriched air pushes away the slag layer on the surface of the bath above the oxygen lance area, exposing the matte to the feeding mouth. The copper-smelting furnace is a horizontal cylindrical vessel, which is geometrically different from a steelmaking ladle or tundish. In addition, the upper layer in the steel ladle or tundish is skinny in comparison to the lower-layer thickness, but the upper slag layer is considerably thick in copper-smelting furnaces. Therefore, the behavior of the plume eye in a copper-smelting furnace is different from that in a steelmaking ladle or tundish. The fundamental studies, including the thermodynamics of the slag and fluid dynamic of the molten bath, have been extensively conducted in recent years to understand and support the new technology [10,25,26]. However, few studies were reported on the plume eye in the bottom-blowing copper-smelting process. Jiang et al. [27] studied the influencing factors of plume eye using the laboratory water model.

This study aims to confirm the existence of plume eye and investigate the reaction mechanisms in a copper bottom-blowing smelting furnace by analyzing the industrial samples collected above the reaction region. The parameters to control the size of the plume eye will be studied using laboratory water-model experiments to provide information for industry operation.

2. Research Methods

2.1. Sampling and Analysis of Industrial Samples

Figure 1 shows a schematic diagram of an oxygen bottom-blowing smelting furnace. The industrial samples were collected from a bottom-blowing copper-smelting furnace (φ3.3 m × 15 m) in Dongying Fangyuan Nonferrous Metals Co., Ltd., Dongying City, China. [10]. A long cold iron rod is stretched into the furnace from the feeding mouth and the sample sprayed above the oxygen lance area stuck on the iron rod and cooled rapidly. The slag sample was collected from the tapping hole using a long cold iron rod dipped into the slag flow. All the quenched samples, including the splash sample and slag sample, were mounted in epoxy resin, polished, and carbon coated for electron probe microanalysis (EPMA). A JXA 8200 Electron Probe Microanalyser (Japan Electron Optics Ltd., Tokyo, Japan) with Wavelength Dispersive Spectroscopy (WDS, Japan Electron Optics Ltd., Tokyo, Japan) was used for the microstructural and compositional analyses. The EPMA was operated on at an accelerating voltage of 15 kV and a probe current of 15 nA. The beam size was set to "0 μm" to accurately measure the composition of an area larger than 1 μm. The ZAF (Z is atomic number correction factor, A is absorption correction factor, and F is fluorescence correction factor) correction procedure was applied for the data analysis. "The standards used for the analysis were from Charles M. Taylor Co. (Stanford, CA, USA): Al_2O_3 for Al, MgO for MgO, Fe_2O_3 for Fe, Cu_2O for Cu, $KAlSi_3O_8$ for K, PbS for Pb, GaAs for As, $CaMoO_4$ for Mo, $CaSiO_3$ for Si and Ca and $CuFeS_2$ for Cu Fe and S, and Micro-Analysis Consultant Ltd.

(Cambridge, UK): ZnO for Zn". The average accuracy of the composition measured using EPMA is within 1 wt%.

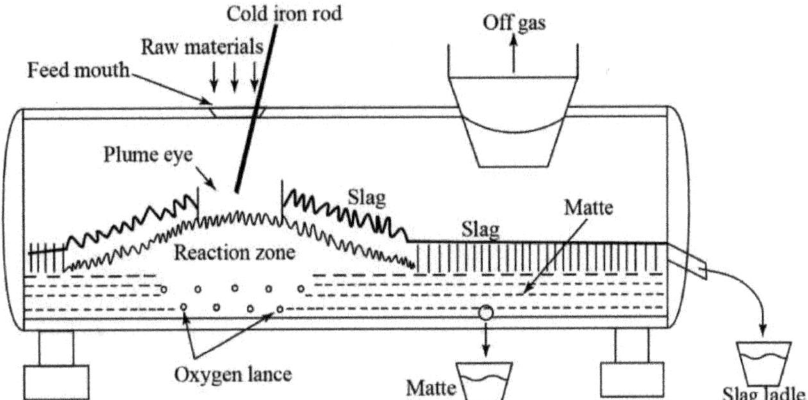

Figure 1. Schematic diagram of oxygen bottom-blowing smelting furnace.

2.2. Water Model Experiment

A horizontal cylindrical reactor was designed according to the ratio of 1:10 of the size of the industrial bottom-blowing copper-smelting furnace. The front view and side view of the reactor are shown in Figure 2a and b, respectively. In this study, high-pressure air was injected into the furnace through the bottom lance. Water and silicone oil at 25 °C have similar kinematic viscosities to matte (8×10^{-7} m^2/s) and slag (6×10^{-6} m^2/s) at 1200 °C [28]. The ratio of water density to silicon oil density is close to that of the matte density to slag density. Water and silicone oil were used to simulate matte and slag, respectively. Low density silicone oil located on the upper part of the molten pool can be a good observation of "plume eye".

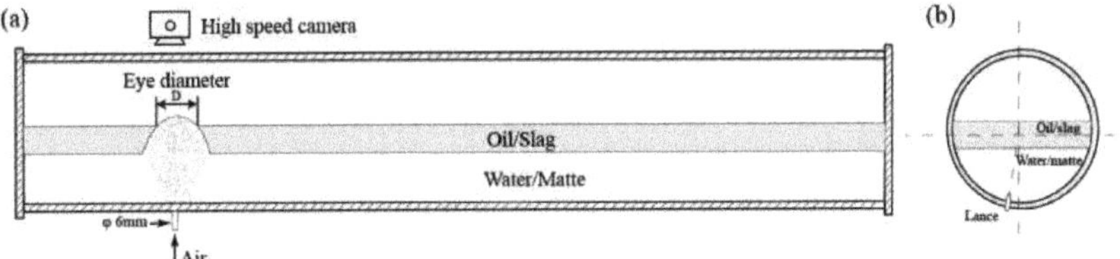

Figure 2. Schematic diagram of water model experiment: (**a**) front view; (**b**) side view.

At the beginning of each experiment, the liquid bath was flashed with high-pressure air for 20 min to stabilize the flow field of the bath in the furnace. The schematic diagram of the plume eye generated in the model furnace is shown in Figure 3. Since the upper liquid is yellow silicone oil, it is easy to distinguish the shape and size of the plume eye. The plume eye was pictured with a high-speed camera and then the diameter Di of the "plume eye" was measured compared to the width of the upper liquid (D5) from the pictures. By measuring the ratio of the real width of the slag to D5, the diameter of the actual plume eye in the water model can be obtained. The diameter D of the plume eye was determined by calculating the average of ten measures in different directions.

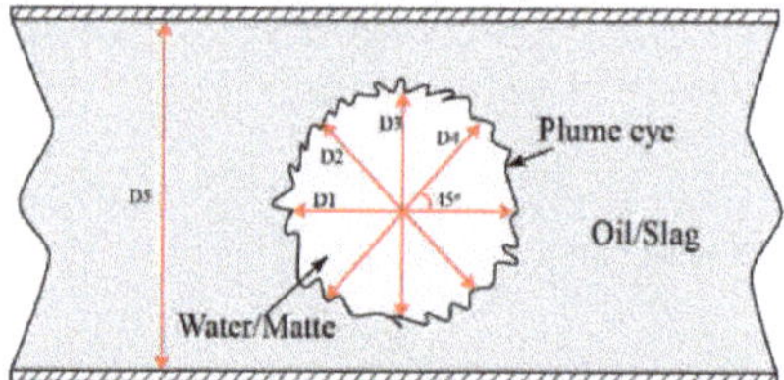

Figure 3. Schematic diagram of measuring the diameter of "plume eye".

3. Results

3.1. Analysis of Industrial Samples

Figure 4 shows a typical microstructure of the quenched slag extracted from the tapping hole. The temperature of the tapping slag was approximately 1180 °C. It can be seen from the figure that liquid, spinel, and matte existed in the quenched slag. Spinel is a primary phase that retained its shapes and compositions on quenching. The presence of the spinel phase indicates that the operating temperature of the furnace (1180 °C) was lower than the liquidus temperature (1350 °C) of the slag, which is the feature of the BBF [29].

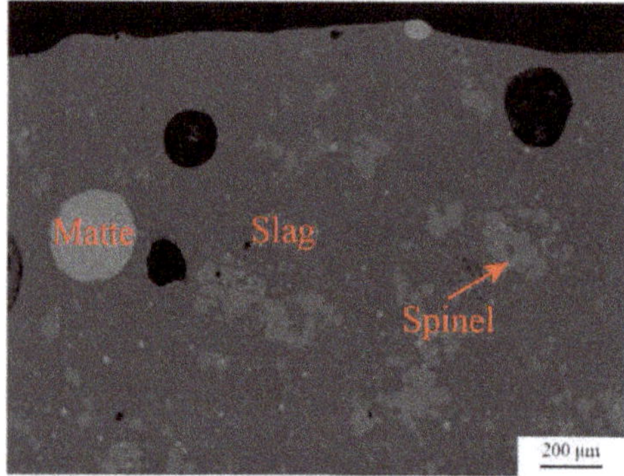

Figure 4. Typical microstructure of a quenched oxygen bottom-blowing slag.

Table 1 provides the compositions of the liquid, spinel, and matte phases measured using EPMA. It can be seen that the spinel mainly contains "FeO" (FeO_x) causing the Fe/SiO_2 ratio in the liquid phase to be lower than that of the bulk slag. In addition to "FeO" and SiO_2, ZnO, Al_2O_3, CaO, MgO, and K_2O also exist in the liquid. It is noted that the dissolved Cu_2O in the liquid phase is only 0.2 wt%. The matte grade entrained in the slag is consistent with the matte grade collected at the matte tapping hole, both of which are 70% Cu.

Table 1. Phase compositions of oxygen bottom-blowing copper-smelting slag measured by EPMA.

Phase	"FeO"	Cu_2O	CaO	SiO_2	Al_2O_3	As_2O_3	MgO	S	PbO	ZnO	MoO_3	K_2O	Fe/SiO_2
Spinel	94.5	0.0	0.0	0.4	2.8	0.0	0.3	0.0	0.0	1.5	0.1	0.0	
Slag	56.9	0.2	1.1	32.1	4.1	0.1	0.8	0.0	0.6	2.5	0.2	1.4	1.38

Phase	Fe	Cu	S	As_2O_3	PbO	ZnO	MoO_3
Matte	2.4	70.1	21.7	0.1	0.1	0.1	0.5

Figure 5 shows the typical microstructures of the splashed samples extracted simultaneously above the lance area. The compositions of all the phases present in the splashed samples are presented in Tables 2 and 3. It can be seen from Figure 5a–c that the matte phase dominates the splashed samples confirming the existence of the plume eye in the BBF.

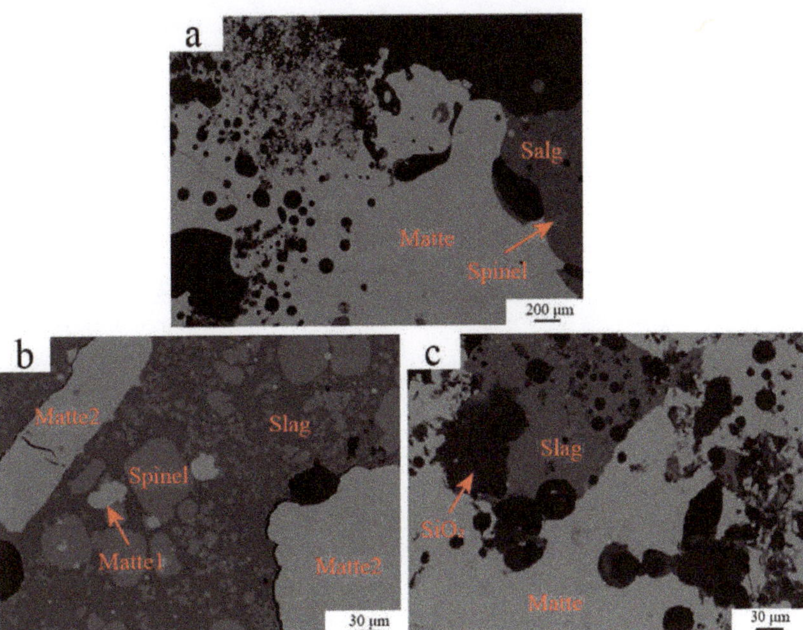

Figure 5. Typical microstructures of the splash samples. (**a**) Sample 1#, (**b**) Sample 2#, and (**c**) Sample 3#.

Table 2. Phase compositions of liquid slag, spinel, and SiO_2 in the splashed samples measured using EPMA.

Sample	Phase	Composition (wt.%)												
		FeO	Cu_2O	CaO	SiO_2	Al_2O_3	As_2O_3	MgO	S	PbO	ZnO	MoO_3	K_2O	Fe/SiO_2
1#	spinel	92.8	0.1	0	0.5	4.1	0	0.3	0	0	1.3	0.2	0	-
	liquid	57.2	0.4	1.2	31.3	4.2	0.1	0.7	0.1	0.4	2.5	0.3	1.3	1.42
2#	spinel	92.9	0.1	0	0.5	3.9	0	0.3	0	0.1	1.5	0	0	-
	liquid	48.6	1.6	1.4	38.1	4.1	0.2	1	0.3	0.5	2.5	0.2	1.5	0.99
3#	SiO_2	4.8	0.5	0.2	93.2	0.4	0	0.2	0	0.1	0.2	0.1	0.3	-
	liquid	47.5	0.4	1.7	40	4.1	0.2	1.1	0.1	0.5	2.8	0.2	1.3	0.92

Table 3. Matte compositions in the splash samples measured using EPMA.

Sample	Phase	Composition (wt.%)						
		Fe	Cu	S	As	Pb	Zn	Mo
1#	matte	2.2	70.4	22.5	0.0	0.1	0.1	0.2
2#	matte 1	2.2	70.6	21.9	0.0	0	0.0	0.2
	matte 2	5.9	64.2	21.4	0.2	0.7	0.1	0.1
3#	matte	4.5	66.6	24.2	0	0.0	0.1	0.2

Figure 5a shows a larger matte together with slag is present. It is seen from Tables 2 and 3 that the Fe/SiO_2 ratio in the liquid slag is 1.42 and the matte grade is 70.4% in the splashed sample 1#. These values are close to those in the final slag shown in Table 1, indicating that the splashed sample 1# is approaching the equilibrium. Figure 5b shows the existence of large and small pieces of matte together with the slag. A large proportion of solid spinel is present in the slag. The Fe/SiO_2 of the liquid slag in the splash sample 2# is 0.99, which is much lower than that in the final slag. In addition, the large matte phase contains 64.2% Cu, which is lower than that in the final matte. Therefore, the splash sample 2# is still in the matte-forming reaction stage. Further oxidation of the low-grade matte will increase the Fe/SiO_2 ratio in the slag and the Cu concentration in the matte. In sample 3#, the undissolved SiO_2 is together with the large matte as shown in Figure 5c. The Fe/SiO_2 in the splash sample 3# is 0.92 and the matte grade is 66.2. It seems that the local SiO_2 concentration in the slag is too high and SiO_2 is the primary phase. Further oxidation of the low-grade matte will increase the "FeO" concentration in the slag and dissolve the SiO_2 phase.

The analyses of the splash samples collected above the lance area indicate that the equilibrated and unequilibrated matte were brought to the surface of the bath by the high-pressure gas. A plume eye exits in this area without slag coverage above the matte. The low-grade matte and low Fe/SiO_2 slag in the splash samples indicate that the smelting reactions are mainly inside the matte layer since the oxygen is injected from the bottom of the bath and is almost completely consumed in the matte.

3.2. The Role of Plume Eye in Copper Bottom-Blowing Smelting

The presence of the plume eye in the bottom-blowing smelting furnace plays an important role to enhance the reactions between the feeding materials and oxygen. It can be seen from Figure 1 that the feeding mouth is directly above the lance area in the BBF. Raw materials such as copper concentrate and SiO_2 flux can directly fall into the molten bath area with gas injection underneath. The reaction mechanisms with and without a plume eye are discussed below.

Figure 6a shows a schematic diagram of the smelting reaction zone of BBF without a plume eye. If the matte is covered by the slag (without plume eye), the copper concentrate and SiO_2 flux fall into the liquid slag. Because the sulfides are immiscible with the slag, which does not contain excess oxygen, the sulfides do not react with the slag directly. The liquid sulfides must pass through the viscus slag layer to enter the matte layer where several phases such as Cu_2O, Cu_2S, and high-grade matte can react with the concentrate (low-grade matte) to form a final matte. The fine sulfide particles melted to small liquid droplets that require a long time to pass through the slag layer. The SiO_2 stays in the slag layer and reacts with liquid slag to form a slag with low Fe/SiO_2.

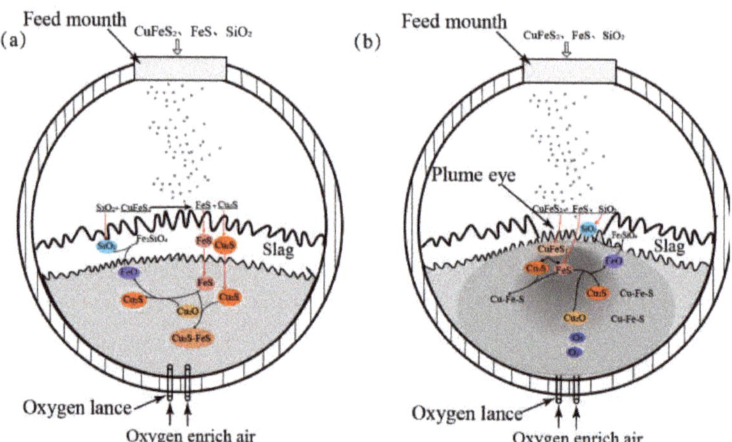

Figure 6. Schematic diagram of smelting reaction zone of oxygen-enriched bottom-blowing furnace (**a**) without "plume eye" and (**b**) with "plume eye".

The schematic diagram of the smelting reaction zone of a BBF with a plume eye is shown in Figure 6b. In this case, the copper concentrate as a low-grade matte falls into the plume eye and contacts with the matte directly. The copper concentrate quickly dissolves into the matte and oxidizes further to form the final matte. The "FeO" formed as a result of the oxidation floats up and reacts with the slag and the SiO_2 around the plume eye to form the final slag. This reaction mechanism can explain why the splash samples contain different grades of matte, as shown in Table 3. The direct dissolution of copper concentrate into the matte through a plume eye significantly increases the overall reaction rate of the copper-concentrate smelting.

3.3. Effects of Operating Conditions on the Plume Eye Diameter

According to the analyses of the industrial samples combined with the study of the reaction mechanism, a plume eye is confirmed to be present and plays an important role in copper bottom-blowing smelting. Controlling the size of the plume eye will provide useful guidelines for industrial operation. The effects of the gas flowrate rate, slag-layer thickness, and matte-layer thickness on the diameter of the plume eye were studied using water-model experiments. Figure 7a shows the influence of the matte layer thickness on the diameter of the plume eye at a fixed slag-layer thickness of 4 cm. It can be seen from the figure that, at a given gas flowrate, the diameter of the plume eye increases with increasing the ratio of matte thickness to slag thickness. The greater difference in thickness between the matte and the slag, the greater the increase in the diameter of the plume eye. This trend is generally consistent with the results of the steel ladle study [7]. On the other hand, the plume eye diameter increases with increasing the gas flowrate.

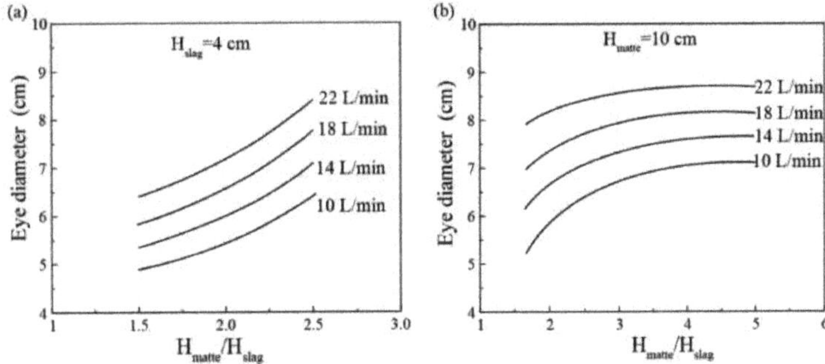

Figure 7. Effect of matte-to-slag ratio on plume eye diameter under different gas flow, (**a**) fixed slag thickness 4 cm and (**b**) fixed matte thickness 10 cm.

Figure 7b shows the influence of slag thickness on the diameter of the plume eye when the thickness of the matte is fixed to 10 cm. At a given gas flowrate, the diameter of the plume eye increases with the ratio of the matte thickness to the slag thickness. However, the increase in the plume eye diameter is not sensitive with a high thickness of the slag layer.

The effect of the gas flowrate on the plume eye diameter is shown in Figure 8 at a fixed ratio of matte thickness to slag thickness. It can be seen that the plume eye diameter is increased linearly with the increase in the gas flowrate at a fixed ratio of water thickness to oil thickness. The gas flowrate per unit area in the laboratory condition is similar to that in the industrial furnace. With a higher-pressure gas injected into the molten bath from the bottom, more dynamic energy will be provided to push the upper layer away.

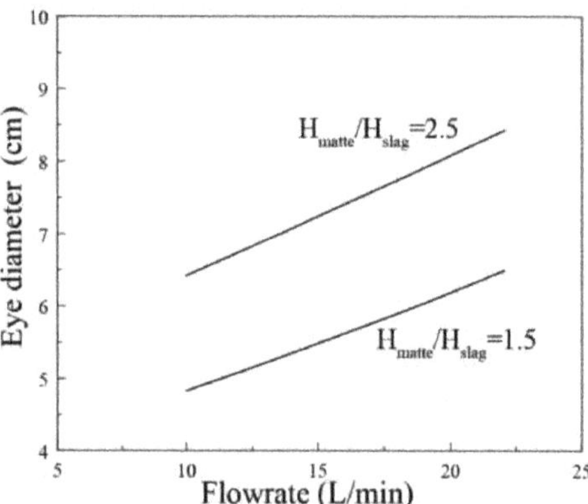

Figure 8. Effect of gas flowrate on the diameter of plume eye at fixed matte-thickness-to-slag-thickness ratio.

4. Discussions

In the copper bottom-blowing smelting furnace, raw materials are fed through a feeding port above the lance area. The presence of a plume eye in the lance area enables the feeds to fall directly into the matte through the plume eye causing the reaction mechanism to be different to previous works. Wang et al. [26] proposed a copper-smelting mechanism in an oxygen bottom-blowing furnace. In their mechanism model, the reaction zone of a BBF is divided into seven functional layers from top to bottom, i.e., gas layer, mineral decomposition transitioning layer, slag layer, slag formation transitioning layer, matte formation transitioning layer, weak oxidizing layer, and strong oxidizing layer. The presence of a plume eye in a BBF improves the understanding of the copper-smelting mechanism. The slag layer is not present above the gas injection zone in a bottom-blowing copper-smelting furnace. Therefore, the slag layer, slag formation transitioning layer, and matte formation transitioning layer proposed by Wang et al. [26] need to be removed and a new reaction mechanism is proposed based on the present study.

The copper concentrate as a low-grade matte rapidly dissolves into the oxygen-rich matte in the lance area. A large plume eye can improve the reaction efficiency and increase the productivity of the furnace. This mechanism explained that the productivity of the first BBF was doubled after a few months' operation [18]. More importantly, with an increased number of copper bottom-blowing smelting furnaces, not all feeding mouths were designed exactly above the gas injection zone. The present study indicates that the location and size of the feeding mouth need to be considered to ensure the raw materials to be dropped directly into the plume eye. The water-model experimental study on the plume eye provides the quantitative effect of different parameters on the size of the plume eye, which will help the copper industry to optimize the operation parameters to control the required plume eye diameter. In the operation of a BBF, the diameter of the plume eye can be increased by increasing the oxygen-enriched air pressure or the thickness of the matte layer. Reducing the thickness of the slag layer can also increase the diameter of the plume eye to a lesser extent. The diameter of the plume eye needs to be adjusted to enable the raw materials from the feeding mouth to be dropped into the matte layer directly as the size of the feeding mouth is fixed.

5. Conclusions

In this study, the existence of plume eyes in a copper bottom-blowing furnace was confirmed by analyzing industrial samples and laboratory water-model experiments. It was found that the splashed samples above the lance zone of the BBF were mainly matte and had a small amount of slag with high SiO_2. The grade of the matte in the splashed sample and the Fe/SiO_2 ratio of the liquid slag are generally lower than those in the final products. The existence of the plume eye enables the copper concentrate to directly dissolve into the matte, which improves the efficiency of the matte-producing reaction and increases the productivity of the bottom-blowing smelting furnace. The water-model experiments show that, when the thickness of the slag layer is constant, the thicker the matte layer, the larger the diameter of the plume eye. When the thickness of the matte layer is constant, the higher the thickness of the slag layer, the smaller the diameter of the plume eye. The plume eye diameter increases significantly with the increase in the gas flowrate. At a given gas flowrate, the diameter of the plume eye increases with the ratio of the matte thickness to the slag thickness. This research is of great significance for further understanding the strengthening smelting mechanism of the BBF and optimizing the industrial operations.

Author Contributions: Formal analysis, J.L.; resources, B.Z. and K.T.; data curation, J.L.; writing—original draft, J.L.; writing—review & editing, B.Z. and K.T.; supervision, B.Z.; project administration, K.T. and B.Z. All authors have read and agreed to the published version of the manuscript.

Funding: This research received no external funding.

Data Availability Statement: Not applicable.

Acknowledgments: The authors would like to acknowledge Dongying Fangyuan Nonferrous Metals Co., Ltd. for providing the financial support and the industrial samples for this research.

Conflicts of Interest: The authors declare no conflict of interest. The funders had no role in the design of the study; in the collection, analyses, or interpretation of data; in the writing of the manuscript; or in the decision to publish the results.

References

1. Conejo, A.N.; Feng, W. Ladle Eye Formation Due to Bottom Gas Injection: A Reassessment of Experimental Data. *Metall. Mater. Trans. B* **2022**, *53*, 999–1017. [CrossRef]
2. Iguchi, M.; Miyamoto, K.; Yamashita, S.; Iguchi, D.; Zeze, M. Spout eye area in ladle refining process. *ISIJ Int.* **2004**, *44*, 636–638. [CrossRef]
3. Krishnapisharody, K.; Irons, G. Modeling of slag eye formation over a metal bath due to gas bubbling. *Metall. Trans. B* **2006**, *37*, 763–772. [CrossRef]
4. Chatterjee, S.; Chattopadhyay, K. Formation of slag 'eye' in an inert gas shrouded tundish. *ISIJ Int.* **2015**, *55*, 1416–1424. [CrossRef]
5. Yonezawa, K.; Schwerdtfeger, K. Spout eyes formed by an emerging gas plume at the surface of a slag-covered metal melt. *Metall. Trans. B* **1999**, *30*, 411–418. [CrossRef]
6. Peranandhanthan, M.; Mazumdar, D. Modeling of slag eye area in argon stirred ladles. *ISIJ Int.* **2010**, *50*, 1622–1631. [CrossRef]
7. Lv, N.; Wu, L.; Wang, H.; Dong, Y.; Su, C. Size analysis of slag eye formed by gas blowing in ladle refining. *JISR Int.* **2017**, *24*, 243–250. [CrossRef]
8. Han, J.; Heo, S.; Kam, D.; You, B.; Pak, J.; Song, H. Transient fluid flow phenomena in a gas stirred liquid bath with top oil layer—Approach by numerical simulation and water model Experiments. *ISIJ Int.* **2001**, *41*, 1165–1172. [CrossRef]
9. Mazumdar, D.; Evans, J. Communications: A model for estimating exposed plume eye area in steel refining ladles covered with thin slag. *Metall. Trans. B* **2004**, *35*, 400–404. [CrossRef]
10. Zhao, B.; Liao, J. Development of bottom-blowing copper smelting technology: A review. *Metals* **2022**, *12*, 190. [CrossRef]
11. Kapusta, J.P.T. Submerged gas jet penetration: A study of bubbling versus jetting and side versus bottom blowing in copper bath smelting. *JOM* **2017**, *69*, 970–979. [CrossRef]
12. Chen, M.; Jiang, Y.; Cui, Z.; Wei, C.; Zhao, B. Chemical degradation mechanisms of magnesia–chromite refractories in the copper smelting furnace. *JOM* **2018**, *70*, 2443–2448. [CrossRef]
13. Xu, L.; Chen, M.; Wang, N.; Gao, S. Chemical wear mechanism of magnesia-chromite refractory for an oxygen bottom-blowing copper-smelting furnace: A post-mortem analysis. *Ceram. Int.* **2021**, *47*, 2908–2915. [CrossRef]
14. Chen, M.; Cui, Z.; Zhao, B. Slag Chemistry of Bottom Blowing Copper Smelting Furnace at Dongying Fangyuan. In Proceedings of the 6th International Symposium on High-Temperature Metallurgical Processing, Orlando, FL, USA, 15–19 March 2015; pp. 257–264.
15. Liang, S. A review of oxygen bottom blowing process for copper smelting and converting. In Proceedings of the 9th International Copper Conference (Copper 2016), Kobe, Japan, 13–16 November 2016; pp. 1008–1014.
16. He, S.; Li, D. Discussion on issues related to Shuikoushan Copper Smelting process. *China Nonferrous Metall.* **2005**, *34*, 19–22. (In Chinese)
17. Chen, Z. The application of oxygen bottom-blowing bath smelting of copper. *China Nonferrous Metall.* **2009**, *5*, 16–22. (In Chinese)
18. Liang, S.; Chen, Z. Application and development of oxygen enriched bottom-blowing copper smelting technology furnace. *Energy Sav. Nonferr. Metall.* **2013**, *4*, 16–19. (In Chinese)

19. Cui, Z.; Wang, Z.; Zhao, B. Features of the bottom blowing oxygen copper smelting technology. In Proceedings of the 4th International Symposium on High-Temperature Metallurgical Processing, San Antonio, TX, USA, 3–7 March 2013; pp. 351–360.

20. Zhao, B.; Cui, Z.; Wang, Z. A new copper smelting technology—Bottom blowing oxygen furnace developed at Dongying Fangyuan Nonferrous Metals. In Proceedings of the 4th International Symposium on High-Temperature Metallurgical Processing, San Antonio, TX, USA, 3–7 March 2013; pp. 3–10.

21. Qu, S.; Li, T.; Dong, Z.; Luan, H. Plant practice of and design discussion on oxygen enriched bottom blowing smelting. *Nonferrous Met. Extr. Metall.* **2012**, *3*, 10–13. (In Chinese)

22. Du, X.; Zhao, G.; Wang, H. Industrial application of oxygen bottom-blowing copper smelting technology. *China Nonferrous Metall.* **2018**, *4*, 4–6. (In Chinese)

23. Wang, H.; Yuan, J.; He, R.; Wang, X. Plant practice of copper oxygen enrichment bottom blowing smelting. *Nonferrous Met. Extr. Metall.* **2013**, *10*, 15–19. (In Chinese)

24. Yan, J. Recent operation of the oxygen bottom-blowing copper smelting and continuous copper converting technologies. In Proceedings of the 9th International Copper Conference (Copper 2016), Kobe, Japan, 13–16 November 2016; pp. 13–16.

25. Song, K.; Jokilaakso, A. Transport phenomena in copper bath smelting and converting processes–A review of experimental and modeling studies. *Miner. Process. Extr. Metall. Rev.* **2020**, *42*, 107–121.

26. Wang, Q.; Guo, X.; Tian, Q. Copper smelting mechanism in oxygen bottom-blowing furnace. *Trans. Nonferrous Met. Soc. China* **2017**, *27*, 946–953. [CrossRef]

27. Jiang, X.; Cui, Z.; Chen, M.; Zhao, B. Study of plume eye in the copper bottom blowing smelting furnace. *Metall. Trans. B* **2019**, *50*, 782–789. [CrossRef]

28. Shui, L.; Cui, Z.; Ma, X.; Rhamdhani, M.A.; Nguyen, A.; Zhao, B. Mixing phenomena in a bottom blowing copper smelter: A water model study. *Metall. Trans. B* **2015**, *46*, 1218–1225. [CrossRef]

29. Liao, J.; Liao, C.; Zhao, B. Comparison of copper smelting slags between flash smelting furnace and bottom-Blowing furnace. In *12th International Symposium on High-Temperature Metallurgical Processing*; Springer: Cham, Switzerland, 2022; pp. 249–259.

Article

Control of Copper Content in Flash Smelting Slag and the Recovery of Valuable Metals from Slag—A Thermodynamic Consideration

Sui Xie [1,2], Xinhua Yuan [3], Fupeng Liu [1] and Baojun Zhao [1,2,*]

[1] Faculty of Materials Metallurgy and Chemistry, Jiangxi University of Science and Technology, Ganzhou 341000, China
[2] International Institute for Innovation, Jiangxi University of Science and Technology, Nanchang 330013, China
[3] Jiangxi Copper Corporation Guixi Smelter, Guixi 335400, China
* Correspondence: bzhao@jxust.edu.cn

Abstract: To determine slag properties and the factors influencing these properties for optimization of operating conditions in the copper flash smelting process, the composition and microstructures of the quenched smelting and converting slags have been analyzed. Thermodynamic software FactSage 8.2 has been used to investigate the effects of matte grade, SO_2 partial pressure, and the Fe/SiO_2 ratio on the liquidus temperature and the copper content of the smelting slag. The possibility to recover valuable metals from the smelting and converting slags through pyrometallurgical reduction by carbon is also discussed. It was found that the flash smelting slag temperature is usually higher than its liquidus temperature and the copper (1.2% Cu) is mainly present in the slag as dissolved copper. In the copper flash smelting process, the copper content in the slag can be decreased by decreasing the Fe/SiO_2 ratio and temperature. In pyrometallurgical slag reduction, most Cu, Mo, and Ni can be recovered as an alloy. The conditions of recovery such as the ratio of smelting slag to converting slag, temperature, and reduction extent have been discussed.

Keywords: smelting slag; liquidus temperature; converting slag; FactSage

Citation: Xie, S.; Yuan, X.; Liu, F.; Zhao, B. Control of Copper Content in Flash Smelting Slag and the Recovery of Valuable Metals from Slag—A Thermodynamic Consideration. *Metals* **2023**, *13*, 153. https://doi.org/10.3390/met13010153

Academic Editor: Mark E. Schlesinger

Received: 27 December 2022
Revised: 9 January 2023
Accepted: 9 January 2023
Published: 11 January 2023

1. Introduction

A total of 80% of copper is extracted from concentrates worldwide by the pyrometallurgical process [1,2]. Flash smelting with the characteristics of copper concentrates directly reacted with oxygen is widely used in the copper matte smelting process as a result of the advantages of high productivity, high automation, and low investment costs [3]. Flash smelting accounts for 43% of all Cu matte smelting [4]. The smelting process is developing toward a high feeding rate, high matte grade, and high oxygen enrichment to increase productivity and SO_2 concentration in the off-gas. However, it is confronted with the challenge of the control of copper content in slag and matte grade [5]. Matte and smelting slag are separated normally in the settler of flash smelting furnaces [6]. If the smelting temperature of the slag is lower than its liquidus temperature, spinel is precipitated from the slag, which increases the viscosity of the slag and restrains the separation of matte and slag [7]. In addition, long-term solid precipitation in the settler decreases its effective volume and separation of matte from the slag [8]. The smelting temperature is generally higher than the liquidus temperature of the slag to avoid the formation of spinel. The liquidus temperature of the slag is related to the composition, matte grade, and SO_2 partial pressure [9,10]. Copper mainly exists in slag as dissolved copper if the smelting temperature is higher than the liquidus temperature [11]. The composition of slag, temperature, matte grade, O_2, and SO_2 partial pressures affect the content of dissolved copper in slag [12]. Chemical analysis of smelting and converting slags indicated that they generally contain valuable metals such as copper, molybdenum, zinc, nickel, and tin [13]. Pyrometallurgical technology can be used to recover copper from smelting and converting slags in general [14]. Deep reduction to recover copper and iron from copper slags was also studied where coal, coke, diesel, hydrogen, and natural gas can be used as the reductants [15]. The reduction of the slag using hydrogen can reduce CO_2 emission and carbon in the Cu-Fe alloy [16]. Nonetheless, the efficiency of the hydrogen utilization is lower and the cost is higher than the solid reductants. However, other valuable metals such as Mo, Ni, and Zn in slags remain unclear during the recovery process. FactSage thermodynamic software contains powerful databases of optimized solutions, pure substances, and matte, and has been successfully used in the processing of copper concentrates [17], as well as the recovery of valuable metals from solid wastes [18].

This study aims to understand the effects of matte grade, SO_2 partial pressure, and Fe/SiO_2 on the liquidus temperature and copper content in smelting slag and explore the likelihood of recovering copper, molybdenum, zinc, nickel, and tin from smelting and converting slags.

2. Research Methodology

2.1. Materials

The smelting and converting slags were provided by Jiangxi Copper Smelter where a flash smelting furnace and PS converter are used for smelting and converting, respectively. The slags were collected from slag tap holes using a cooled iron bar, which maintained the microstructures and compositions of the slags at high-temperature conditions. The slag samples were mounted in epoxy resin, polished, and examined using optical and scanning electron microscopy. An MLA 650L scanning electron microscope (SEM-EDS), FEI, Hillsboro, OR, USA, was employed to determine the microstructures of the slags. The bulk compositions of the smelting and converting slags were measured by XRF (Axios max, produced by PANalytical B.V, Almelo, The Netherlands).

2.2. Thermodynamic Predictions

The thermodynamic software, FactSage 8.2, which integrates a substantially optimized oxide, pure substance, and matte database, is widely used in pyrometallurgical processing [19,20]. The databases selected included "FactPs", "Ftmisc", and "Ftoxide". The solution phases selected in the smelting of copper concentrate included "FTmisc-MATT", "FToxid-SLAGA", FToxid-SPINC", FToxid-MeO", "FToxid-OlivA", and "FToxid-ZNIT". The solution phases selected in pyrometallurgical reduction included "FTmisc-FeLQ", "FTmisc-CuLQ", "FTmisc-FeCu", "FTmisc-MATT", "FToxid-SLAGA", "FToxid-SPINC", FToxid-MeO", "FToxid-cPyrA", "FToxid-WOLLA", "FToxid-Bred", "FToxid-bC2SA", "FToxid-AC2S", "FToxid-Mel_A", "FToxid-OlivA", "FToxid-Mull", "FToxid-CORU", and "FToxid-ZNIT". Effects of matte grade, SO_2 partial pressure, and Fe/SiO_2 on liquidus temperature and copper content of slag have been systematically calculated. In addition, the effects of carbon addition on the liquidus temperature of slag and the reduction level of valuable metals in pyrometallurgical reduction have also been calculated.

3. Results and Discussion

3.1. Characterization of Slags

The composition of the smelting and converting slags analyzed by XRF are shown in Table 1. As can be seen from Table 1, Fe/SiO_2 values in the smelting and converting slags are 1.3 and 2.0, respectively. The content of copper in the smelting and converting slags is 1.3% and 5.2% Cu_2O, respectively. Moreover, the smelting slag contains 0.4% MoO_3 and 1.5% ZnO, and the converting slag contains 0.3% NiO, 0.2% SnO, and 2.2% ZnO. Mo, Ni, Sn, and Zn in the slags are valuable elements that should be recovered.

Table 1. Composition of quenched smelting and converting slags (wt.%).

Processes	"FeO"	SiO_2	Al_2O_3	CaO	MgO	Cu_2O	S	PbO	ZnO	As_2O_3	Na_2O	K_2O	MoO_3	NiO	SnO	TiO_2
Smelting	53.2	31.8	4.9	2.0	1.2	1.3	0.5	0.3	1.5	0.6	0.5	1.2	0.4	0.0	0.0	0.5
Converting	64.0	24.9	1.1	0.2	0.3	5.2	0.2	0.6	2.2	0.1	0.2	0.2	0.1	0.3	0.2	0.1

Typical microstructures of the quenched smelting and converting slags are shown in Figure 1. As can be seen from Figure 1a, the slag does contain large solid and matte phases indicating that the smelting temperature was higher than the liquidus temperature of the slag and the separation between the slag and the matte was efficient. The fine matte and dendritic crystals were precipitated from the slag during cooling. The microstructure of the quenched smelting slag confirms that all copper present in the slag was chemically dissolved. The converting slag shown in Figure 1b contains large spinel and matte indicating that the converting temperature (1250 °C) was lower than the liquidus temperature of the slag (1420 °C). The presence of an excessive solid phase in the slag increased the viscosity and affected the separation of the matte droplets from the slag. The copper content in the converting slag (5.2% Cu_2O) includes chemically dissolved and entrained matte droplets.

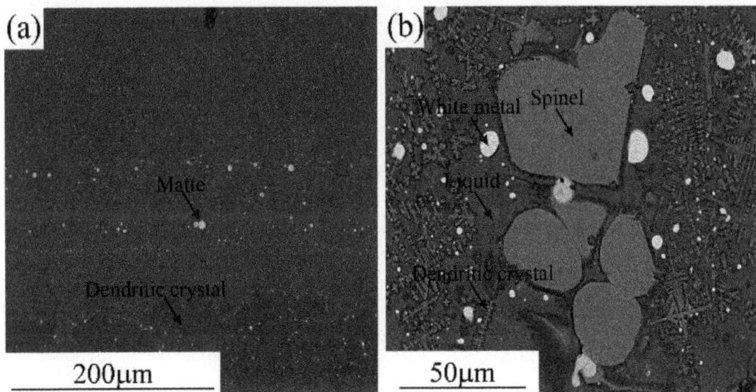

Figure 1. Typical microstructures of the quenched smelting and converting slags: (**a**) Smelting slag and (**b**) Converting slag.

3.2. Control of Liquidus Temperature of Flash Smelting Slag

The liquidus temperature of a smelting slag is directly related to its composition and oxygen partial pressure, which determines the FeO/Fe_2O_3 ratio in the slag [21]. The oxygen partial pressure in the system cannot be measured directly but correlates to the matte grade, SO_2 partial pressure, and temperature [22,23]. The matte grade in the smelting process can be controlled by the composition of copper concentrate and the ratio of oxygen to the concentrate. The SO_2 partial pressure in the off-gas is determined by the oxygen enrichment in the input gas mixture. Figure 2 shows the effect of matte grade and SO_2 partial pressure on the liquidus temperature of the smelting slag whose composition is based on Table 1. FactSage calculations show that the composition of the smelting slag shown in Figure 2 is in the spinel primary phase field. It can be seen that the liquidus temperatures of the smelting slag increase with the increasing matte grade and SO_2 partial pressure. At low-oxygen enrichment, P_{SO2} is approximately 0.3 atm. It can be seen from Figure 2 that the liquidus temperature increases from 1150 to 1215 °C when the matte grade is increased from 55 to 75. On the other hand, the liquidus temperature can be increased by approximately 12 °C when P_{SO2} is increased from 0.3 to 0.6 atm at a fixed matte grade.

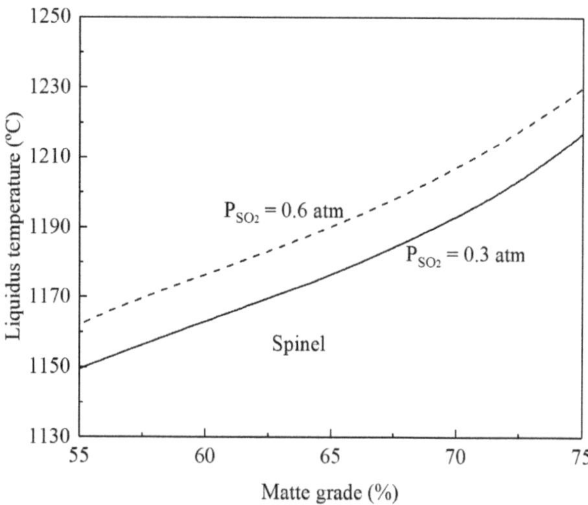

Figure 2. Effects of matte grade and SO$_2$ partial pressure on liquidus temperature of smelting slag, predicted by FactSage 8.2.

SiO$_2$ is usually added as a flux in the copper smelting process, and the Fe/SiO2 ratio can represent the composition of a copper smelting slag. It can be seen from Figure 3 that at fixed matte grade and P$_{SO2}$, the liquidus temperatures of a copper smelting slag increase with an increasing Fe/SiO$_2$ ratio, i.e., with a decreasing SiO$_2$ addition. At a fixed operating temperature and production of high-grade matte or high-SO$_2$ off-gas, more SiO$_2$ flux needs to be added to keep the liquidus temperature below the operating temperature. The slope of the liquidus temperature is decreased with the increase in Fe/SiO$_2$. For example, the temperature increase is 30 °C when the Fe/SiO$_2$ ratio increases from 1.3 to 1.5. When the Fe/SiO$_2$ ratio is increased from 1.5 to 1.7, the increment of the liquidus temperature is only 20 °C at SO$_2$ partial pressure 0.6 atm and matte grade 70%.

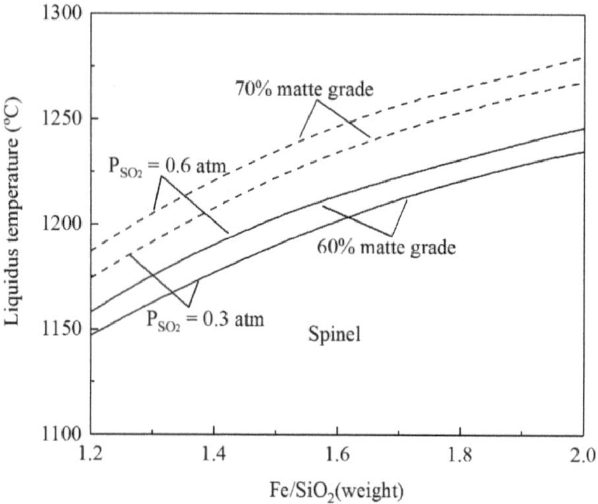

Figure 3. Effects of the Fe/SiO$_2$ ratio on liquidus temperature of smelting slag, predicted by FactSage 8.2.

In summary, low matte grade, low SO_2 partial pressure, and high SiO_2 addition can decrease the liquidus temperature of the copper smelting slag. The modern smelting process tends to increase productivity by producing high-grade matte with high-oxygen enrichment. The intensified smelting can treat more copper concentrate and reduce the volume of the off-gas capacity for easy production of sulfuric acid. Consequently, increased SiO_2 addition seems to be the efficient direction to control the liquidus temperature of the smelting slag.

3.3. Control of Copper Content in Copper Smelting Slag

The microstructure of the quenched flash smelting slag shown in Figure 1a indicates that the copper in the smelting slag is present as dissolved copper. The effects of matte grade on copper content in the smelting slag are predicted by FactSage 8.2 and shown in Figure 4. The copper content in smelting slag decreases with the increasing matte grade up to 70% Cu. With a matte grade greater than 70%, the copper content in the slag increases with the increasing matte grade. This trend is confirmed by the experimental studies of Fallah-Mehrjardi et al. [24,25] but is different from the findings reported by Shimpo et al. [26]. There is a general understanding [27] that the sulfidic dissolution of copper in copper smelting decreases with the increasing matte grade. However, the oxidic dissolution of copper in the slag increases with the increasing matte grade. As a result, the copper content in the smelting slag is not a simple function of the matte grade. A total of 70% Cu in the matte seems to be the optimum grade to reduce the copper loss in the slag to 0.65%. It can be seen from the figure that SO_2 partial pressure has little effect on the copper content in the slag. This is a good indication that increased oxygen enrichment does not increase the copper content in the smelting slag.

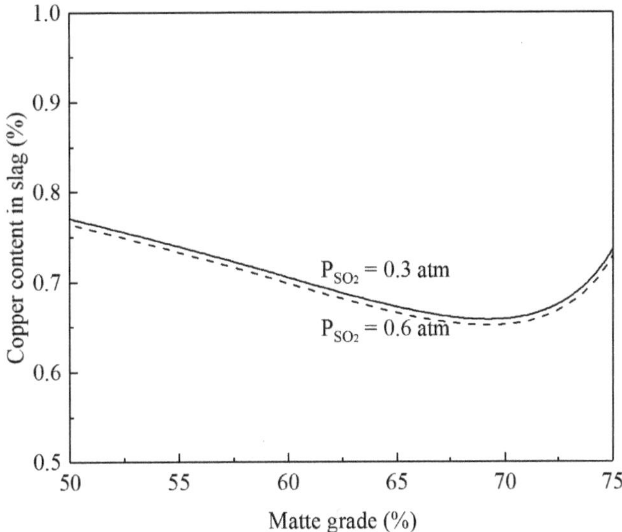

Figure 4. Effects of matte grade and SO_2 partial pressure on copper content in smelting slag at 1250 °C and Fe/SiO_2 1.3, predicted by FactSage 8.2.

The effects of temperature and matte grade on copper content in the smelting slag are shown in Figure 5 at a fixed SO_2 partial pressure of 0.3 atm and a Fe/SiO_2 ratio of 1.3. It can be seen that the copper content in the smelting slag increases rapidly with the increasing temperature. Nearly 0.3% Cu is increased in the slag if the temperature is increased from 1200 to 1300 °C. This is a significant copper loss as the volume of the smelting slag is very high. An increase in the matte grade from 60 to 70% can decrease the copper solubility in the slag.

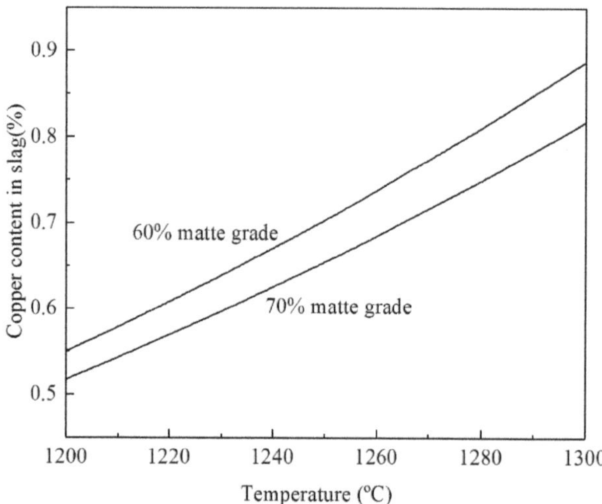

Figure 5. Effects of temperature and matte grade on copper content in smelting slag at a fixed SO_2 partial pressure of 0.3 atm and a Fe/SiO_2 ratio of 1.3, predicted by FactSage 8.2.

Figure 6 shows the copper content in the smelting slag as a function of the Fe/SiO_2 ratio in the slag. It can be seen that copper content in the smelting slag increases with the increasing Fe/SiO_2 ratio in the slag. However, the increment is different at different matte grades. When 70% matte is produced, the copper content in the slag increases slowly with the increasing Fe/SiO_2 ratio in the slag. An increase in the Fe/SiO_2 ratio from 1.2 to 1.7 only increases the copper content in slag by 0.05%. In contrast, an increase in the Fe/SiO_2 ratio from 1.2 to 1.7 can increase the copper content in slag by 0.2% with 60% matte production. The increase in Fe/SiO_2 means a decrease in SiO_2 addition and slag volume. Copper loss in the slag needs to consider both Cu% in the slag and slag volume. The selection of the Fe/SiO_2 ratio in the copper smelting slag needs to balance liquidus temperature, copper content in slag, and slag volume.

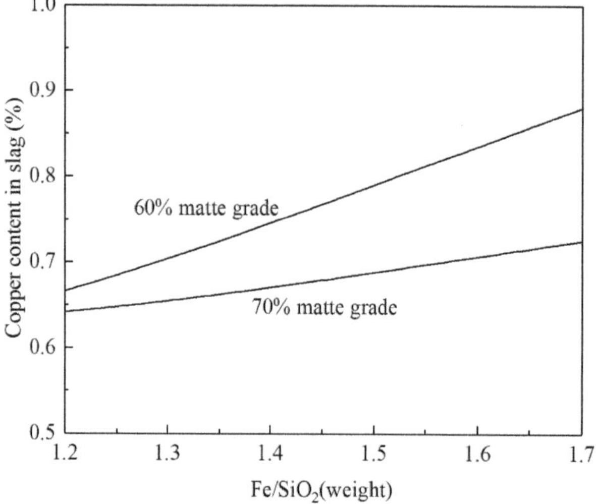

Figure 6. Effect of the Fe/SiO_2 ratio on copper content in smelting slag at 1250 °C and a fixed SO_2 partial pressure of 0.3 atm, calculated by FactSage 8.2.

Copper in the smelting slag can be recovered by a pyrometallurgical process or flotation process. The flotation process can only recover matte and metal particles. Fine matte and copper metal may precipitate from the smelting slag on slow cooling. However, the slag needs to be ground very finely to liberate the copper matte and metal from hard oxides, which increases the cost of copper recovery. Therefore, the selection of an optimum condition including an optimized matte grade, low temperature, and optimized Fe/SiO_2 ratio can reduce the copper loss in the flash smelting slag.

3.4. Recovery of Valuable Metals from Smelting and Converting Slags

It can be seen from Table 1 that the smelting and converting slags provided by the smelter contain significant amounts of valuable elements such as copper, molybdenum, zinc, nickel, and tin. Copper can be effectively recovered by the flotation process. However, it is unclear whether other elements can be recovered in the flotation process. The possibility of recovering all valuable elements in the pyrometallurgical process is evaluated by the thermodynamic calculations of FactSage 8.2. Smelting and converting slags can be treated individually or together depending on the economic evaluation. The composition values given in Table 1 are not sufficient for the calculation of reduction reactions. According to the smelting temperature and matte grade provided by the smelter, the oxygen partial pressure and Fe^{2+}/Fe^{3+} ratio in the smelting slag were calculated by FactSage 8.2 and are shown in Table 2. Similarly, the oxygen partial pressure and Fe^{2+}/Fe^{3+} ratio in the converting slag were also calculated by assuming the slag was in equilibrium with Cu_2S. Using the information given in Tables 1 and 2, the reduction of the smelting and converting slags by carbon were predicted by FactSage 8.2 and discussed below.

Table 2. Fe^{2+}/Fe^{3+} ratio in the smelting and converting slags calculated by FactSage 8.2.

Processes	Temperature (°C)	P_{O2} (atm)	FeO (wt%)	Fe_2O_3 (wt%)	Fe^{2+}/Fe^{3+}
Smelting	1300	10^{-7}	47.9	5.6	9.55
Converting	1250	10^{-5}	26.18	39.24	0.74

At high temperature reduction, the oxides of copper, molybdenum, zinc, nickel, iron, and tin in the smelting and converting slags can be reduced with carbon addition. The liquidus temperatures of the slag will change accordingly with the reduction. The effect of carbon addition on liquidus temperatures when treating individual slag or mixed slags (80% smelting slag and 20% converting slag) is shown in Figure 7. It can be seen that spinel is the primary phase at low carbon addition and the liquidus temperatures decrease with increasing the carbon percentage. More carbon addition brings the slags to the olivine primary phase field where the liquidus temperatures are lower than 1200 °C. The smelting slag enters into the SiO_2 primary phase field with further addition of carbon. The liquidus temperatures rapidly increase with the increasing carbon in the SiO_2 primary phase field. In a copper smelter, the ratio of the smelting slag to converting slag is approximately 4:1. It can be seen from Figure 7 that treatment of the mixed smelting and converting slags is possible. The liquidus temperatures can be controlled below 1200 °C without flux addition.

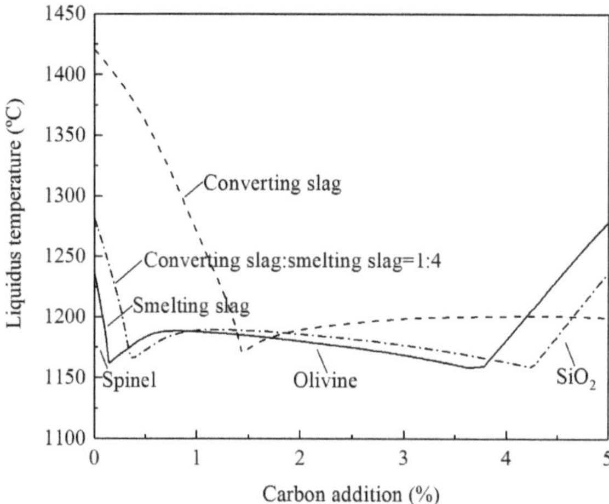

Figure 7. Effect of carbon addition on the liquidus temperature of slag, calculated by FactSage 8.2.

The effect of carbon addition on the recovery of copper, iron, zinc, and molybdenum from the smelting slag at 1250 °C is shown in Figure 8. It can be seen that copper is first reduced followed by molybdenum. With less than 1% carbon addition, 100% molybdenum oxide has been reduced. When 5% carbon is added, the recovery of copper, zinc, and iron is 88%, 58%, and 51%, respectively. It can be seen from Figure 7 that the liquidus temperature is above 1250 °C when 5% carbon is added to the smelting slag. To add more carbon to increase the recovery of copper and zinc, flux such as CaO needs to be added to lower the liquidus temperature of the slag.

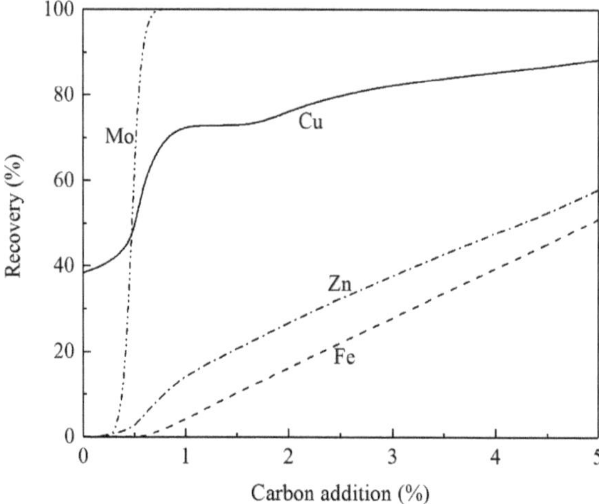

Figure 8. Effect of carbon addition on the recovery of copper, iron, zinc, and molybdenum from smelting slag at 1250 °C, calculated by FactSage 8.2.

Figure 9 shows the effect of carbon addition on the recovery of copper, iron, zinc, nickel, and tin from converting slag at 1250 °C. Note that the converting slag contains less sulfur than the smelting slag which causes different reduction behavior. With a 5% carbon addition, over 95% of copper and nickel have been recovered from the slag. Less than 50% of the zinc and 10% of the tin are reduced from the slag. Deep reduction is required to fully reduce the oxides of zinc and tin.

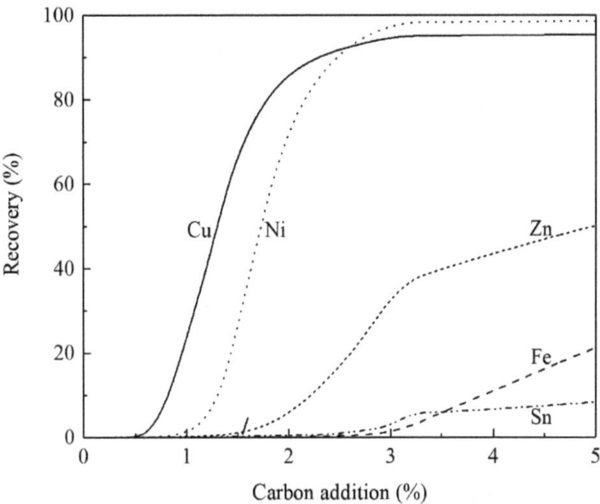

Figure 9. Effect of carbon addition on the recovery of copper, iron, zinc, nickel, and tin in converting slag at 1250 °C, calculated by FactSage 8.2.

It is shown in Figure 7 that the liquidus temperatures of the reduced slag mixture can be controlled below 1250 °C. Figure 10 shows the effect of carbon addition on the recovery of copper, iron, zinc, molybdenum, nickel, and tin from the mixed slag (smelting slag:converting slag = 4:1) at 1250 °C. It can be seen that almost all nickel and molybdenum can be recovered from the mixed slag with a 5% carbon addition. In addition, 85% of the copper can also be recovered. However, only part of the zinc oxide and tin oxide is reduced at the same condition. Considering the liquidus temperature of the reduced slag and recovery of multi-elements, it is more beneficial to treat the smelting slag and converting slag together.

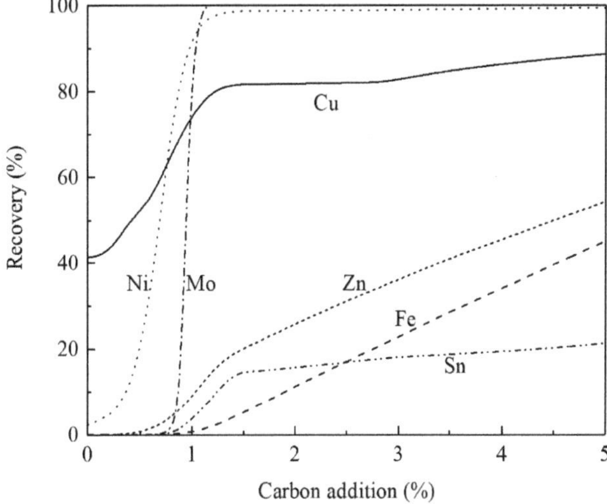

Figure 10. Effect of carbon addition on the recovery of copper, zinc, molybdenum, nickel, iron, and tin from the mixed slag (smelting slag:converting slag = 4:1) at 1250 °C, calculated by FactSage 8.2.

4. Conclusions

Quenched flash smelting slag and converting slag have been analyzed as a base for thermodynamic predictions. The control of liquidus temperatures and copper content in the smelting slag and the recovery of valuable elements from the slags have been calculated by FactSage 8.2. It was found that the smelting temperature in the flash furnace was higher than the liquidus temperature of the slag and all copper was present in the flash smelting slag as dissolved copper. High matte grade, SO_2 partial pressure, and Fe/SiO_2 can cause high liquidus temperature of the smelting slag. Optimum matte grade, low temperature, and low Fe/SiO_2 are beneficial to decrease the copper content in the smelting slag. Copper, nickel, and molybdenum can be recovered in alloy by reduction of the mixed smelting and converting slags. Recovery of zinc and tin from the slags requires deep reduction and flux addition.

Author Contributions: Methodology, B.Z.; software, B.Z.; resources, X.Y.; writing – original draft, S.X.; writing – review & editing, X.Y., F.L. and B.Z. All authors have read and agreed to the published version of the manuscript.

Funding: This research received no external funding.

Institutional Review Board Statement: Not applicable.

Informed Consent Statement: Not applicable.

Data Availability Statement: Not applicable.

Acknowledgments: The authors would like to thank Guixi Smelter of Jiangxi Copper for providing the smelting and converting slags.

Conflicts of Interest: The authors declare no conflict of interest.

References

1. Potysz, A.; Hullebusch, E.; Kierczak, J.; Grybos, M.; Lens, P.; Guibaud, G. Copper metallurgical slags–current knowledge and fate: A review. *Crit. Rev. Environ. Sci. Technol.* **2015**, *45*, 2424–2488. [CrossRef]
2. Moskalyk, R.; Alfantazi, A. Review of copper pyrometallurgical practice: Today and tomorrow. *Miner. Eng.* **2003**, *16*, 893–919. [CrossRef]
3. Zhao, B.; Liao, J. Development of Bottom-blowing copper smelting technology: A review. *Metals* **2022**, *12*, 190. [CrossRef]
4. Wan, X.; Shen, L.; Jokilaakso, A.; Eriç, H.; Taskinen, P. Experimental Approach to Matte-Slag Reactions in the Flash Smelting Process. *Miner. Process. Extr. Metall. Rev.* **2021**, *42*, 231–241. [CrossRef]
5. Kojo, I.; Jokilaakso, A.; Hanniala, P. Flash smelting and converting furnaces: A 50 year retrospect. *JOM* **2000**, *52*, 57–61. [CrossRef]
6. Fagerlund, K.; Jalkanen, H. Microscale simulation of settler processes in copper matte smelting. *Metall. Mater. Trans. B* **2000**, *31*, 439–451. [CrossRef]
7. Liao, J.; Liao, C.; Zhao, B. Comparison of Copper Smelting Slags Between Flash Smelting Furnace and Bottom-Blowing Furnace. In Proceedings of the 12th International Symposium on High-Temperature Metallurgical Processing, Orlando, FL, USA, 15–18 March 2021.
8. Zhou, J.; Chen, Z.; Zhou, P.; Yu, J.; Liu, A. Numerical simulation of flow characteristics in settler of flash furnace. *Trans. Nonferrous Met. Soc. China* **2012**, *22*, 1517–1525. [CrossRef]
9. Zhao, B.; Xie, S.; Zhou, L. Study on slag-matte-gas equilibrium in copper flash smelting process. *Nonferrous Met. (Extr. Metall.)* **2022**, *3*, 38–43. (In Chinese)
10. Sridhar, R.; Toguri, J.; Simeonov, S. Copper losses and thermodynamic considerations in copper smelting. *Metall. Mater. Trans. B* **1997**, *28*, 191–200. [CrossRef]
11. Bacedoni, M.; Moreno-Ventas, I.; Ríos, G. Copper flash smelting process balance modelling. *Metals* **2020**, *10*, 1229. [CrossRef]
12. Cornejo, K.; Chen, M.; Zhao, B. Control of Copper Loss in Flash Smelting Slag. In *Materials Engineering—From Ideas to Practice: An EPD Symposium in Honor of Jiann-Yang Hwang*; Springer: Orlando, FL, USA, 2021.
13. Zhang, H.; Wang, Y.; He, Y.; Xu, S.; Hu, B.; Cao, H.; Zhou, J.; Zheng, G. Efficient and safe disposition of arsenic by incorporation in smelting slag through copper flash smelting process. *Miner. Eng.* **2021**, *160*, 106661. [CrossRef]
14. Tian, H.; Guo, Z.; Pan, J.; Zhu, D.; Yang, C.; Xue, Y.; Li, S.; Wang, D. Comprehensive review on metallurgical recycling and cleaning of copper slag. *Resour. Conserv. Recy.* **2021**, *168*, 105366. [CrossRef]
15. Qu, G.; Wei, Y.; Li, B.; Wang, H.; Yang, Y.; McLean, A. Distribution of copper and iron components with hydrogen reduction of copper slag. *J. Alloy Compd.* **2020**, *824*, 153910. [CrossRef]
16. Zhang, B.; Zhang, T.; Zheng, C. Reduction Kinetics of Copper Slag by H_2. *Minerals* **2022**, *12*, 548. [CrossRef]

17. Xu, F.; Weng, T.; Tan, K.; Liao, J.; Zhao, B.; Xie, S. Distribution and Control of Arsenic during Copper Converting and Refining. *Metals* **2023**, *13*, 85. [CrossRef]
18. Yu, Y.; Wang, H.; Hu, J. Co-treatment of electroplating sludge, copper slag, and spent cathode carbon for recovering and solidifying heavy metals. *J. Hazard. Mater.* **2021**, *417*, 126020.
19. Bale, C.; Bélisle, E.; Chartrand, P.; Decterov, S.; Eriksson, G.; Gheribi, A.; Hack, K.; Jung, I.; Kang, Y.; Melançon, J.; et al. FactSage thermochemical software and databases, 2010–2016. *Calphad* **2016**, *54*, 35–53. [CrossRef]
20. Bale, C.; Chartrand, P.; Degterov, S.; Eriksson, G.; Hack, K.; Ben Mahfoud, R.; Melançon, J.; Pelton, A.; Petersen, S. FactSage thermochemical software and databases. *Calphad* **2002**, *26*, 189–228. [CrossRef]
21. Zhao, B.; Hayes, P.; Jak, E. Effects of CaO, Al_2O_3 and MgO on liquidus temperatures of copper smelting and converting slags under controlled oxygen partial pressures. *J. Min. Metall. B* **2013**, *49*, 153. [CrossRef]
22. Jak, E. Integrated experimental and thermodynamic modelling research methodology for metallurgical slags with examples in the copper production field. In Proceedings of the IX International Conference on Molten Slags, Fluxes and Salts, Beijing, China, 27–30 May 2012.
23. Yazawa, A. Thermodynamic evaluations of extractive metallurgical processes. *Metall. Mater. Trans. B* **1979**, *10*, 307–321. [CrossRef]
24. Fallah-Mehrjardi, A.; Hidayat, T.; Hayes, P.; Jak, E. Experimental investigation of gas/slag/matte/tridymite equilibria in the Cu-Fe-O-S-Si system in controlled gas atmospheres: Experimental results at 1473 K (1200 °C) and $P(SO_2)$ = 0.25 atm. *Metall. Mater. Trans. B* **2017**, *48*, 3017–3026. [CrossRef]
25. Fallah-Mehrjardi, A.; Hayes, P.; Jak, E. The effect of CaO on gas/slag/matte/tridymite equilibria in fayalite-based copper smelting slags at 1473 K (1200 °C) and $P(SO_2)$ =0.25 atm. *Metall. Mater. Trans. B* **2018**, *49*, 602–609. [CrossRef]
26. Shimpo, R.; Goto, S.; Ogawa, O.; Asakura, L. A study on the equilibrium between copper matte and slag. *Can. Metall. Quart.* **1986**, *25*, 113–121. [CrossRef]
27. Furuta, S.; Tanaka, S.; Hamamoto, M.; Inada, H. Analysis of copper loss in slag in Tamano type flash smelting furnace. In *Sohn International Symposium: International Symposium of Sulfide Smelting*; Wiley-TMS: San Diego, CA, USA.

Article

Distribution and Control of Arsenic during Copper Converting and Refining

Feiyan Xu [1], Tao Weng [1], Keqin Tan [2], Jinfa Liao [3], Baojun Zhao [3] and Sui Xie [3,*]

1 Faculty of Materials Metallurgy and Chemistry, Jiangxi University of Science and Technology, Ganzhou 341000, China
2 Dongying Fangyuan Nonferrous Metals, Dongying 257091, China
3 International Institute for Innovation, Jiangxi University of Science and Technology, Nanchang 330013, China
* Correspondence: 7120210006@mail.jxust.edu.cn

Abstract: Arsenic content in copper concentrates is continuously increasing worldwide. It is desirable to remove arsenic from copper in the earlier stages of copper making due to the deposition of arsenic to cathode copper during the electrorefining process. Effects of temperature, flux, and oxygen on the distribution of arsenic during copper converting and fire refining processes were studied using FactSage 8.2. The results showed that arsenic can be effectively removed by proper selection of converting and refining slags. The decrease in Fe/SiO_2 or Fe/CaO ratio in the converting slag is favorable for arsenic distributed to slag. CaO is more effective than SiO_2 in decreasing the liquidus temperature of the slag and arsenic content in the blister copper during the converting process. Na_2O or CaO as a flux is effective to remove arsenic in the fire refining process.

Keywords: arsenic distribution; arsenic control; FactSage; blister copper; anode copper

1. Introduction

Most copper is extracted from concentrates from pyrometallurgical processes which include smelting, converting, and fire and electro refining steps to obtain high-purity cathode copper [1,2]. The Outokumpu flash furnace and bath smelting furnaces are widely used in copper matte smelting processes [3]. The Peirce–Smith batch converting process dominates the production of blister copper from matte. However, the Peirce–Smith batch converting process has the disadvantages of low SO_2 utilization, gas fugitive emissions, and discontinuous operation. The flash converting and bath converting technologies have been developed to overcome the disadvantages [4,5]. Copper fire refining is mostly carried out in a rotary refining furnace, blowing gas, such as O_2, CO_2, and N_2, for the removal of sulfur and adding reductants for the removal of oxygen [6,7].

Arsenic is a harmful impurity present in copper concentrates [8]. In copper smelting, converting, and fire refining processes, arsenic can be present in the condensed phases and gas [8]. Arsenic-containing materials in the off gas is usually collected in the form of dust [9]. In the condensed phases, the arsenic can be distributed between matte and slag in the smelting process, between the blister copper and slag in the converting process, and between the anode copper and slag in the fire refining process [10]. With the high-quality copper concentrates exhausted worldwide, the content of arsenic in the copper concentrates is increasing continuously [11]. If arsenic is not handled properly in copper production, it will cause difficulty for copper extraction and endanger the environment [12,13]. Arsenic from anode copper can accumulate in the electrolyte during copper electrorefining. Deposit of the arsenic on the cathode copper significantly affects its quality and production rate during the copper electrorefining [14]. Removal of arsenic before the electrorefining process is important to produce high-quality cathode copper.

Swinbourne and Kho [15] studied the distribution of arsenic during copper flash converting using HSC Chemistry 7.1 software. It was found that the distribution of ar-

Citation: Xu, F.; Weng, T.; Tan, K.; Liao, J.; Zhao, B.; Xie, S. Distribution and Control of Arsenic during Copper Converting and Refining. *Metals* **2023**, *13*, 85. https://doi.org/10.3390/met13010085

Academic Editor: Daniel Assumpcao Bertuol

Received: 19 November 2022
Revised: 14 December 2022
Accepted: 26 December 2022
Published: 29 December 2022

senic among copper, slag, and waste gas depends on their thermodynamic properties. Swinbourne and Kho reported that one-half of arsenic is distributed in blister copper due to the arsenic activity coefficient being lower than in slag. Li et al. [16,17] studied the distribution behavior of arsenic during copper flash converting with the multi-phase equilibrium mathematical model. It was found that arsenic is mainly present in blister copper (51.4%) and slag (41.9%). All of these thermodynamics studies only used properties of pure substances and did not consider solution phases. Yuan and Liu [18] and Liu [19] studied the distribution of arsenic in the bottom-blowing smelting and converting processes. It was found that increase in SiO_2 in slag was beneficial to reduce the arsenic in blister copper. Guo et al. [20] established a mass balance model and substance flow charts of arsenic in the bottom-blowing smelting and Peirce–Smith converting processes. It was found that arsenic mainly existed in both smelting and converting dusts. Zhou et al. [21] investigated the distribution of arsenic in ISA smelting, Peirce–Smith converting, and rotary refining processes in detail. It was found that the addition of alkaline fluxes was efficient to remove the arsenic from blister copper during the refining process. Kucharski [6] studied the effect of Na_2CO_3 on the removal of arsenic from blister copper during the refining process. It was found that arsenic from the blister copper reacted with Na_2CO_3 to produce Na_3AsO_4. Hidayat [22] summarized the experimental research of Kaur et al. [23], Nagamori and Mackey [24], and Avarmaa et al. [25], and determined the distributions of Arsenic between liquid copper and silicate slag at tridymite saturation. It was found that arsenic preferentially dissolved into copper instead of slag. Moreover, the arsenic in the slag was increased with increasing the oxygen partial pressure.

Most of the studies focused on the distribution of arsenic between different phases. Few studies discussed the control of arsenic by slag in copper-converting and -refining processes. Iron silicate is the major component of copper-smelting slags with variable Fe/SiO_2 ratios. It is not considered to use slag to control arsenic distribution in the smelting process. However, different fluxes can be used in the converting and fire refining processes. It is possible to evaluate the ability of different slags for the removal of arsenic from blister copper and anode copper. Recently updated FactSage 8.2 with extensive slag and alloy solution phases are used to investigate the optimal conditions to fix the arsenic in the copper-converting and -refining slags [26]. This study aims to understand the distribution of arsenic in matte converting and blister copper refining from a thermodynamics aspect.

2. Thermodynamic Predictions

FactSage 8.2 software was widely used in pyrometallurgical processing of different minerals [27–29]. Propper selection of the databases and phases is important to obtain reliable results. In the present study, the databases selected included "FactPs", "Ftmisc", and "Ftoxide". The solution phases selected in the smelting of copper concentrate included "FTmisc-PYRRB", "Ftmisc-MeS2", "Ftmisc-BCCS", "Ftmisc-FCCS", "Ftoxid-SLAGA", "Ftoxid-SPINA", and "FToxid-Oliv". The solution phases selected in the matte converting included "FTmisc-CuLQ", "FTmisc-FeCu", "FToxid-SLAGA", "FToxid-SPINA", "FToxid-MeO", and "FToxid-Oliv". The solution phases selected in fire refining of blister copper included "FTmisc-CuLQ", "FToxid-SLAGA", "FToxid-SPINA", and "FToxid-MeO."

3. Results and Discussion

3.1. Converting of Copper Matte

Peirce–Smith converter is still the major converting technology in copper smelters due to its advantages of easy operation and high adaptability of complex feeds. However, the P–S converter has the disadvantages of low SO_2 concentration, gas fugitive emissions, and discontinuous operation. Several continuous converting technologies have been developed in recent years, including flash converting and bottom-blowing converting processes. These new technologies have good automation and more efficient capture and utilization of SO_2. The published data on arsenic distribution on different technologies during converting are shown in Table 1. It can be seen from the table that more arsenic turns to the gas phase in the P–S converting process because of the strong stirring. In the condensed phases, arsenic tends to stay in the blister copper with a small proportion of removal by the slag. Guo et al. [20] simulated the distribution of arsenic in the converting process. They indicated that up to 77% could go to gas with less than 10% arsenic left in the blister. It seems that all plant data are far away from the simulation. Comparing with the P–S converter, flash-converting and bottom-blowing-converting furnaces retain more than 95% arsenic in the condensed phases where higher arsenic enters the blister.

Table 1. The published data on arsenic distribution during converting.

Authors	Technology	Distribution (wt%)		
		Blister Copper	**Slag**	**Gas**
Li [16]	Flash	52	42	6
Swinbourne [15]	Flash	60	37	3
Liu [19]	Bottom blowing	65.7	31.4	2.9
Guo [20]	Peirce-smith	12.9	9.4	77.7
Zhou [21]	Peirce-smith	39.0	17.0	44.0
Zhang [30]	Peirce-smith	28.0	13.0	58.0
Wang [31]	Peirce-smith	66.4	5.4	28.2
Wang [32]	Peirce-smith	59.9	5.1	34.9
Vogt J [33]	Peirce-smith	50.0	32.0	18.0

Typical composition of copper concentrate, matte, and blister copper from a smelting plant were used in the present study as shown in Table 2 [21]. A total of 50% copper in matte was produced in the smelting stage and the slag had Fe/SiO_2 1.2. The slag temperature was 1200 °C and O_2 content in the gas was 22.83%. It can be seen from Table 2 that the arsenic content in the concentrate, matte, and blister copper was 0.30, 0.37, and 0.43%, respectively.

Table 2. Typical compositions of copper concentrate, matte, and blister copper.

	Cu	Fe	S	As	O	Ca	Al_2O_3	CaO	MgO
Concentrate	29.50	32.60	31.80	0.30	0.00	0.00	2.60	2.30	0.90
Matte	50.01	23.70	24.45	0.37	1.18	0.28	0.00	0.00	0.00
Blister copper	98.50	0.00	0.73	0.43	0.34	0.00	0.00	0.00	0.00

Matte converting was conducted at 1290 °C with a conventional slag containing Fe/SiO_2 ratio of 1.7. The effect of copper content in blister copper on arsenic distribution during the converting is shown in Figure 1. Oxygen partial pressures during the converting are also shown in the figure. Note that Figure 1 only shows the end of the converting stage where the matte phase has been converted to the blister copper. It can be seen that the P_{O_2} increases with increasing copper content in the blister. At the same time, the arsenic increases in the slag and decreases in the blister. According to the thermodynamic predictions, arsenic in the gas phase is neglected as shown in Figure 1. It seems that the converting reactions in the flashing and bottom furnaces approached equilibrium as indicated in Table 1.

Figure 2 shows the phase proportion as a function of copper content in blister copper under the same conditions as Figure 1. It can be seen that the proportion of gas remains constant when the Cu in the blister is increased from 97 to 99%. However, the proportion of the slag increases slowly initially and then rapidly with increasing copper content in the blister. In contrast, the proportion of the blister decreases slowly initially and then rapidly with increasing copper content in the blister. The reason is that high Cu in blister corresponds to high oxygen partial pressure as shown in Figure 1. Cu in the blister is oxidized to Cu_2O which is dissolved into the slag resulting in copper loss in the converting process. Arsenic in the blister can be reduced from 63 to 20% when the Cu in the blister is increased from 97 to 99% according to Figure 1. However, it can be seen from Figure 2 that the proportion of the blister is reduced from 30 to 20%. The direct recovery of copper in the converting process is decreased by 33% if a low-As and high-Cu blister is produced. More effort will be required to recover the copper from the converting slag and the productivity of the converting furnace is significantly affected. It is a choice of the operator to balance the purity and productivity of the blister during the converting process.

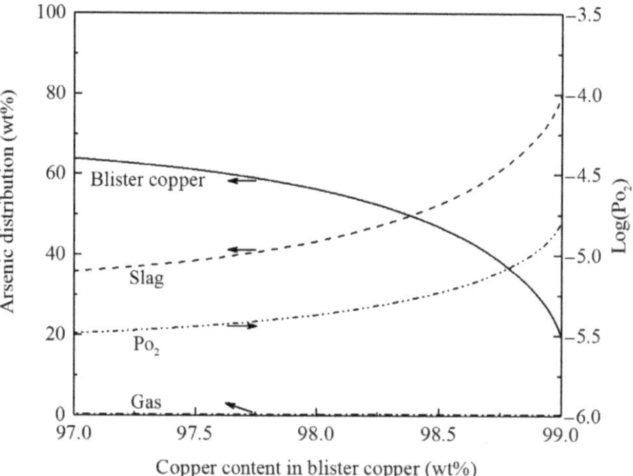

Figure 1. Effect of copper content in blister copper on arsenic distribution at 1290 °C with a slag Fe/SiO$_2$ ratio of 1.7.

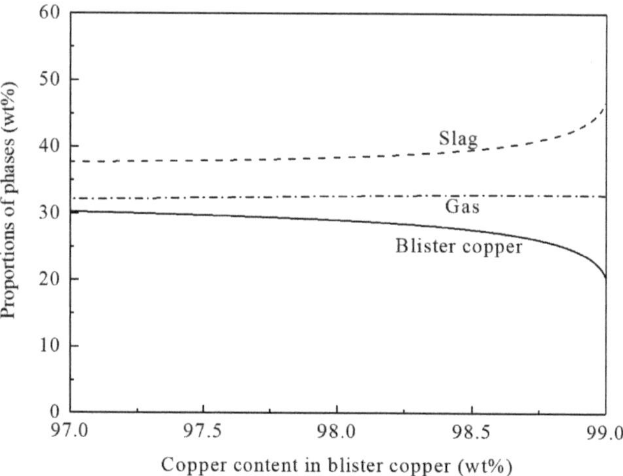

Figure 2. Effect of copper content in blister copper on proportions of the phases at 1290 °C with a slag Fe/SiO$_2$ ratio of 1.7.

Figure 3 shows the effect of temperature on arsenic distribution during converting at a slag Fe/SiO$_2$ 1.7 and 98.5% Cu in blister copper. Little arsenic is present in the gas phase. The arsenic is decreased slightly in the slag and increased slightly in the blister with increasing temperature. Thermodynamically, a high converting temperature is not beneficial to remove the arsenic from the blister.

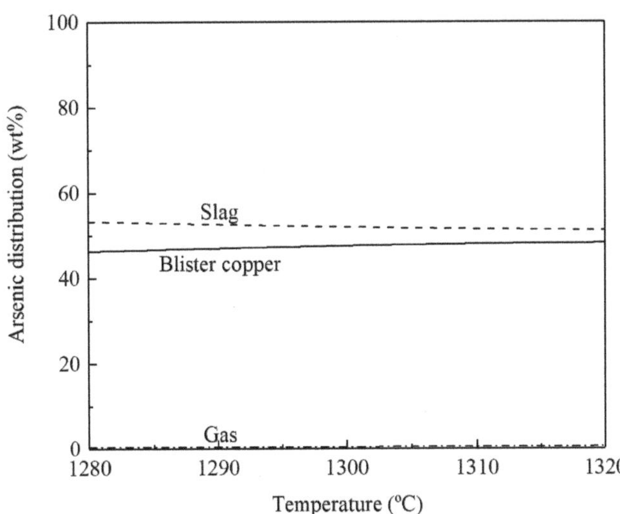

Figure 3. Effect of temperature on arsenic distribution at slag Fe/SiO$_2$ 1.7 and 98.5% Cu in blister copper.

Both SiO$_2$ and CaO can be used in the copper converting process as a flux. The primary role of a flux is to form a liquid slag with the impurities. Effects of SiO$_2$ and CaO on liquidus temperature of the converting slag are shown in Figure 4 at a fixed copper content 98.5% in blister copper. Fe/SiO$_2$ or Fe/CaO ratio is usually used in the operation to control the addition of the flux and the composition of the converting slag. Figure 4 shows that, when SiO$_2$ is used as a flux, spinel is the only primary phase in the Fe/SiO$_2$ ratios of 1.5 to 4.0. The liquidus temperatures increase with increasing Fe/SiO$_2$ ratio. Ca$_2$Fe$_2$O$_5$ and spinel are the primary phases when CaO is used as a flux. In the Ca$_2$Fe$_2$O$_5$ primary phase field, liquidus temperatures decrease with increasing the Fe/CaO ratio. On the other hand, the spinel is the primary phase when the Fe/CaO ratio is hgher than 2.65. The liquidus temperatures in the spinel primary phase field increase with increasing the Fe/CaO ratio in the slag. It is obvious that CaO is a more efficient flux to lower the liquidus temperature of the converting slag. Much less CaO is required to control the liquidus temperature. As a result, more scrap can be treated in the converting furnace.

Figure 5 illustrates the arsenic distribution as a function of Fe/SiO$_2$ or Fe/CaO ratio at 1290 °C and copper content in blister copper 98.5%. In both cases the arsenic in the gas phase is neglected. The arsenic always increases in the slag and decreases in the blister copper with decreasing Fe/SiO$_2$ or Fe/CaO ratio. This means that both SiO$_2$ and CaO can fix the arsenic in the slag and decrease arsenic in the blister copper. Comparing with SiO$_2$, more arsenic can be fixed in the slag by adding CaO. Fe/SiO$_2$ and Fe/CaO are usually controlled between 1–2 and 3–4, respectively, in the converting process [34]. It can be seen from Figures 4 and 5 that, the liquidus temperature is 1264 °C and 52% arsenic is dissolved in the blister copper when the Fe/SiO$_2$ ratio is 1.5. In contrast, the liquidus temperature is 1217 °C and only 34% arsenic is dissolved in the blister copper when the Fe/CaO ratio is 3.0.

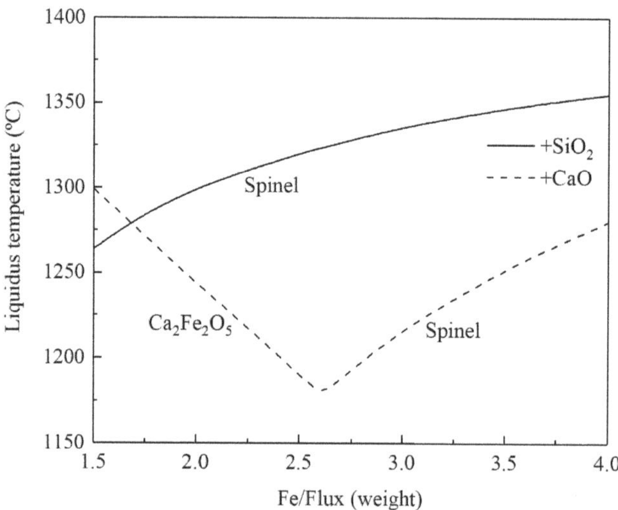

Figure 4. Effect of SiO$_2$ or CaO on liquidus temperatures of converting slag at a fixed copper content 98.5% in blister copper.

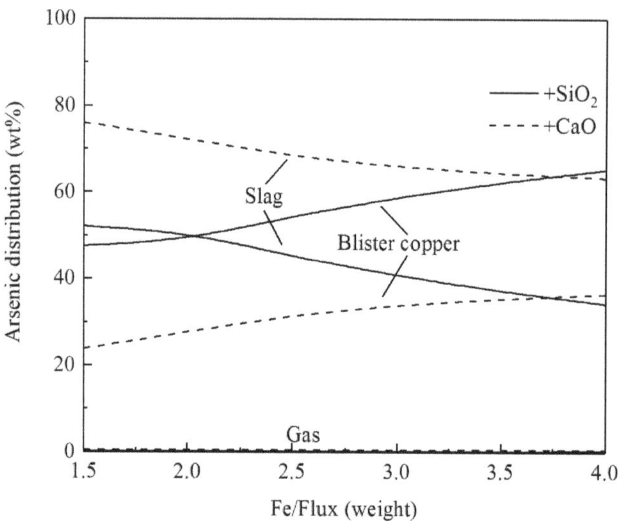

Figure 5. Effect of Fe/Flux ratio on arsenic distribution at 1290 °C and 98.5% Cu in the blister copper.

3.2. Fire refining of Blister Copper

After the converting process, blister copper containing 98.5% Cu is taken as an example for refining. It can be seen from Table 2 that arsenic, sulfur, and oxygen are the major impurities in blister copper and their concentrations are 0.43%, 0.73%, and 0.34%, respectively.

All blister copper needs to be fire refined to produce anode copper for electrorefining. The purpose of fire refining is to produce liquid copper which can be cast with minimal gas porosity, since sulfur and oxygen can disrupt the downstream electrorefining process. The conventional fire refining process usually includes an oxidation stage to remove sulfur and a reduction stage to remove oxygen. It can be seen from Figure 6 that in the oxidation stage, sulfur content in the anode copper decreases rapidly at the beginning of oxygen blowing and then slowly after the sulfur content is lower than 0.1%. On the other hand, oxygen content in the anode copper increases slowly with increasing oxygen blowing. However, continuous blowing of oxygen will increase the oxygen in copper quickly, which will require more reductant and time to remove oxygen in the reduction stage.

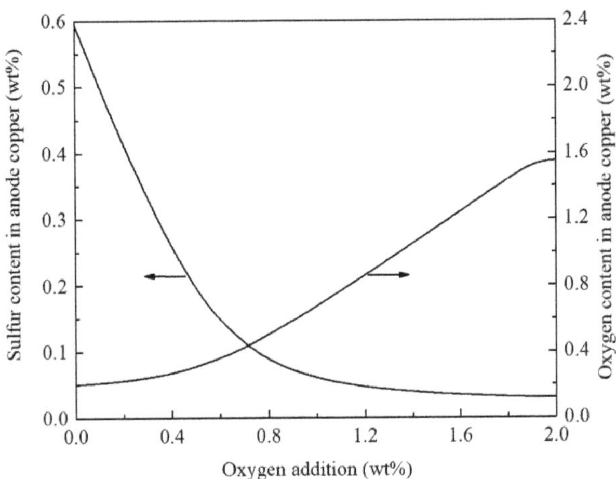

Figure 6. Effect of O_2/blister on sulfur and oxygen contents in anode copper at 1200 °C.

With increased arsenic content in the copper concentrates, it can be seen from Tables 1 and 2 that arsenic cannot be completely removed in the smelting and converting process. Arsenic in the anode copper will be brought into the electrorefining process which will affect the quality of the cathode copper and the efficiency of the process. The electrolyte needs to be purified frequently which will reduce the productivity of the electrorefining process. Flux is not usually used in the conventional fire refining process because the removal of sulfur and oxygen does not require a flux. It can be seen from Table 3 that over 80% arsenic is present in the anode copper after fire refining [20,21]. When Na_2CO_3 was used as a flux, the arsenic in the anode copper can be reduced to 1% [6]. It is therefore necessary to add a flux in the fire refining process to control arsenic in the anode copper.

Table 3. The published data on arsenic distribution during fire refining.

Authors	Flux	Mass Fraction (%)		
		Copper Anode	Slag	Gas
Zhou [21]	-	83.5%	-	-
Guo [20]	-	96.9	3.1	-
Kucharski [6]	Na_2CO_3	1	-	-

It can be seen from Figure 7 that if only oxygen is used in the refining process, arsenic in the anode copper is almost constant. In contrast, the addition of Na_2O or CaO can significantly reduce arsenic in the anode copper. Na_2O is more effective than CaO for removing arsenic from the copper.

The main reactions between arsenic and flux (CaO or Na_2O) are

$$3CaO + 2[As] + 5[O] = Ca_3(AsO_4)_2 \tag{1}$$

$$3Na_2O + 2[As] + 5[O] = 2Na_3AsO_4 \tag{2}$$

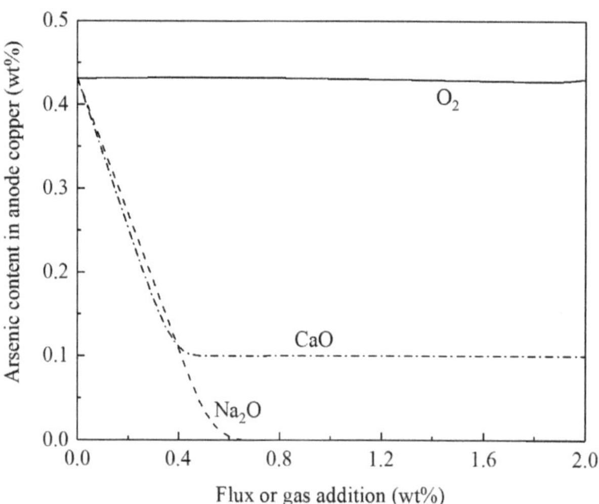

Figure 7. Effect of flux and oxygen on arsenic content in anode copper at 1200 °C.

It can be seen from the reactions that oxygen plays an important role in the removal of arsenic. Figure 8 shows the effect of oxygen on arsenic content in anode copper with 0.5% flux addition. It can be seen that oxygen does not affect the arsenic content in anode copper when Na_2O is a flux. However, the arsenic content in anode copper decreases significantly with increasing oxygen/blister ratio. When the oxygen addition is higher than 0.6%, CaO is a more effective flux than Na_2O to remove arsenic. When the oxygen addition is lower than 0.6%, it can be seen from Figure 6 that 0.15–0.6% S is present in the copper, which has a competing reaction with As to react with CaO. The addition of more oxygen removes sulfur from copper to enable the arsenic to react with CaO. On the other hand, the same weight of Na_2O has less mole% than CaO due to its higher molecular mass. Na_2O can react with both of sulfur and arsenic in the copper, which consumed all of the 0.5% Na_2O added without oxygen addition. More CaO in mole can remove more arsenic from copper when sulfur in copper is low.

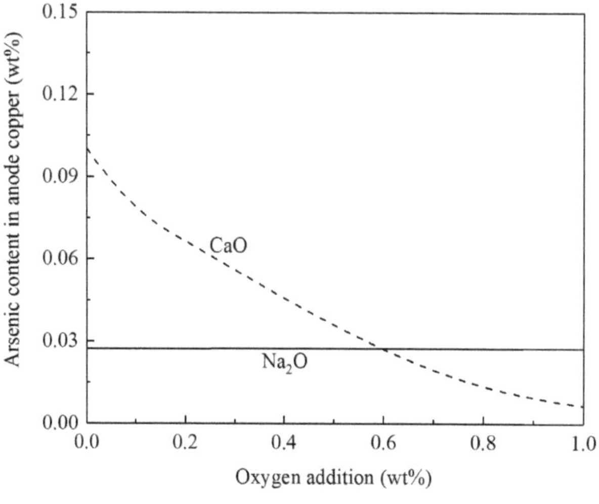

Figure 8. Effect of O_2/blister on arsenic content in anode copper with 0.5% flux/blister.

Effect of temperature on arsenic content in anode copper is shown in Figure 9 with 0.2% flux addition. It can be seen that at lower temperature, the arsenic content in anode copper can be reduced with increasing the temperature. The effect of temperature on arsenic content in anode copper is not significant at high temperatures and high flux additions.

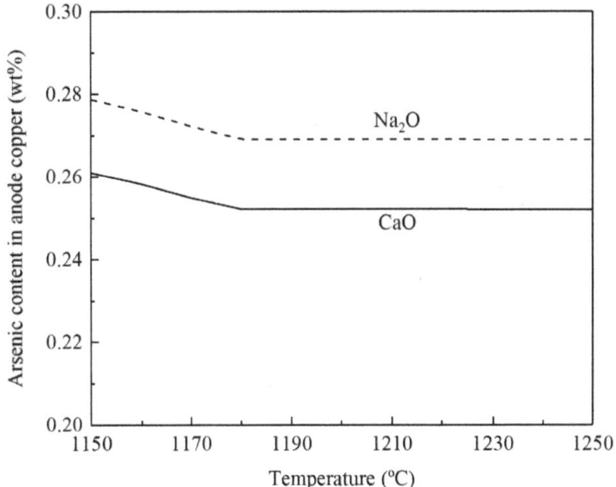

Figure 9. Effect of temperature on arsenic content in anode copper with 1.5% O$_2$/blister and 0.2% flux/blister.

Figure 10 shows the effect of flux addition on arsenic distribution at 1200 °C with O$_2$ addition 1.5%. It can be seen that CaO is a better flux than Na$_2$O in the removal of arsenic if oxygen is added which is the condition of oxidation stage in the fire refining process. With 0.55% CaO or Na$_2$O addition, over 99% arsenic brought from blister can be removed in the slag phase and a clean anode copper is obtained. The molecular mass of Na$_2$O (61.98) is larger than the molecular mass of CaO (56.08). When the same weight of Na$_2$O or CaO is added to the copper refining furnace with sufficient oxygen, CaO is a better flux than Na$_2$O to remove arsenic from the anode copper.

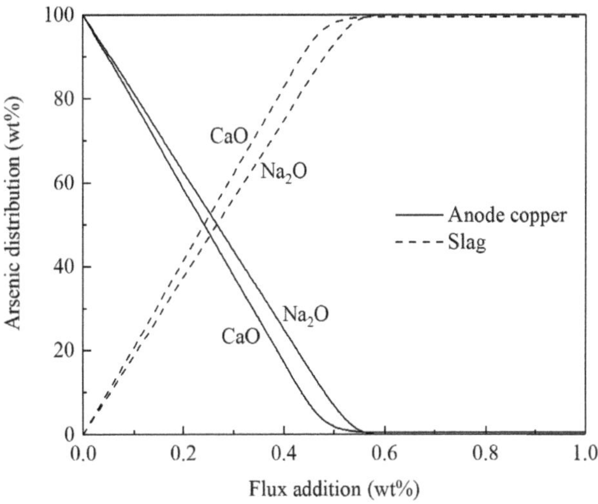

Figure 10. Effect of flux on arsenic distribution with 1.5% O$_2$/blister at 1200 °C.

4. Conclusions

Control of arsenic in the copper production is becoming to be a more and more important issue. Most of the studies were focused on the smelting process. The distribution and control of arsenic during copper converting and refining have been investigated by a review of the industrial data and thermodynamic calculations. Conventional converting and fire refining processes cannot remove arsenic efficiently. CaO is found to be a more effect flux in decreasing the liquidus temperature of the slag and arsenic content in the blister copper during the converting process. Both CaO and Na_2O can fix arsenic in the slag during the fire refining process. When oxygen is present, CaO is a better flux than Na_2O to remove arsenic from copper. Arsenic in anode copper can be slightly decreased with increasing temperature when the flux is added. More oxygen is beneficial for the removal of arsenic from anode copper with CaO addition.

Author Contributions: Conceptualization, B.Z. and S.X.; methodology, B.Z.; software, B.Z.; validation, F.X., T.W. and J.L.; formal analysis, J.L. and K.T.; investigation, F.X. and T.W.; data curation, F.X. and T.W.; writing—original draft preparation, F.X. and T.W.; writing—review and editing, K.T., B.Z. and S.X.; visualization, J.L. All authors have read and agreed to the published version of the manuscript.

Funding: This research received no external funding.

Institutional Review Board Statement: Not applicable.

Informed Consent Statement: Not applicable.

Data Availability Statement: Data is contained within the article.

Conflicts of Interest: The authors declare no conflict of interest.

References

1. Schlesinger, M.; Sole, K.; Davenport, W.; Alvear, G. Chapter 1—Overview. In *Extractive Metallurgy of Copper*, 6th ed.; Elsevier: Oxford, UK, 2021; pp. 1–18. ISBN 9780128218754.
2. Zhang, H.; Wang, Y.; He, Y.; Xu, S.; Hu, B.; Cao, H. Efficient and safe disposition of arsenic by incorporation in smelting slag through copper flash smelting process. *Miner. Eng.* **2021**, *160*, 106661. [CrossRef]
3. Zhou, H.; Liu, G.; Zhang, L.; Zhou, C. Mineralogical and morphological factors affecting the separation of copper and arsenic in flash copper smelting slag flotation beneficiation process. *J. Hazard. Mater.* **2021**, *401*, 123293. [CrossRef] [PubMed]
4. Zhao, B.; Liao, J. Development of Bottom-blowing copper smelting technology: A review. *Metals* **2022**, *12*, 190. [CrossRef]
5. Gao, D.; Peng, X.; Song, Y.; Zhu, Z.; Dai, Y. Mathematical modeling and numerical optimization of particle heating process in copper flash furnace. *Trans. Nonferrous Met. Soc. China* **2021**, *31*, 1506–1517. [CrossRef]
6. Kucharski, M. Arsenic removal from blister copper by soda injection into melts. *Scand. J. Metall.* **2002**, *31*, 246–250. [CrossRef]
7. Schlesinger, M.; Sole, K.; Davenport, W.; Alvear, G. Chapter 12—Fire refining (S and O removal) and anode casting. In *Extractive Metallurgy of Copper*, 6th ed.; Elsevier: Oxford, UK, 2021; pp. 313–329. ISBN 9780128218754.
8. Long, G.; Peng, Y.; Bradshaw, D. A review of copper–arsenic mineral removal from copper concentrates. *Miner. Eng.* **2012**, *36*, 179–186. [CrossRef]
9. Zhou, H.; Liu, G.; Zhang, L.; Zhou, C.; Mian, M.; Cheema, A. Strategies for arsenic pollution control from copper pyrometallurgy based on the study of arsenic sources, emission pathways and speciation characterization in copper flash smelting systems. *Environ. Pollut.* **2021**, *270*, 116203. [CrossRef]
10. Nagamori, M.; Mackey, P.; Tarassoff, P. Copper solubility in FeO- Fe_2O_3- SiO_2- Al_2O_3 slag and distribution equilibria of Pb, Bi, Sb and As between slag and metallic copper. *Metall. Mater. Trans. B* **1975**, *6*, 295–301. [CrossRef]
11. Flores, G.; Risopatron, C.; Pease, J. Processing of complex materials in the copper industry: Challenges and opportunities ahead. *JOM* **2020**, *72*, 3447–3461. [CrossRef]
12. Yang, W.; Qian, L.; Jin, B.; Feng, Q.; Li, L.; He, K.; Yang, J. Leaching behaviors of copper and arsenic from high-arsenic copper sulfide concentrates by oxygen-rich sulfuric acid leaching at atmospheric pressure. *J. Environ. Chem. Eng.* **2022**, *10*, 107358. [CrossRef]
13. Wang, Q.; Guo, X.; Tian, Q.; Chen, M.; Zhao, B. Reaction mechanism and distribution behavior of arsenic in the bottom blown Copper smelting process. *Metals* **2017**, *7*, 302. [CrossRef]
14. Zheng, Y.; Peng, Y.; Ke, L.; Chen, W. Separation and recovery of Cu and As from copper electrolyte through electrowinning and SO_2 reduction. *Trans. Nonferrous Met. Soc. China* **2013**, *23*, 2166–2173. [CrossRef]
15. Swinbourne, D.; Kho, T. Computational thermodynamics modeling of minor element distributions during copper flash converting. *Metall. Mater. Trans. B* **2012**, *43*, 823–829. [CrossRef]

16. Li, M.; Zhou, J.; Zhang, W.; Li, H.; Tong, C. Thermodynamics analysis of distribution behavior of impurity elements during copper flash converting. *Trans. Nonferrous Met. Soc.* **2017**, *27*, 1952–1958. (In Chinese)

17. Li, M.; Zhou, J.; Zhang, W.; Li, H.; Tong, C. Multiphase equilibrium thermodynamics analysis of copper flash converting process. *Trans. Nonferrous Met. Soc.* **2017**, *27*, 1494–1503. (In Chinese)

18. Yuan, Y.; Liu, S. Distribution and removal of arsenic in the process of bottom-blowing continuous copper smelting. *China Nonferrous Metall.* **2020**, *2*, 37–40. (In Chinese)

19. Liu, S. Operation practice of copper matte bottom blowing continuous converting. *Nonferrous Met. Extr. Metall.* **2016**, *12*, 17–19. (In Chinese)

20. Guo, X.; Chen, Y.; Wang, Q.; Wang, S.; Tian, Q. Copper and arsenic substance flow analysis of pyrometallurgical process for copper production. *Trans. Nonferrous Met. Soc. China* **2022**, *32*, 364–376. [CrossRef]

21. Zhou, Y.; Li, J.; Li, R.; Wu, F. Distribution of impurity elements in copper pyrometallurgical process and its influence on the recovery. *China Nonferrous Metall* **2019**, *4*, 9–16. (In Chinese)

22. Hidayat, T.; Chen, J.; Hayes, P.C.; Jak, E. Distributions of As, Pb, Sn and Zn as minor elements between iron silicate slag and copper in equilibrium with tridymite in the Cu-Fe-O-Si system. *Int. J. Mater. Res.* **2021**, *112*, 178–188. [CrossRef]

23. Kaur, R.; Swinbourne, D.; Nexhip, C. Nickel, lead and antimony distributions between ferrous calcium silicate slag and copper at 1300 °C. *Miner. Process. Extr. Metall. Rev.* **2009**, *118*, 65–72. [CrossRef]

24. Nagamori, M.; Mackey, P. Distribution equilibria of Sn, Se and Te between $FeO\text{-}Fe_2O_3\text{-}SiO_2\text{-}Al_2O_3\text{-}CuO_{0.5}$ slag and metallic copper. *Metall Mater Trans. B.* **1977**, *8*, 39–46. [CrossRef]

25. Avarmaa, K.; Yliaho, S.; Taskinen, P. Recoveries of rare elements Ga, Ge, In and Sn from waste electric and electronic equipment through secondary copper smelting. *Waste Manag.* **2018**, *71*, 400–410. [CrossRef]

26. Bale, C.; Bélisle, E.; Chartrand, P.; Decterov, S.; Eriksson, G.; Gheribi, A.; Hack, K.; Jung, I.; Kang, Y.; Melançon, J.; et al. FactSage thermochemical software and databases, 2010–2016. *Calphad* **2016**, *54*, 35–53. [CrossRef]

27. Liao, C.; Xie, S.; Zhao, B.; Cai, B.; Wang, L. Thermodynamic and experimental analyses of the carbothermic reduction of tungsten slag. *JOM* **2021**, *73*, 1853–1860. [CrossRef]

28. Yu, W.; Tang, Q.; Chen, J.; Sun, T. Thermodynamic analysis of the carbothermic reduction of a high-phosphorus oolitic iron ore by FactSage. *J. Miner. Met. Mater.* **2016**, *23*, 1126–1132. [CrossRef]

29. Nezhad, M.; Zabett, A. Thermodynamic analysis of the carbothermic reduction of electric arc furnace dust in the presence of ferrosilicon. *Calphad* **2016**, *52*, 143–151. [CrossRef]

30. Zhang, Z.; Yuan, L.; Huang, L.; Xu, Z. Trend and recovery of arsenic, antimony and bismuth in copper smelting. *Nonferrous Met. Sci. Eng.* **2019**, *10*, 13–19. (In Chinese)

31. Wang, S. Arsenic Distribution Surveying in Copper Smelting Process of Jinlong Copper Co., Ltd. *Nonferrous Met. Eng.* **2016**, *6*, 83–86. (In Chinese)

32. Wang, Q.; Zhong, S.; Guo, X.; Chen, J. Arsenic challenge and control in zijin copper smelter. In Proceedings of the 58th Annual Conference of Metallurgists (COM) Hosting the 10th International Copper Conference 2019, Vancouver, BC, Canada, 18–21 August 2019; p. 16.

33. Schlesinger, M.; Sole, K.; Davenport, W.; Alvear, G. Chapter 8—Converting of copper matt. In *Extractive Metallurgy of Copper*, 6th ed.; Elsevier: Oxford, UK, 2021; pp. 187–229. ISBN 9780128218754.

34. Yang, X.; Han, Z.; Guo, Y.; Xu, M. Study on lead impurity movement and removal in copper flash converting. *China Nonferrous Metall.* **2018**, *6*, 9–11. (In Chinese)

metals

Article

Recovery of Cu-Fe Alloy from Copper Smelting Slag

Yi Qu [1,2], Keqin Tan [3], Baojun Zhao [1,2] and Sui Xie [1,2,*]

1   Faculty of Materials Metallurgy and Chemistry, Jiangxi University of Science and Technology, Ganzhou 341000, China
2   International Institute for Innovation, Jiangxi University of Science and Technology, Nanchang 330013, China
3   Dongying Fangyuan Nonferrous Metals Co., Ltd., Dongying 257091, China
*   Correspondence: xiesui21@126.com

Abstract: Copper smelting slag usually contains 1–6 wt% copper, which can be recovered by pyrometallurgical and flotation processes. However, the tailing slags still consist of 0.3–0.7 wt% Cu and 35–45 wt% Fe equivalents to those in the copper and iron ores, respectively. Most of the research was focused on the recovery of iron from the tailing slags. Copper can increase the mechanical strength, corrosion resistance and antibacterial property of some steels. A new process to recover copper and iron directly and fully from hot copper smelting slag is proposed to produce Cu-Fe alloy for steel production. Effects of flux, temperature, reaction time, reductant type and addition on the recovery of copper and iron were investigated by high-temperature experiments and thermodynamic calculations. It was found that, with 5% CaO and 13–16% carbon additions, most of the copper and iron can be recovered from copper smelting slag at 1350–1400 °C. The copper and iron contents of the reduced slag are lower than 0.1% and 0.5%, respectively, at optimum condition. The new process has the advantages of low energy consumption, low flux addition and high recovery of copper and iron.

Keywords: copper smelting slag; smelting reduction; Cu-Fe alloy; FactSage

Citation: Qu, Y.; Tan, K.; Zhao, B.; Xie, S. Recovery of Cu-Fe Alloy from Copper Smelting Slag. *Metals* **2023**, *13*, 271. https://doi.org/10.3390/met13020271

Academic Editor: Nebojša Nikolić

Received: 31 December 2022
Revised: 27 January 2023
Accepted: 28 January 2023
Published: 29 January 2023

1. Introduction

80% to 90% of copper is produced through a pyrometallurgical process from concentrates around the world [1]. Copper smelting slag is a by-product of the pyrometallurgical process. 2–3 tons slag is generated for producing per ton of copper [2]. Approximately 35.2 million tons were produced globally in 2022 [3]. In the modern smelting process, copper smelting slag contains 1–6% copper and 35–45% iron [1]. Most of the copper in the slag is present as entrained matte and can be recovered by the cleaning process [2,4]. Pyrometallurgical cleaning [5–9] and flotation cleaning [10–13] are the main methods to recover copper from the smelting slag. The tailing slag contains 0.7% Cu from the pyrometallurgical cleaning [5–9] and 0.3% copper from the flotation cleaning. In addition, 35–45% iron in the cleaning slag is not recovered.

The contents of copper (0.3–0.7%) and iron (35–45%) in the cleaning slags are close to the grades of copper ores and iron ores, respectively. Extensive studies have been conducted to recover iron from the cleaning slags through reduction process [14–22]. For example, a tailing copper smelting slag containing 0.24% Cu and 38.23% Fe was used to recover iron at 1100–1300 °C [15]. 16% CaO and 20% coke fines were mixed with the tailing slag and pelletized for reduction. After reduction, the sample was ground, and the alloy powder was recovered by magnetic separation. The alloy contained 91% Fe and 0.18% Cu together with 0.1–0.6% Al_2O_3, CaO, MgO, Na_2O and SiO_2. This process used tailing slag powder for low-temperature reduction, which saved ground and heating energy. However, the maximum recovery of iron was only 92%, and the alloy cannot be directly used to produce steel due to the high impurities. Smelting separation at 1500 °C was used to remove the impurities, which increased the cost of the alloy [21]. The smelting reduction at 1500 °C was undertaken with additions of CaO and MgO. The maximum recovery of 90% Fe was achieved with 20% CaO and 10% MgO [14]. Copper recovery was not mentioned in these studies as the Cu content in the slags was low. The smelting reduction can separate the alloy from the oxide slag completely to obtain a clean product. However, this process did not consider the recovery of copper, and the smelting conditions such as flux addition and temperature need to be optimized to increase the recovery of the metals. All copper smelters have a slag cleaning process to recover copper from the slags. Most of the research was focused on the recovery of iron from the cold tailing slags. The recovered alloy was not suitable for general steel production because the copper in the alloy is considered an impurity in most of the steels. All proposed processes to recover iron from copper cleaning slags are not commercialized due to the high cost and high level of impurities.

In most of the steels, copper is considered an impurity that induces severe microstructural distortions in the steel and decreases its tensile strength, impact energy and hardness [23]. On the other hand, it was found that 2.5–4% copper in steel increases the mechanical strength, corrosion resistance and antibacterial property of steel [24–30]. Copper metal is usually used to produce Cu-containing steels. However, the cost is high to use high-purity copper, and the distribution of copper in the steel is not uniform, which limits the applications of the Cu-containing steels. Hao et al. studied the possibility of producing antibacterial stainless steel from copper cleaning slag via smelting reduction [31]. The optimum condition was found at 1450 °C with 20% CaO and 13% carbon additions. The overall metal recovery was 91.76% at optimum condition. The contents of Cu and Fe in the reduced slag were 0.015% and 0.29%, respectively. The copper content in the alloy was 1.65%, which is lower than that required for antibacterial stainless steel. Extra copper metal needs to be added to the steel production. A new process to recover iron and copper directly from copper smelting slag is developed in the present study. The molten slag as a feed can be used in the reduction process to save energy. The new technology omits the cleaning process and produces a Cu-Fe alloy with high Cu, which can be directly used as a feed for the production of Cu-containing steels. Effects of the parameters including flux, temperature, reaction time, reductant type and addition on the recovery of copper and iron from copper smelting slag are investigated in this study.

2. Materials and Methods

2.1. Materials

The copper smelting slag was provided by a smelter, and the composition of the slag measured by an XRF (X-ray Fluorescence Spectrometer) is shown in Table 1. The iron in the slag includes both Fe^{2+} and Fe^{3+}. For presentation purposes, all iron is expressed as "FeO". It can be seen that "FeO" and SiO_2 are the main components in the slag. 6.83% Cu_2O and 8.52% ZnO are also present in the slag. In the high temperature experiments, coke and graphite were used as reductants and CaO (analytically pure) was used as a flux.

Table 1. Composition of the smelting slag measured by XRF.

"FeO"	MgO	CaO	SiO_2	Al_2O_3	ZnO	Cu_2O	S
54.67	2.53	3.22	20.54	2.23	8.52	6.83	1.46

2.2. Experiments

The equipment used in this study is shown in Figure 1. The experiments were carried out in a vertical tube furnace heated by $MoSi_2$ heating elements. A Pt/Pt-13%Rh thermocouple inside an alumina sheath was placed in the hot zone of the furnace to accurately measure the temperature of the sample. The sample, consisting of the required slag, reductant and flux was mixed in an agate mortar for 30 min and pelletized. The pelletized mixture was placed in a corundum crucible and suspended at the bottom of the furnace. The furnace was properly sealed, and ultrahigh purity argon was flushed for 30 min to remove the air inside. The crucible with the mixture was raised from the bottom to the hot zone of the reaction tube for the required reaction time. The sample was quenched in water after the reduction was finished. The experimental conditions are shown in Table 2. Alloy and slag were separated after the reduction and quenching. The compositions of the alloy and slag were measured by ICP-OES (Inductively Coupled Plasma Optical Emission Spectrometer; Agilent 5900, Palo Alto, CA, USA), and the microstructures of the quenched slag were observed by SEM (Scanning electron microscopy; FEI, Hillsboro, OR, USA).

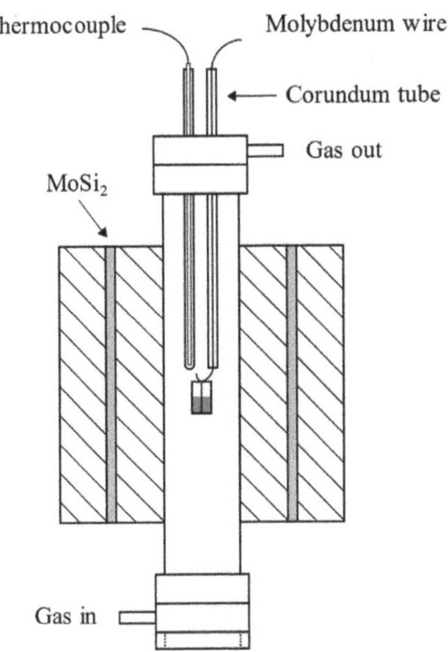

Figure 1. Schematic diagram of the experimental system used in this study.

Table 2. Experimental conditions.

Exp	Slag/g	CaO/g	Carbon/g	C Type	Temperature/°C	Time/min
1	10	0.5	1.3	Graphite	1250	120
2	10	0.5	1.3	Graphite	1350	120
3	10	0.5	1.3	Graphite	1450	120
4	10	0.5	1	Graphite	1400	120
5	10	0.5	1.3	Graphite	1400	120
6	10	0.5	1.6	Graphite	1400	120
7	10	0.5	1.4	Graphite	1350	120
8	10	0.5	1.3	Graphite	1400	60
9	10	0.5	1.3	Graphite	1400	120
10	10	0.5	1.3	Graphite	1400	180
11	10	0.5	1	Coke	1400	120
12	10	0.5	1.3	Coke	1400	120
13	10	0.5	1.6	Coke	1400	120

2.3. Thermodynamic Predictions

The thermodynamic software FactSage has been widely used in the fields of materials and metallurgy [32]. The latest version, FactSage 8.2 (Thermfact/CRCT and GTT-Technologies), was used in the present study to predict the reactions and plan the experiments. The databases "FactPS", "FToxid" and "FTmisc" were used in the "Equilib" module. The solution phases selected in the calculations included "FTmisc-FeLQ", "FTmisc-CuLQ", "FTmisc-FeCu", "FTmisc-MATT", "FToxid-SLAGA", "FToxid-SPINC", FToxid-MeO", "FToxid-cPyrA", "FToxid-WOLLA", "FToxid-bC2SA", "FToxid-aC2S", "FToxid-Mel_A", "FToxid-OlivA".

3. Results and Discussion

The overall reaction in the present study to recover valuable metals from the copper smelting slag includes the following:

$$\text{Slag1}(Cu_2O, FeOx, SiO_2, Al_2O_3, MgO)(l) + CaO(s) + C(s) \rightarrow \text{Cu-Fe}(l) +$$
$$\text{Slag2}(Cu_2O, FeOx, ZnO, SiO_2, Al_2O_3, MgO)(l)$$

Slag1, in the previous studies, is a tailing slag which after cleaning contained little copper. In the present study, Slag1 is a smelting slag containing relatively more copper. CaO is a flux that can be added as $CaCO_3$ or CaO. C is a reductant to be added as graphite, coal or coke. The Cu_2O and FeOx in Slag1 can be reduced to form the Cu-Fe alloy. Other components from Slag1 react with the flux to form Slag2, which is a tailing slag. The variables in the reaction include the ratios of flux and reductant to Slag1, temperature and reaction time, which will be discussed in the following sections.

3.1. Control of Liquidus Temperature of Reduced Slag

When the copper smelting slag reacts with carbon, the oxides of copper, iron and zinc can be reduced to form the Cu-Fe-C alloy, and zinc is vaporized from the condensed system. Most of sulfur in the slag is also present in the alloy. After reduction, the remaining slag contains only CaO, MgO, Al_2O_3 and SiO_2. The liquidus temperature of the alloy Cu-Fe-C is low. It is essential to keep the reduced slag fully liquid to separate the alloy from the slag completely. The effect of flux addition on the liquidus temperature of the reduced slag is calculated by FactSage 8.2 and shown in Figure 2. It can be seen that the liquidus temperature of the reduced slag without flux is 1560 °C. Both CaO and MgO can decrease the liquidus temperature of the reduced slag. 5% CaO can rapidly decrease the liquidus temperature to 1270 °C. If the CaO addition is more than 7%, the slag will enter the $CaSiO_3$ primary phase field and the liquidus temperatures will increase. MgO can only decrease the liquidus temperatures to 1380 °C, showing that CaO is a more effective flux to control the liquidus temperature of the reduced slag.

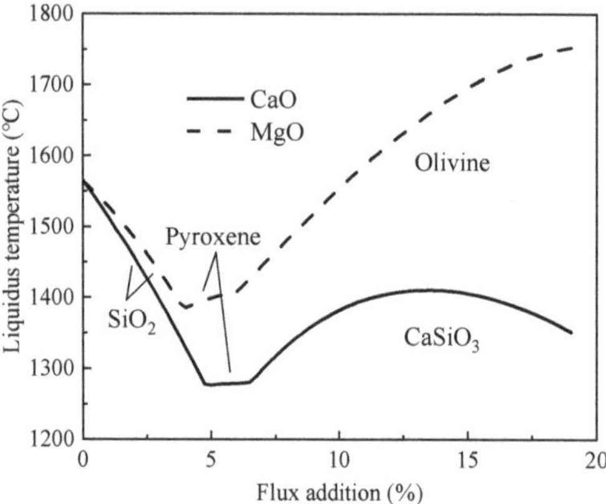

Figure 2. Effect of flux addition on the liquidus temperature of slag.

Figure 3 shows the appearance of the alloy from experiments 1 to 6 in Table 2. It can be seen from the figure that, after the reduction at 1250 °C (Exp 1), both alloy and slag were not liquid and they were not separated. As the reduction temperature was increased to 1350 °C (Exp 2), the alloy was separated from the slag. However, there are still slag and pores present inside the alloy, indicating that it was not completely melted. A dense alloy was obtained from experiments 3–6 indicating that the reduction temperature of 1400 °C is suitable to produce a dense alloy.

Figure 3. Appearance of the alloy after reduction: (**1**) 1250 °C, 13% C; (**2**) 1350 °C, 13% C; (**3**) 1450 °C, 13% C; (**4**) 1400 °C, 10% C; (**5**) 1400 °C, 13% C; (**6**) 1400 °C, 16% C.

Figure 4a shows the typical microstructure of the quenched sample after reduction at 1250 °C. The black phase shown in Figure 4a is graphite, the grey phase is slag, and the white phase is alloy. It seems that the alloy was not liquid at 1250 °C because it did not separate from the graphite and slag. Figure 4b shows the typical microstructure of the quenched sample after reduction at 1350 °C. It can be seen that the slag was fully liquid at a high temperature and converted to a homogeneous glass on quenching. The microstructures shown in Figure 4 confirm the observations shown in Figure 3 and the prediction shown in Figure 2. With the support of thermodynamic predictions, high temperature experiments show that only a 5% addition of CaO can decrease the liquidus temperature of the reduced slag to approximately 1300 °C. This result provides the possibility of low-temperature smelting reduction compared to those (1450–1500 °C) previously reported in the literature [14,31]. 5% CaO addition not only decreases the flux cost but also reduces the volume of the reduced slag by approximately 30% compared to the previous studies [14,31] where 20% CaO was added.

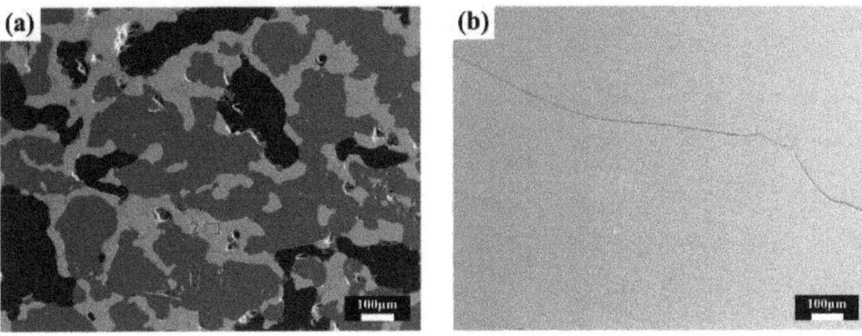

Figure 4. Typical microstructures of the quenched slags after reduction, (**a**) 1250 °C; (**b**) 1350 °C.

3.2. Effect of Reductant on the Recovery of Copper and Iron

The effect of carbon addition on amounts of the phases during the reduction of the copper smelting slag was calculated by FactSage 8.2, and the results are shown in Figure 5. 100 g of copper smelting slag was used as a base, and the temperature and CaO addition were 1400 °C and 5%, respectively, for the calculations. It can be seen that the proportion of the slag phase decreases with increasing carbon addition as "FeO", Cu_2O and ZnO in the slag were reduced from the slag. As a result, Fe and Cu form an alloy, and Zn vaporizes as a gas. Zinc can be collected in flue gas as a dust which is an extra advantage of this technology. According to the prediction, ZnO in the slag is completely reduced with the

2.5% carbon addition. All iron and copper oxides are reduced with a 14% carbon addition. Further addition of carbon causes the formation of graphite.

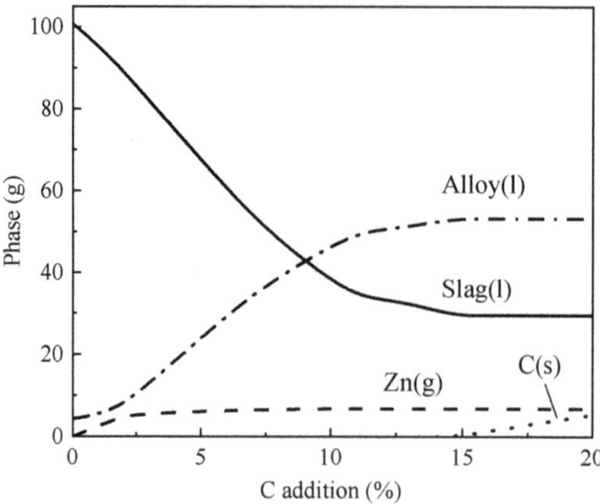

Figure 5. Effect of carbon addition on the amounts of the phases during reduction at 100g slag, 1400 °C and 5% CaO addition.

Figure 6 shows the experimental results on the copper and iron contents remaining in the reduced slag after reduction at 1400 °C with 5% CaO addition. As can be seen from Figure 6, after reduction by 10% graphite, the copper and iron contents in the remining slag are 0.14% and 16%, respectively. The copper and iron contents of the remining slag decrease with increasing carbon addition. As the carbon addition increased to 16%, the copper and iron in the reduced slag decreased to 0.06% and 0.5%, respectively. The experimental results confirmed the reductant ratio predicted by FactSage and that reported by Hao et al. [31]. The exact requirement of the reductant depends on the proportions of the reducible oxides in the slag, the types of the reductants and the reaction temperature.

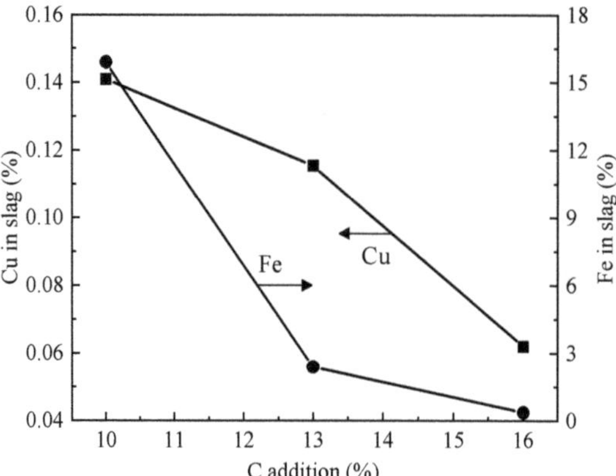

Figure 6. Experimentally determined copper and iron contents in the reduced slag at 1400 °C with 5% CaO addition.

Figures 7 and 8 show the effect of carbon type on the copper and iron contents in the reduced slags after reduction at 1400 °C and 5% CaO addition. It can be seen from Figure 7 that the copper content in the reduced slag decreases with increasing graphite and coke additions. The coke (which contains 85% carbon) is a more effective reductant than the graphite due to its higher reactivity and surface area. With 13% graphite and coke (13% caron equivalent) addition, the copper content in the reduced slag is 0.12 and 0.053%, respectively. To decrease the copper content to 0.06%, 16% graphite is required. It can be seen from Figure 8 that 13% carbon equivalent coke addition can reduce the iron in the slag to 0.2%. The same amount of graphite addition only reduced the iron content in the slag to 2.5%. Similarly, more than 16% graphite is required to reduce the iron content in the slag to 0.2%.

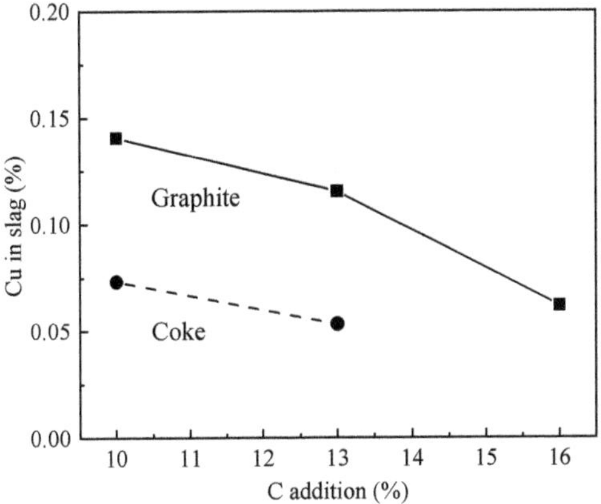

Figure 7. Effect of carbon type on copper content in the reduced slags after reduction at 1400 °C and 5% CaO addition.

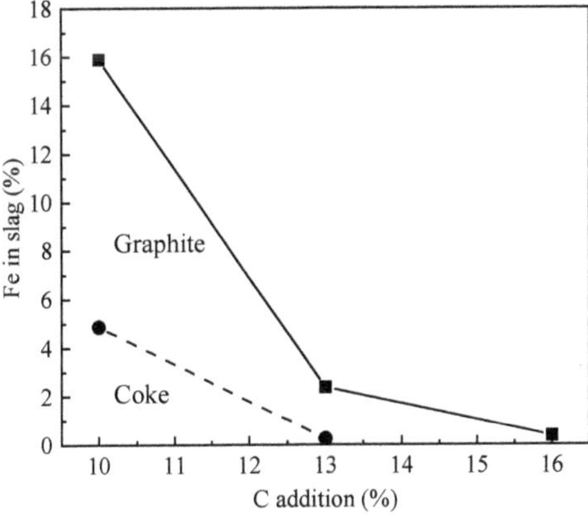

Figure 8. Effect of carbon type on iron content in the reduced slags after reduction at 1400 °C and 5% CaO addition.

3.3. Effect of Temperature on the Recovery of Copper and Iron

The effect of temperature on the copper and iron contents in the reduced slag, predicted by FactSage 8.2, is shown in Figure 9. From 1350 to 1410 °C, the amount of iron in the reduced slag decreases slowly with increasing temperature. Above 1410 °C, the iron content of the reduced slag decreases rapidly with increasing temperature. In contrast, copper in the reduced slag increases with decreasing temperature below 1410 °C. Above 1410 °C, the copper content of the reduced slag decreases rapidly with increasing temperature. The increase of copper in the slag can be explained by the sulfur in the slag. It can be seen from Figure 9 that sulfur in the slag first increases and then decreases with increasing temperature. When the sulfur is high in the slag, the chemical bond Cu-S is strong, which affects the reduction of copper from the slag.

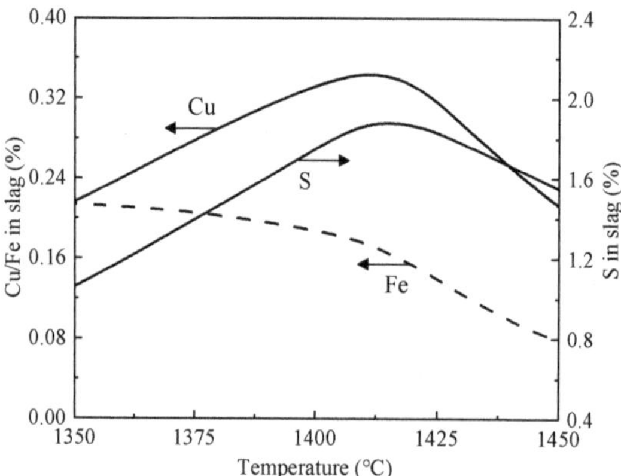

Figure 9. Effect of temperature on the copper and iron contents in the reduced slag, predicted by FactSage 8.2: 13% carbon and 5% CaO.

Figure 10 shows the experimental results about the effect of temperature on the copper and iron contents in the reduced slag with 5% CaO and 13% graphite. It can be seen that the iron content of the slag decreases slowly at lower temperatures and fast at higher temperatures, which shows the same trend as that predicted by FactSage. In the temperature range investigated (1350–1450 °C), copper in the slag continuously increases, which is different from that predicted by FactSage as shown in Figure 9. An optimum temperature should be selected to maximize the recovery of copper by considering the liquidus temperatures of the alloy and slag and the copper content in the reduced slag. Hao et al. [31] reported that the copper content in the reduced slag decreased from 1400 to 1450 °C and then increased from 1450 to 1500 °C, followed by a decrease from 1500 to 1550 °C. They also reported that the iron content of the reduced slag increased from 1400 to 1500 °C, followed by a decrease from 1500 to 1550 °C, but the reasons for the observed trends were not mentioned.

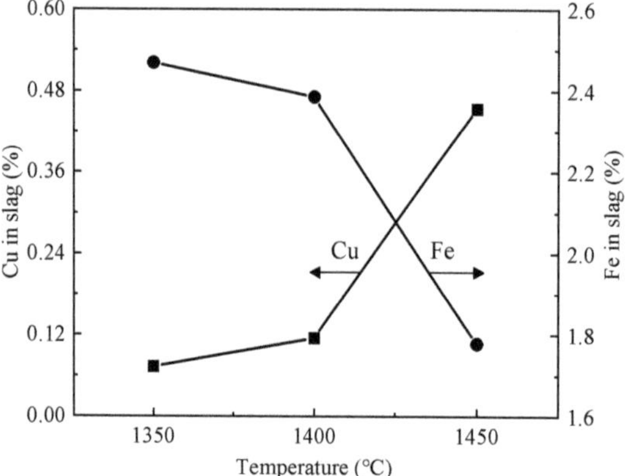

Figure 10. Effect of temperature on the copper and iron contents in the reduced slag by experiments, 5% CaO and 13% graphite.

3.4. Effects of Reduction Time on the Recovery of Copper and Iron

The effect of reduction time on the copper and iron contents in the reduced slag was studied by experiments at 1400 °C with 5% CaO addition and 13% graphite. As can be seen from Figure 11, the copper and iron contents of the slag decrease with increasing reduction time. It seems that the reduction of copper approached equilibrium because thre was only a small decrease in the copper content from 120 min to 180 min. On the other hand, the iron content in the slag continuously decreases, indicating that the reduction of iron did not reach equilibrium in 180min. A proper reaction time is determined economically, as a long reaction time increases the fuel cost and reduces productivity. Heo et al. [14] reported that the iron content in the reduced slag decreased linearly with the reaction time up to 60 min which showed the same trend as the present study. However, in the study of Hao et al. [31], the contents of copper and iron in the reduced slag showed a fluctuating upward trend with the reaction time between 15 and 60 min.

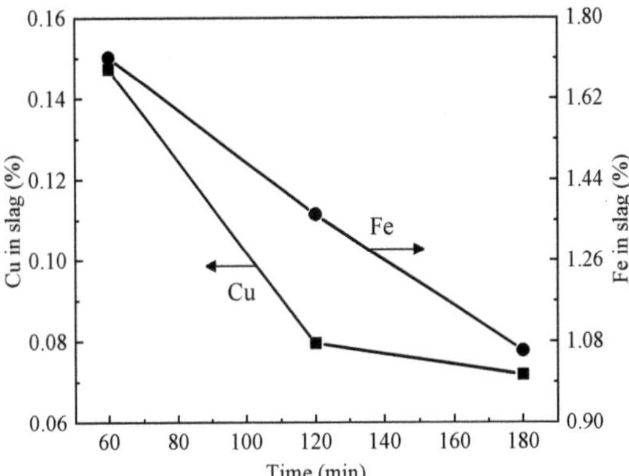

Figure 11. Effect of reduction time on the copper and iron contents in the reduced slag at 1400 °C, 5% CaO and 13% graphite.

Laboratory research is a systematic study to identify the optimum conditions to maximize the recovery of valuable metals from the copper smelting slag with a minimum cost. Copper smelting slag is around 1200 °C when it is tapped from the smelting furnace. In most of the smelters, the slag is slow-cooled and ground to recover copper through a flotation process. The previous research was focused on the recovery of metals (mainly iron) from the tailing slag. In the present study, a novel process is proposed to recover valuable metals directly from the hot copper smelting slag. This process does not need a flotation plant and fully uses the heat from the molten slag. In one single step, copper, iron and zinc can be recovered through smelting and reduction. The present study demonstrates that the optimum conditions to achieve a good recovery of the metals from the copper smelting slag include 5% CaO and 13–16% carbon additions, 1350–1400 °C, and a 60 min reaction. These conditions are specifically for the as-received copper smelting slag and available flux and reductants. In an industrial operation, maximum recovery is not the target, and economic factors need to be taken into account. For example, higher temperatures and longer reaction times can increase the recovery but also increase the cost of the fuel and refractory. The optimum conditions in the laboratory study may not be the best when they are applied to an industrial operation. The research is only used as a guideline to support the industry in adjusting the operating parameters to maximize the economic benefits.

4. Conclusions

Cu-Fe alloy has wide applications in the production of Cu-containing steels. A new process to produce Cu-Fe alloy directly from hot copper smelting slag. Effects of flux, temperature, reaction time, reductant type, and amount on the recovery of copper and iron were studied by thermodynamic calculations and high temperature experiments. Optimal flux and reductant additions have been determined. High temperatures decrease the iron content but increases the copper content in the slag. Coke is a better reductant than graphite to reduce copper and iron from the slag. In optimal conditions, over 99% of copper and iron can be recovered from the copper smelting slag. Zinc in the slag can also be recovered from the dust. The proposed process has the potential to replace the existing slag cleaning plant in the copper smelter to produce a Cu-Fe alloy suitable for steel production.

Author Contributions: Methodology, B.Z.; software, B.Z.; resources, K.T.; writing—original draft preparation, Y.Q.; writing—review and editing, K.T., B.Z. and S.X. All authors have read and agreed to the published version of the manuscript.

Funding: The authors would like to thank Dongying Fangyuan Nonferrous Metals Co., Ltd. and Jiangxi University of Science and Technology for providing financial support for this research.

Conflicts of Interest: The authors declare no conflict of interest.

References

1. Zuo, Z.L.; Feng, Y.; Dong, X.J.; Luo, S.; Ren, D.D.; Wang, W.H.; Wu, Y.X.; Yu, Q.B.; Lin, H.; Lin, X.Q. Advances in Recovery of Valuable Metals and Waste Heat from Copper Slag. *Fuel Process. Technol.* **2022**, *235*, 107361. [CrossRef]
2. Tian, H.Y.; Guo, Z.Q.; Pan, J.; Zhu, D.Q.; Yang, C.C.; Xue, Y.X.; Li, S.W.; Wang, D.Z. Comprehensive Review on Metallurgical Recycling and Cleaning of Copper Slag. *Resour. Conserv. Recycl.* **2021**, *168*, 105366. [CrossRef]
3. U.S. Geological Survey. *Mineral Commodity Summaries 2022*; U.S. Geological Survey: Reston, VA, USA, 2022. [CrossRef]
4. Zhou, W.T.; Liu, X.; Lyu, X.J.; Gao, W.H.; Su, H.L.; Li, C.M. Extraction and Separation of Copper and Iron from Copper Smelting Slag: A Review. *J. Clean. Prod.* **2022**, *368*, 133095. [CrossRef]
5. Zhou, S.W.; Wei, Y.G.; Shi, Y.; Li, B.; Wang, H. Characterization and Recovery of Copper from Converter Copper Slag Via Smelting Separation. *Metall. Mater. Trans. B* **2018**, *49*, 2458–2468. [CrossRef]
6. Zhang, J.; Qi, Y.H.; Yan, D.L.; Xu, H.C. A New Technology for Copper Slag Reduction to Get Molten Iron and Copper Matte. *J. Iron Steel Res. Int.* **2015**, *22*, 396–401. [CrossRef]
7. Zhang, H.P.; Li, B.; Wei, Y.G.; Wang, H.; Yang, Y.D. Effect of CaO on Copper Loss and Phase Transformation in Copper Slag. *Metall. Mater. Trans. B* **2022**, *53*, 1538–1551. [CrossRef]
8. Guo, Z.Q.; Pan, J.; Zhu, D.Q.; Zhang, F. Green and Efficient Utilization of Waste Ferric-Oxide Desulfurizer to Clean Waste Copper Slag by the Smelting Reduction-Sulfurizing Process. *J. Clean. Prod.* **2018**, *199*, 891–899. [CrossRef]
9. Zuo, Z.L.; Yu, Q.B.; Wei, M.Q.; Xie, H.Q.; Duan, W.J.; Wang, K.; Qin, Q. Thermogravimetric Study of the Reduction of Copper Slag by Biomass: Reduction Characteristics and Kinetics. *J. Therm. Anal. Calorim.* **2016**, *126*, 481–491. [CrossRef]
10. Shang, X.K.; Lan, Z.Y.; Zhang, Q.F.; Li, T.T. A Comprehensive Recovery of Copper-Iron Experimental Research in the Copper Smelting Slag. *Adv. Mater. Res.* **2014**, *926*, 4197–4200. [CrossRef]
11. Roy, S.; Datta, A.; Rehani, S. Flotation of Copper Sulphide from Copper Smelter Slag Using Multiple Collectors and Their Mixtures. *Int. J. Miner. Process.* **2015**, *143*, 43–49. [CrossRef]
12. Meng, X.Y.; Li, Y.; Wang, H.Y.; Yang, Y.D.; Mclean, A. Effects of Na$_2$O Additions to Copper Slag on Iron Recovery and the Generation of Ceramics from the Non-Magnetic Residue. *J. Hazard. Mater.* **2020**, *399*, 122845. [CrossRef] [PubMed]
13. Liu, R.Q.; Zhai, Q.L.; Wang, C.; Li, X.; Sun, W. Optimizing the Crystalline State of Cu Slag by Na$_2$CO$_3$ to Improve Cu Recovery by Flotation. *Minerals* **2020**, *10*, 820. [CrossRef]
14. Heo, J.H.; Kim, B.-S.; Park, J.H. Effect of CaO Addition on Iron Recovery from Copper Smelting Slags by Solid Carbon. *Metall. Mater. Trans. B* **2013**, *44*, 1352–1363. [CrossRef]
15. Long, H.; Meng, Q.; Chun, T.; Wang, P.; Li, J. Preparation of Metallic Iron Powder from Copper Slag by Carbothermic Reduction and Magnetic Separation. *Can. Metall. Q.* **2016**, *55*, 338–344. [CrossRef]
16. Guo, Z.Q.; Zhu, D.Q.; Pan, J.; Yao, W.J.; Xu, W.Q.; Chen, J.N. Effect of Na$_2$CO$_3$ Addition on Carbothermic Reduction of Copper Smelting Slag to Prepare Crude Fe-Cu Alloy. *JOM* **2017**, *69*, 1688–1695. [CrossRef]
17. Guo, Z.Q.; Zhu, D.Q.; Pan, J.; Zhang, F. Innovative Methodology for Comprehensive and Harmless Utilization of Waste Copper Slag via Selective Reduction-Magnetic Separation Process. *J. Clean. Prod.* **2018**, *187*, 910–922. [CrossRef]
18. Wang, J.-P.; Erdenebold, U. A Study on Reduction of Copper Smelting Slag by Carbon for Recycling into Metal Values and Cement Raw Material. *Sustainability* **2020**, *12*, 1421. [CrossRef]
19. Łabaj, J.; Blacha, L.; Jodkowski, M.; Smalcerz, A.; Fröhlichová, M.; Findorak, R. The Use of Waste, Fine-Grained Carbonaceous Material in the Process of Copper Slag Reduction. *J. Clean. Prod.* **2021**, *288*, 125640. [CrossRef]
20. Du, J.L.; Zhang, F.X.; Hu, J.H.; Yang, S.L.; Liu, H.L.; Wang, H. Direct Reduction of Copper Slag Using Rubber Seed Oil as a Reductant: Iron Recycling and Thermokinetics. *J. Clean. Prod.* **2022**, *363*, 132546. [CrossRef]
21. Pan, L.T.; Zhu, D.Q.; Guo, Z.Q.; Pan, J. Preparation of a Master Fe–Cu Alloy by Smelting of a Cu-Bearing Direct Reduction Iron Powder. *Metals* **2019**, *9*, 701. [CrossRef]
22. Gorai, B.; Jana, R.K. Premchand Characteristics and Utilisation of Copper Slag—A Review. *Resour. Conserv. Recycl.* **2003**, *39*, 299–313. [CrossRef]
23. Sekunowo, O.I.; Durowaye, S.I.; Gbenebor, O.P. Effect of Copper on Microstructure and Mechanical Properties of Construction Steel. *Int. J. Chem. Mol. Nucl. Mater. Metall. Eng.* **2014**, *8*, 839–843.

24. Takaki, S.; Fujioka, M.; Aihara, S.; Nagataki, Y.; Yamashita, T.; Sano, N.; Adachi, Y.; Nomura, M.; Yaguchi, H. Effect of Copper on Tensile Properties and Grain-Refinement of Steel and Its Relation to Precipitation Behavior. *Mater. Trans.* **2004**, *45*, 2239–2244. [CrossRef]
25. Hong, I.T.; Koo, C.H. Antibacterial Properties, Corrosion Resistance and Mechanical Properties of Cu-Modified SUS 304 Stainless Steel. *Mater. Sci. Eng. A* **2005**, *393*, 213–222. [CrossRef]
26. Vaynman, S.; Fine, M.E.; Chung, Y.-W.; Hahin, C. *Low-Carbon, Cu-Precipitation-Strengthened Steel*; Association for Iron & Steel Technology: Warrendale, PA, USA, 2011; pp. 181–188.
27. Yokoi, T.; Maruyama, N.; Takahashi, M.; Sugiyama, M. Application of Controlled Cu Nano-Precipitation for Improvement in Fatigue Properties of Steels. *Nippon Steel Tech. Rep.* **2005**, *91*, 49–55.
28. Ren, L.; Nan, L.; Yang, K. Study of Copper Precipitation Behavior in a Cu-Bearing Austenitic Antibacterial Stainless Steel. *Mater. Des.* **2011**, *32*, 2374–2379. [CrossRef]
29. Yin, Y.; Zhang, X.; Wang, D.; Nan, L. Study of Antibacterial Performance of a Type 304 Cu Bearing Stainless Steel against Airborne Bacteria in Real Life Environments. *Mater. Technol.* **2015**, *30*, B104–B108. [CrossRef]
30. Zhang, X.R.; Yang, C.G.; Yang, K. Contact Killing of Cu-Bearing Stainless Steel Based on Charge Transfer Caused by the Microdomain Potential Difference. *ACS Appl. Mater. Interfaces* **2020**, *12*, 361–372. [CrossRef]
31. Hao, J.; Dou, Z.H.; Zhang, T.A.; Jiang, B.C.; Wang, K.; Wan, X.Y. Manufacture of Wear-Resistant Cast Iron and Copper-Bearing Antibacterial Stainless Steel from Molten Copper Slag via Vortex Smelting Reduction. *J. Clean. Prod.* **2022**, *375*, 134202. [CrossRef]
32. Bale, C.W.; Bélisle, E.; Chartrand, P.; Decterov, S.A.; Eriksson, G.; Gheribi, A.E.; Hack, K.; Jung, I.-H.; Kang, Y.-B.; Melançon, J.; et al. Reprint of: FactSage thermochemical software and databases, 2010–2016. *Calphad* **2016**, *55*, 1–19. [CrossRef]

Article

Liquid Formation in Sinters and Its Correlation with Softening Behaviour

Vaishak Kamireddy [1], Dongqing Wang [2], Wen Pan [2], Shaoguo Chen [2], Tim Evans [3], Fengqiu Tang [1,*], Baojun Zhao [1,4] and Xiaodong Ma [1,*]

[1] Sustainable Minerals Institute, The University of Queensland, Brisbane 4072, Australia; v.kamireddy@uq.edu.au (V.K.); baojun@uq.edu.au (B.Z.)
[2] Shougang Research Institute of Technology, Beijing 100041, China; wangdongqing1980@163.com (D.W.); panwen@shougang.com.cn (W.P.); chenshaoguo@shougang.com.cn (S.C.)
[3] Rio Tinto Iron Ore, Perth 6000, Australia; tim.evans@riotinto.com
[4] REEM International Institute for Innovation, Jiangxi University of Science and Technology, Ganzhou 341000, China
* Correspondence: f.tang@uq.edu.au (F.T.); x.ma@uq.edu.au (X.M.)

Abstract: Modern blast furnaces with extensive operational volume demand better-quality iron agglomerates as feed for stable operation. Sinter is the principal feed used in blast furnaces across Asia. Liquid generated during the sintering process plays an essential role in the coalescence of the sinter blend and in sinter quality. Therefore, an estimation of liquid properties at peak bed conditions during sintering helps manage sintering liquid behaviour, leading to better control of final sinter properties. In this study, three different iron sinters were reheated to sinter bed conditions, followed by quenching. Electron probe X-ray microanalysis (EPMA) was used to identify the resultant phases and quantify their chemical compositions. The impact of sinter bulk compositions was analysed, especially on sintering liquid properties. Furthermore, experiments were conducted to study the softening and melting behaviour of the sinters, and the cohesive range of the sinters was identified. Finally, the effect of the sinter bulk compositions on sintering liquid properties and softening behaviour is detailed.

Keywords: sinter; sintering liquid; EPMA; sinter softening behaviour

Citation: Kamireddy, V.; Wang, D.; Pan, W.; Chen, S.; Evans, T.; Tang, F.; Zhao, B.; Ma, X. Liquid Formation in Sinters and Its Correlation with Softening Behaviour. *Metals* **2022**, *12*, 885. https://doi.org/10.3390/met12050885

Academic Editor: Jean-François Blais

Received: 14 April 2022
Accepted: 21 May 2022
Published: 23 May 2022

Publisher's Note: MDPI stays neutral with regard to jurisdictional claims in published maps and institutional affiliations.

1. Introduction

Steel is an important alloy used in our daily lives, spanning households to space technology. The predominant means of steel production are through basic oxygen furnaces (BOF). The blast furnace route produces the principal amount of the total hot metal used as raw material in BOF [1]. Efficient blast furnaces with extensive operational volume are competitive, mainly because of the economies of scale, operational continuity, and no colossal electricity consumption. Moreover, as the blast furnaces are becoming taller and broader with increased operating volume, the demand for good quality feed withstanding the harsh environment inside the blast furnace is rising, leading to increased adoption of agglomeration techniques for preparing blast furnace feed [2]. Agglomerates, such as sinters and pellets, have become significant, mainly due to their mechanical reliability and suitability for large-scale production [1,2].

Despite its competitive operation, the blast furnace unit is the foremost contributor of CO_2 equivalent emissions in a steel plant. Therefore, in addition to pulverised coal injection and other techniques to increase the ore/coke ratio, there is also a significant effort to control the slag volume. As sinters are the predominant feed for blast furnaces across Asia, sinters generating low-FeO containing slags on reduction are increasingly preferred so as to have a higher softening start temperature and a narrow cohesive zone, and slags containing low-FeO and higher basicity have reduced adhesion to the iron surface, leading to the latter's increased carburisation. In addition, the increased gangue content in sinters increases the melt-down temperature, expanding the cohesive zone [3].

To achieve lower slag content, when Al_2O_3 content and FeO content were reduced in sinters, an improvement in shaft efficiency of the blast furnace was observed using a blast furnace inner reaction simulator [4]. Here, the silica content was also reduced to compensate for the excess melt generated during sintering due to the reduced alumina content. The silica content is vital in melt/liquid formation during sintering. The increase in silica content helps reduce the liquid formation temperature and thereby increases the amount of liquid formed when the peak sinter bed temperature is reached [5,6]. However, higher silica content increases the FeO content in the final sinter as more of $2FeO.SiO_2$ is formed, blocking FeO reoxidation [7]. The challenge for sinter plants is to produce low-silica and low-FeO sinters without compromising final sinter properties. Reduction properties of the sinter can be improved by controlling the liquid formation during sintering as the sintering liquid not only agglomerates the blend but also plays an essential role in sinter mineralogy and pore structure [4]; better control of sintering liquid properties helps the sinter plants to produce high-quality sinters, even when using economic and low-grade ores.

Before preparing the above-mentioned new sinters, the behaviour of current sinters with respect to their bulk compositions needs to be assessed, especially, the amount of sintering liquid generated at peak bed conditions should be estimated. The research published so far primarily focused on the impact of sinter blend properties on final sinter properties using micro-sintering experiments. There are only a few publications on liquid formation during sintering [8–12], however, the details regarding the sintering liquid properties published so far were qualitative, focusing on changes in height or the cross-sectional area of the sintered sample upon temperature variance to calculate the liquid properties. The data regarding these phases and phase compositions present at peak bed conditions were simulated using FactSage [6,12,13].

The current study obtained three sinters with a stable blast furnace operation from a steel plant. These sinters were reheated to peak sinter bed conditions in a tube furnace, followed by quenching to analyse the phases and phase compositions present and calculate the sintering liquid properties. Apart from sintering liquid properties, assessing the softening and melting behaviour of the above sinters is also needed to ascertain or judge the new sinter behaviour in the blast furnace, as sinters with very high reducibility tend to coagulate, leading to increased melting temperatures [14]. Therefore, the softening and melting experiments of the three sinters were conducted, and the softening start and end temperatures and the melting temperature were calculated based on the displacement changes in the sandwich bed with the rise in temperature [15]. This study helps understand the effect of sinter bulk compositions on sintering liquid generated at peak bed conditions. Understanding and controlling sintering liquid properties help control the required final sinter compositions. In addition, analyses of softening and melting behaviour of the current sinters with stable blast furnace operation help determine or adjust the feed-ratio and operational conditions of the blast furnace to prevent coagulation and have a permeable cohesive layer while using new sinters generating low-slag volume.

2. Materials and Methods

2.1. Reheating of Sinters

The sinters were reheated to peak sinter bed conditions in a vertical tube furnace heated by $LaCrO_3$ elements. The temperature in the pre-determined hot zone of the furnace was relatively stable, with a variation of less than 2 K (2 °C). The details of the tube furnace, including a schematic diagram, are shown elsewhere [16]. Based on industry data, the peak sintering temperature was set to 1300 °C, and the oxygen partial pressure was maintained at approximately 0.01 atm by passing pure argon mixed with 1% O_2 gas for the reheating experiments. Sinters used in these experiments were sourced from different sinter plants. The sinter samples were named S1, S2, and S3. In sinters, ferrous iron content was determined by titration, and the rest of the compositions were determined by X-ray fluorescence. Fe_2O_3 was calculated by removing ferrous iron from total iron. The significant compositions were normalised and detailed in Table 1.

Table 1. Sinter samples used in the present study along with their bulk compositions.

Sinter	Composition (wt.%)					
	Fe_2O_3	FeO	CaO	SiO_2	Al_2O_3	MgO
S1	74.7	7.8	9.0	5.0	1.9	1.6
S2	72.3	8.4	10.6	5.4	1.8	1.5
S3	67.8	8.6	12.6	6.8	2.2	2.0

The sinter samples were powdered and mixed using mortar and pestle for uniformity, and then the powder was pressed into pellets around $5 \times 5 \times 5$ mm^3 by applying pressure. Next, the sample was placed in a basket made of Kanthal wire and raised to the hot zone of the furnace. The sinters were held in the hot zone at a temperature of 1300 °C for 10 min and then quenched directly into the water container by displacing the removable glass end at the bottom of the tube furnace. After drying, the quenched sample was mounted with epoxy resin and then polished and carbon-coated for electron probe X-ray microanalysis (EPMA). To further understand the sintering liquid behaviour, the peak temperature was varied from 1250 °C to 1350 °C, and the holding time was varied from 10 min to 60 min for sinter S2.

2.2. Sinter Softening Experiments

The softening experiments were carried out by applying a specific load on the bed under controlled atmospheres in a Pyrox furnace with $LaCrO_3$ heating elements, which is shown in Figure 1. The quantifying of the softening behaviours were determined by the bed contraction rate, which is the ratio of the sample bed thickness on heating to its original thickness [15]. The load was selected to be 1 kg/cm^2, and a gas mixture of Ar and CO (30% CO + 70% Ar) was used. The samples were heated at a constant ramping rate of 300 °C per hour, with isothermal holding at 950 °C for 10 min to simulate the blast furnace thermal reserve zone.

The furnace tube and the upper chamber were sealed in a gas-tight condition. Two independent gas flow circuits were used to suppress heat transfer and protect the laser displacement sensor in the upper chamber from high temperatures. A graphite crucible with holes drilled in the bottom was placed on a supporting alumina platform. An alumina tube was used to support the crucible to be located within the hot zone of the furnace, and a B-type thermocouple was inserted into the tube to monitor the temperature continuously.

Within the graphite crucible, the sandwich bed structure comprised two layers of 2 cm coke at the bottom and top, with 4 cm iron ore samples in the middle. The displacement changes to the sandwich bed (sample bed) were recorded with the help of the laser. The laser displacement sensor was fixed on a movable platform in the upper chamber. The change in sample bed height containing only coke that was 4 cm thick coke layer without any iron feeds was considered as the baseline curve for the softening experiments.

2.3. Characterization

The sinter phase and composition were characterised using a JXA 8200 electron probe X-ray microanalyser (EPMA) (from Japan Electron Optics Ltd., Tokyo, Japan) with wavelength dispersive detectors (WDD). An accelerating voltage of 15 kV and probe current of 15 nA was used. The standards used are hematite (Fe_2O_3), wollastonite ($CaSiO_3$), and spinel ($MgAl_2O_4$). The measurement time for Fe was 60 s on peak and 15 s on background, and other elements (Ca, Si, Mg, Al) were measured for 40 s on peak and 10 s on background. Metal cation measurements were adjusted to selected oxidation states. Under experimental conditions in this study, there may be both Fe^{2+} and Fe^{3+} present in the samples, but all iron concentrations are univocally presented as Fe^{3+} for a clear-cut explanation.

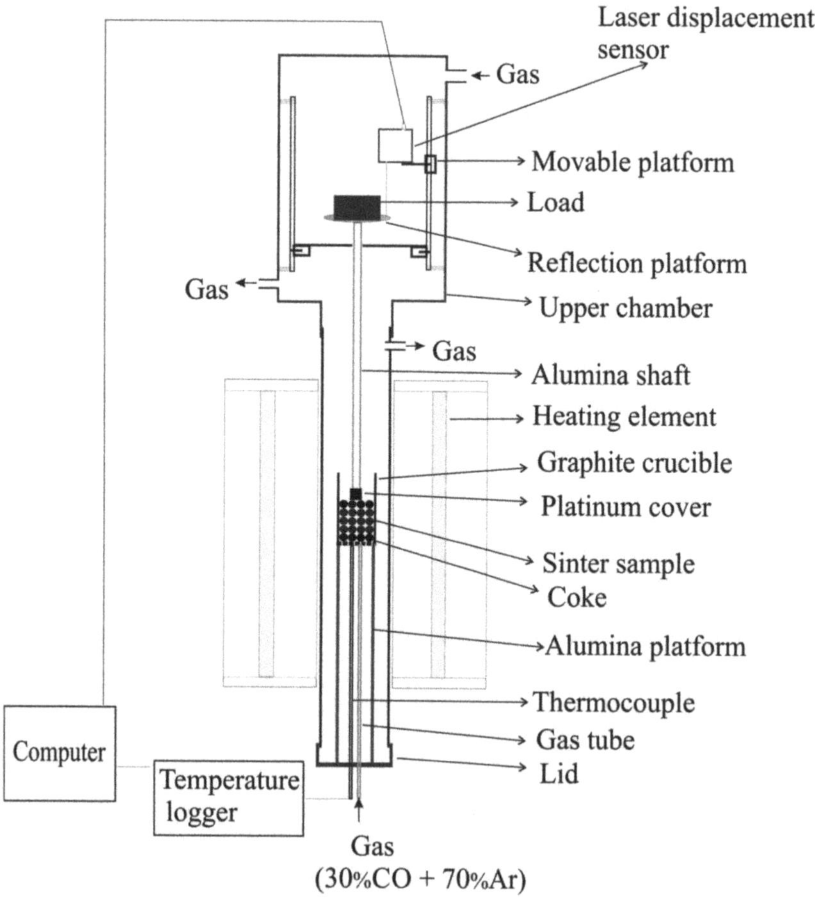

Figure 1. Schematic diagram of Pyrox furnace used for softening and melting test along with the sandwich bed containing coke and sinters, and laser displacement sensor.

3. Results & Discussion

3.1. Reheated and Quenched Sinter Analysis

3.1.1. Analysis of Phases and Phase Composition of the Three Sinters

Phase compositions were assessed at 80 points for each phase in the reheated and quenched sample. The average composition of each phase is presented in Table 2. Typical microstructures of the reheated sinter samples quenched from 1300 °C are shown in Figure 2. The two phases present at the peak temperature are indicated, where 'L' is the liquid phase, and 'H' is hematite. We could see that a more liquid phase appears from S1 to S3 because of more SiO_2 and CaO content in the sinter composition. The compositions of the liquid phase and hematite are presented in Table 2. There is no silica content present in the hematite. The phase proportions are calculated as per Equation (1); the liquid basicity is the ratio of

CaO/SiO$_2$, and the viscosity of the liquid at 1300 °C is estimated by FactSage 7.3, based on liquid compositions from EPMA, and is included in Table 2.

$$\mathrm{Liquid\,phase\,proportion} = \frac{\mathrm{Silica\,content\,in\,sinter\,bulk}}{\mathrm{Silica\,content\,in\,sinter\,liquid}} \times 100\% \tag{1}$$

Table 2. Phases and phase compositions present upon reheating the three sinters at 1300 °C for 10 min and Po$_2$ of 0.01 atm, followed by quenching and EPMA.

Sinter	Phase	Composition (wt.%)					Phase Proportion (%)	Liquid Basicity	Viscosity (Poise)
		Fe$_2$O$_3$	CaO	SiO$_2$	Al$_2$O$_3$	MgO			
S1	Sintering liquid	38.4	31.3	28.9	1.3	0.1	17.3	1.1	2.43
	Hematite	96	2.2	0	1.4	0.4	82.7	N/A	N/A
S2	Sintering liquid	49.7	31.9	14.8	3.3	0.3	36.5	2.2	1.03
	Hematite	95.7	2.3	0	1.6	0.4	63.5	N/A	N/A
S3	Sintering liquid	49.7	32.7	13.5	3.7	0.4	50.4	2.4	0.99
	Hematite	94.9	2.6	0	1.7	0.8	49.6	N/A	N/A

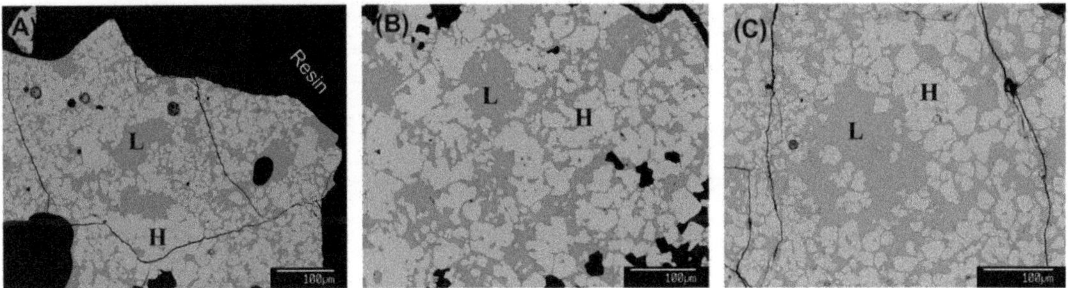

Figure 2. Microstructural BSE images of various reheated sinters quenched from 1300 °C. Microstructure (**A**) S1, (**B**) S2, and (**C**) S3. All samples comprise the hematite (H) and liquid phase (L).

The sintering liquid basicity and liquid proportion increased, and the liquid viscosity decreased from S1 to S3, as shown in Figure 3. In comparison to the bulk properties of the three sinters, the sinter basicity increased from S1 to S2 and then decreased, whereas the silica content and ferrous iron content increased from sinters S1 to S3. The rise in silica, FeO, and the lowering of alumina in sinter bulk (compositions detailed in Table 1) contributed to a more than doubling of liquid proportion from S1 to S2. Despite the rise in alumina content from S2 to S3, the significant increase in silica content contributed to the rise in liquid proportion from S2 to S3.

3.1.2. Effect of Peak Temperature

Experiments were carried out to analyse the impact of variation of the peak operating temperature on sinter liquid. The microstructures of reheated S2 samples are presented in Figure 4. The phases present and their compositions upon reheating and quenching are shown in Table 3.

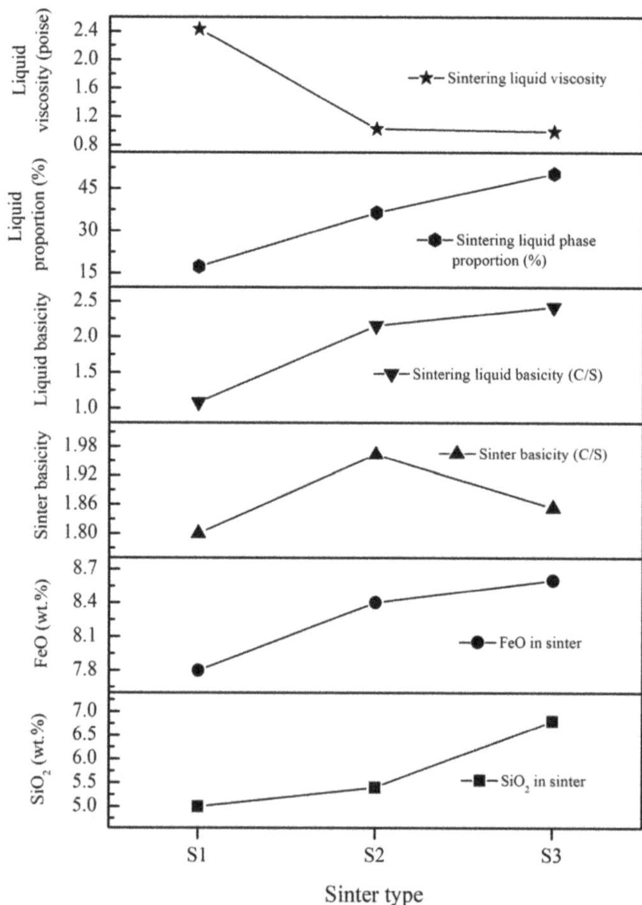

Figure 3. Sintering liquid properties of the three sinters upon reheating and quenching with respect to sinter bulk compositions.

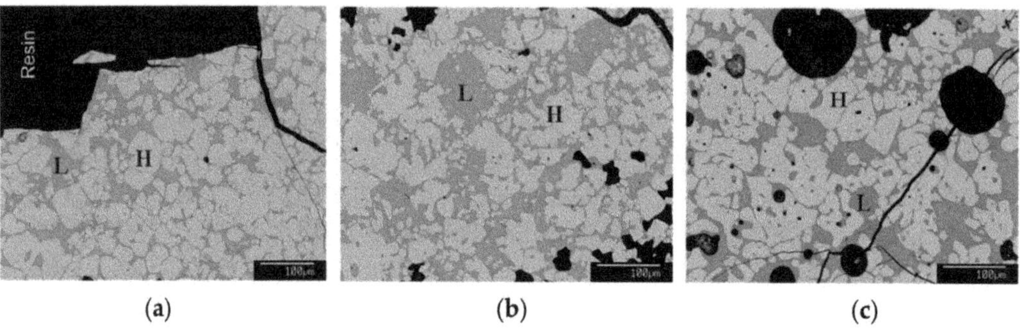

Figure 4. Microstructural BSE images of reheated sinter 'S2' at varied peak temperatures. (**a**) 1250 °C, (**b**) 1300 °C, and (**c**) 1350 °C.

Table 3. Phases and phase compositions present in sinter 'S2' upon reheating at 1250 °C, 1300 °C, and 1350 °C, followed by quenching and EPMA.

Peak Temperature (°C)	Holding Time (min)	Phase	Composition of Liquid (wt.%)					Phase Proportion (wt.%)	Liquid Basicity	Viscosity (Poise)
			Fe_2O_3	CaO	SiO_2	Al_2O_3	MgO			
1250	10	Sintering liquid	44.3	32.0	20.5	3.2	0.1	26.4	1.6	1.90
		Hematite	97.4	1.1	0.0	1.3	0.2	73.6	N/A	N/A
1300	10	Sintering liquid	49.7	31.9	14.8	3.3	0.3	36.5	2.2	1.03
		Hematite	95.7	2.3	0	1.6	0.4	63.5	N/A	N/A
1350	10	Sintering liquid	51.2	31.8	13.2	3.7	0.2	41.0	2.4	0.74
		Hematite	94.0	2.9	0.0	1.4	1.6	59.0	N/A	N/A

Figure 4 shows the presence of sintering liquid and hematite upon reheating and quenching at 1250 °C, 1300 °C, and 1350 °C. The microstructures show an increase in the liquid phase and further deformation of hematite with the temperature rise. Analyses of liquid and hematite compositions in Table 3 show the increase of Fe_2O_3 in sinter liquid and a decrease of hematite with a temperature rise from 1250 to 1350 °C. In contrast, the SiO_2 amount decreased in the liquid phase with the temperature rise. Therefore, the calculated liquid proportion and liquid basicity increased as the temperature increased, whereas the liquid viscosity decreased, as shown in Figure 5. This varied peak temperature and liquid proportion estimation help control the liquid proportion while preparing the low-silica and low-FeO sinters.

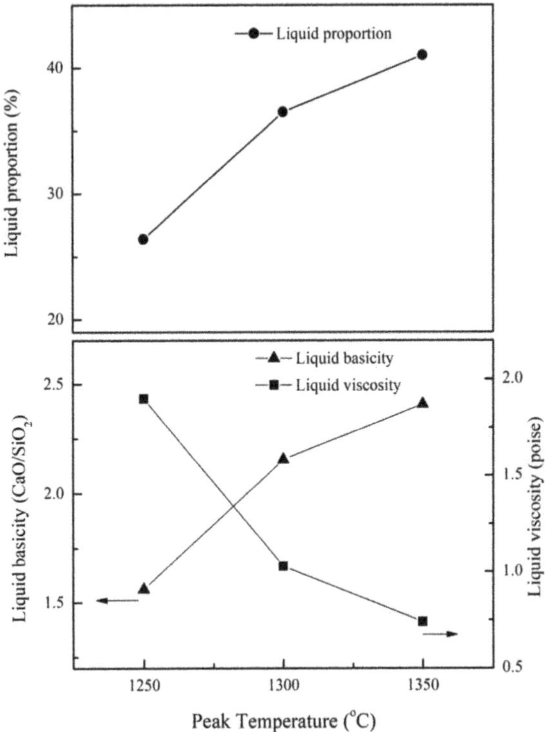

Figure 5. Variation of sinter liquid properties with varied reheating temperatures.

3.1.3. Effect of Holding Time

Experiments were carried out to analyse the impact of varied holding times on sinter liquid. The microstructures of reheated and quenched S2 samples are presented in Figure 6. The phases present and their compositions are shown in Table 4.

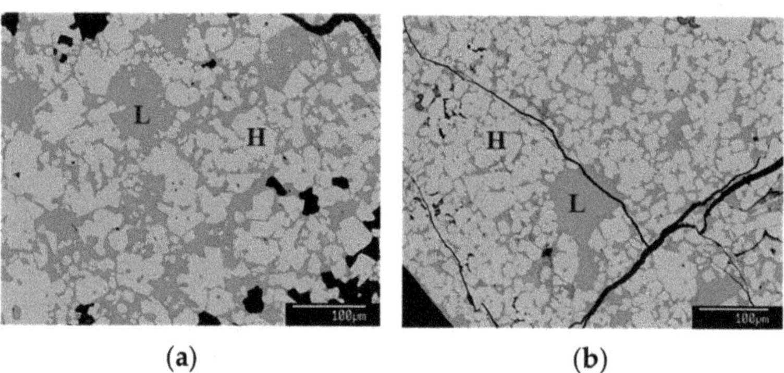

(a) **(b)**

Figure 6. Microstructural BSE images of reheated S2 sinters at different holding times. (**a**) 10 min, and (**b**) 60 min.

Table 4. Phases present and phase compositions in sinter 'S2' upon reheating for two different holding times of 10 min and 60 min at 1300 °C.

Peak Temperature (°C)	Holding Time (min)	Phase	Composition of Liquid (wt.%)					Phase Proportion (wt.%)	Liquid Basicity	Viscosity (Poise)
			Fe_2O_3	CaO	SiO_2	Al_2O_3	MgO			
1300	10	Sintering liquid	49.7	31.9	14.8	3.3	0.3	36.5	2.2	1.03
		Hematite	95.7	2.3	0	1.6	0.4	63.5	N/A	N/A
1300	60	Sintering liquid	49.0	32.0	15.8	3.1	0.1	34.3	2.0	1.07
		Hematite	96.1	2.2	0.0	1.2	0.5	65.7	N/A	N/A

On varying the holding time from 10 min to 60 min at 1300 °C, while the liquid viscosity increased from 1.03 poise to 1.07 poise, which is a minor change, the Fe_2O_3 content decreased from 49.7% to 49% in the liquid phase, and the liquid basicity decreased from 2.2 to 2.0. The variation with increased holding time on liquid properties is very minimal, showing the sample was in near equilibrium for a holding time of 10 min at 1300 °C.

3.2. Softening Behaviour of Sinters

As the temperature increased, the sample bed initially expanded during the softening tests, and then there was a continuous contraction. The present study defines the temperature at 0% displacement contraction as the softening starting temperature (Ts). At 40% displacement contraction, the temperature is defined as the softening ending temperature (Te). Finally, the temperature at 100% displacement contraction is defined as the melting temperature (Tm) [15]. The softening and melting temperatures are represented in Graph A in Figure 7 by considering the S1 curve. The three sinters' and coke's displacement contraction curves are shown in Graph B in Figure 7.

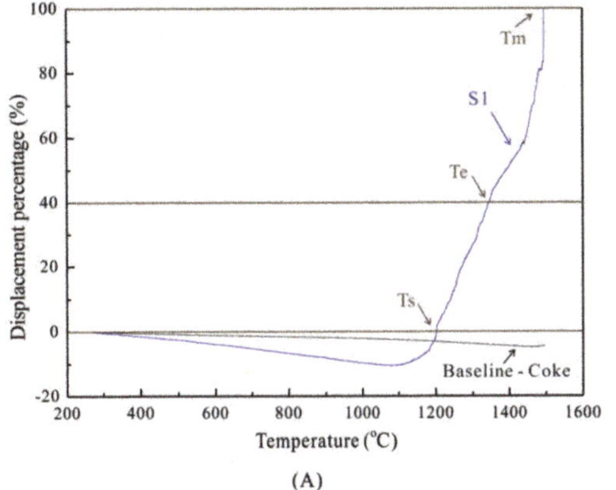
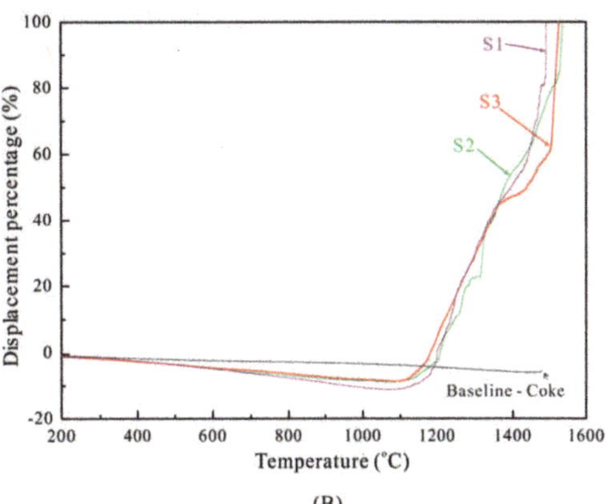

(A)   (B)

Figure 7. Graph (**A**) indicates the softening and melting temperature at the respective displacement percentage for sinter S1. Graph (**B**) shows the sinter softening and melting behaviour with initial expansion followed by continuous contraction for the three sinters, S1, S2 and S3.

The softening temperature range is the difference between the softening ending temperature and softening starting temperature (Te-Ts). The cohesive zone range is the difference between the melting and softening starting temperatures (Tm-Ts). Therefore, as the sample expands and then contracts back to the initial position, then the temperatures of Ts followed by Te and Tm are calculated at the corresponding displacement percentage. The softening and cohesive range for the three different sinters are detailed in Table 5.

Table 5. Softening and melting temperatures of the sinters S1, S2, and S3 along with softening and cohesive range.

Sample	Softening Starting Temperature (Ts, °C)	Softening Ending Temperature (Te, °C)	Melting Temperature (Tm, °C)	Softening Temperature Range (Te-Ts, °C)	Cohesive Range (Tm-Ts, °C)
S1	1205	1333	1479	128	274
S2	1209	1335	1527	126	318
S3	1178	1334	1518	128	340

The sinters should have a higher softening starting temperature and a lower melting temperature, resulting in a narrow cohesive zone for the efficient functioning of the blast furnace. Although the three sinters are currently used as blast furnace feed, on conducting the softening and melting experiments and measuring the cohesive range among these sinters, S1 has the narrowest cohesive range, while S3 has the broadest. In addition, S3, the sinter with the highest FeO content and silica content among the three sinters, has the lowest softening start temperature and the highest melting temperature.

3.3. Correlation of Sinter Liquid Properties to Sinter Softening Properties

Understanding sinter bulk compositions on sinter softening behaviour is essential in preparing sinters with higher softening start temperature, low-FeO containing slags on reduction, and a narrow cohesive zone. In addition, the estimation of phases present and phase compositions at peak sinter bed conditions is vital, as the sintering liquid phase properties play a crucial role in the final sinter morphology. Once the sintering liquid properties were determined by reheating and quenching the sinters, it was followed by the sinters' softening and melting behaviour analyses. Finally, the impact of sinter bulk compositions and gangue content ($(CaO + MgO + SiO_2 + Al_2O_3)/TFe$) [3] on sinter cohesive zone behaviour was analysed and correlated to the sinter liquid behaviour in Figure 8.

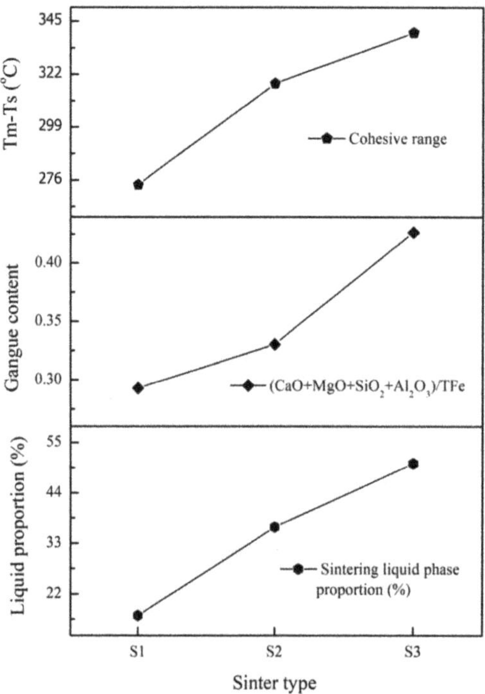

Figure 8. Correlation of sintering liquid proportion and sinters gangue content with the sinters cohesive range.

The liquid proportion, gangue content, and cohesive range have increased from S1 to S3. Though the rise in gangue content from S1 to S2 was lower compared to the increase from S2 to S3, as shown in Figure 8, the liquid phase proportion more than doubled from S1 to S2, and the increased silica, calcia, and FeO contents along with decreased alumina in the sintered bulk, as detailed in Table 1, contributed to this. Despite an increase in the softening starting point from S1 to S2, as shown in Table 5, the significant rise in liquid proportion might have led to an increased glass phase in the sinters, resulting in the generation of viscous slags containing higher FeO, causing coagulation, and therefore leading to an increase in the melting point of S2. Additional increases in silica, calcia, and FeO contents in S3 showed a further rise in the liquid phase proportion. In S3, the liquid phase proportion was nearly 50% at the peak temperature of 1300 °C. It is far beyond the optimal liquid phase required; this resulted in a lower softening starting temperature and further broadening of the cohesive range from S2 to S3. This correlation shows the necessity of controlling the sintering liquid behaviour to achieve better sinter properties.

4. Conclusions

The sinters' reheating and quenching experiments determined the phases and phase compositions present at the peak sinter bed conditions. The liquid basicity and the liquid proportion increased with an increase in silica content in sinter bulk. Apart from that, the liquid phase proportion increased on increasing the peak operational temperature.

As the gangue content increases in the sinters, broadening/widening of the cohesive range is observed, detrimental to blast furnace efficacy. This suggests that though an optimal liquid phase is needed to bind the blend, the excess liquid raises the gangue content in sinters, resulting in high-FeO containing slags on reduction, apart from increasing the glass phase and increased sinter returns generation.

This research is a primary step in exploring the relationship between sintering liquid properties, sinter bulk properties, and softening properties. This study helps better understand the sintering liquid behaviour at peak temperature, which impacts the final sinter morphology, thereby the reduction behaviour of the sinters.

Author Contributions: Conceptualization, V.K., T.E., F.T., B.Z. and X.M.; methodology, B.Z. and X.M.; validation, V.K., B.Z., F.T. and X.M.; formal analysis, V.K.; investigation, V.K.; resources, D.W., W.P., S.C., T.E., B.Z. and X.M.; data curation, V.K. and F.T.; writing—original draft preparation, V.K.; writing—review and editing, V.K., F.T. and X.M.; visualization, V.K., D.W., W.P., S.C., F.T. and X.M.; supervision, X.M.; project administration, F.T. and X.M.; funding acquisition, F.T., B.Z. and X.M. All authors have read and agreed to the published version of the manuscript.

Funding: This research was funded by Shougang Group and Rio Tinto Iron Ore.

Data Availability Statement: Not applicable.

Acknowledgments: The authors would like to thank Shougang Group and Rio Tinto for providing the sinter samples and their financial support. The authors acknowledge the facilities, and the scientific and technical assistance, of the Australian Microscopy and Microanalysis Research Facility at the Centre for Microscopy and Microanalysis, the University of Queensland.

Conflicts of Interest: The authors declare no conflict of interest.

References

1. Fernández-González, D.; Ruiz-Bustinza, I.; Mochón, J.; González-Gasca, C.; Verdeja, L.F. Iron Ore Sintering: Process. *Mineral Process. Extr. Metall. Rev.* **2017**, *38*, 215–227. [CrossRef]
2. Ghosh, A.; Chatterjee, A. *Iron Making and Steelmaking: Theory and Practice*; PHI Learning Pvt. Ltd.: New Delhi, India, 2008.
3. Shatokha, V.; Velychko, O. Study of Softening and Melting Behaviour of Iron Ore Sinter and Pellets. *High Temp. Mater. Processes* **2012**, *31*, 215–220. [CrossRef]
4. Higuchi, K.; Takamoto, Y.; Orimoto, T.; Takehiko, S.; Koizumi, F.; Kazuyuki, S.; Furuta, H. Quality Improvement of Sintered Ore in Relation to Blast Furnace Operation. *Nippon Steel Tech. Rep.* **2006**, *96*, 36–41.
5. Lv, X.; Bai, C.; Deng, Q.; Huang, X.; Qiu, G. Behavior of liquid phase formation during iron ores sintering. *ISIJ Int.* **2011**, *51*, 722–727. [CrossRef]
6. Wu, S.; Zhai, X. Factors influencing melt fluidity of iron ore. *Metall. Res. Technol.* **2018**, *115*, 505. [CrossRef]
7. Wu, S.-L.; Wang, Q.-F.; Bian, M.-L.; Zhu, J.; Long, F.-Y. Influence of iron ore characteristics on FeO formation during sintering. *J. Iron Steel Res. Int.* **2011**, *18*, 5–10. [CrossRef]
8. Murao, R.; Kimura, M. Investigation on reaction schemes of iron ore sintering process by high temperature in-situ x-ray diffraction and micro-texture observation. *Nippon. Steel Sumitomo Met. Tech. Rep.* **2018**, *118*, 59–64.
9. Zhang, G.-L.; Wu, S.-L.; Chen, S.-G.; Su, B.; Que, Z.-G.; Hou, C.-G. Influence of gangue existing states in iron ores on the formation and flow of liquid phase during sintering. *Int. J. Miner. Metall. Mater.* **2014**, *21*, 962–968. [CrossRef]
10. Kasai, E.; Sakano, Y.; Kawaguchi, T.; Nakamura, T. Influence of properties of fluxing materials on the flow of melt formed in the sintering process. *ISIJ Int.* **2000**, *40*, 857–862. [CrossRef]
11. Okazaki, J.; Higuchi, K.; Hosotani, Y.; Shinagawa, K. Influence of Iron Ore Characteristics on Penetrating Behavior of Melt into Ore Layer. *ISIJ Int.* **2003**, *43*, 1384–1392. [CrossRef]
12. Van den Berg, T.; de Villiers, J.; Cromarty, R. Variation of the redox conditions and the resultant phase assemblages during iron ore sintering. *Int. J. Mineral Process.* **2016**, *150*, 47–53. [CrossRef]
13. Nicol, S.; Chen, J.; Qi, W.; Mao, X.; Jak, E.; Hayes, P.C. Measurement of Process Conditions Present in Pilot Scale Iron Ore Sintering. *Minerals* **2019**, *9*, 374. [CrossRef]
14. Nishimura, T.; Higuchi, K.; Naito, M.; Kunitomo, K. Evaluation of Softening, Shrinking and Melting Reduction Behavior of Raw Materials for Blast Furnace. *ISIJ Int.* **2011**, *51*, 1316–1321. [CrossRef]
15. Wang, D.; Chen, M.; Zhang, W.; Zhao, Z.; Zhao, B. Softening Behaviors of High Al_2O_3 Iron Blast Furnace Feeds. In Proceedings of the 10th CSM Steel Congress & 6th Baosteel Biennial Academic Conference, Shanghai, China, 21–23 October 2015.
16. Chen, M.; Zhao, B. Phase Equilibrium Studies of "Cu_2O"–SiO_2–Al_2O_3 System in Equilibrium with Metallic Copper. *J. Am. Ceram. Soc.* **2013**, *96*, 3631–3636. [CrossRef]

metals

Article

Phase Equilibria Studies in the CaO-MgO-Al₂O₃-SiO₂ System with Al₂O₃/SiO₂ Weight Ratio of 0.4

Jinfa Liao [1], Gele Qing [2] and Baojun Zhao [1,3,*]

1   International Institute for Innovation, Jiangxi University of Science and Technology, Nanchang 330013, China
2   Shougang Research Institute of Technology, Beijing 100043, China
3   Sustainable Minerals Institute, University of Queensland, Brisbane, QLD 4072, Australia
*   Correspondence: bzhao@jxust.edu.cn

Abstract: With the raw materials for ironmaking becoming increasingly complex, more accurate phase equilibrium information on the slag is needed to refine the blast furnace operation to reduce the energy cost and CO_2 emissions. CaO-SiO_2-Al_2O_3-MgO is a basic system of ironmaking slag in which CaO and MgO mainly come from the flux, SiO_2 and Al_2O_3 are mainly from raw materials. The effect of flux additions on the phase equilibrium of the slag can be described by a pseudo-ternary system CaO-MgO-(Al_2O_3+SiO_2) at a fixed Al_2O_3/SiO_2 ratio of 0.4. Liquidus temperatures and solid solutions in the CaO-MgO-Al_2O_3-SiO_2 system with Al_2O_3/SiO_2 weight ratio of 0.4 have been experimentally determined using high temperature equilibration and quenching techniques followed by electron probe microanalysis. Dicalcium silicate (Ca_2SiO_4), cordierite ($2MgO\cdot2Al_2O_3\cdot5SiO_2$), spinel ($MgO\cdot Al_2O_3$), merwinite ($3CaO\cdot MgO\cdot2SiO_2$), anorthite ($CaO\cdot Al_2O_3\cdot2SiO_2$), mullite ($Al_2O_3\cdot SiO_2$), periclase ($MgO$), melilite ($2CaO\cdot MgO\cdot2SiO_2$-$2CaO\cdot Al_2O_3\cdot SiO_2$) and forsterite ($Mg_2SiO_4$) are the major primary phases in the composition range investigated. A series of pseudo-binary phase diagrams have been constructed to demonstrate the application of the phase diagrams on blast furnace operation. Composition of the solid solutions corresponding to the liquidus have been accurately measured and will be used for the development of the thermodynamic database.

Keywords: phase equilibria; liquidus temperature; CaO-MgO-Al_2O_3-SiO_2 system; blast furnace slag; basicity

Citation: Liao, J.; Qing, G.; Zhao, B. Phase Equilibria Studies in the CaO-MgO-Al₂O₃-SiO₂ System with Al₂O₃/SiO₂ Weight Ratio of 0.4. *Metals* **2023**, *13*, 224. https://doi.org/10.3390/met13020224

Academic Editors: Fernando Castro and Mark E. Schlesinger

Received: 7 December 2022
Revised: 13 January 2023
Accepted: 19 January 2023
Published: 25 January 2023

1. Introduction

Due to the shortage of high-grade iron ore resources and the increasingly fierce competition in the steel market, many steel enterprises face the use of low-quality raw materials [1]. However, a variety of negative problems, including incomplete separation of slag and hot metal, difficult tapping of the slag from the blast furnace and inefficient reaction between slag and hot metal, can happen when using low-quality raw materials [2–4]. On the other hand, a modern blast furnace with automatic operating system can adjust the parameters accurately if the required information is available. Hence, more accurate slag phase equilibrium information should be provided for modern blast furnaces to treat complex raw materials [5–7]. The oxide system CaO-MgO-Al_2O_3-SiO_2 forms a base for ironmaking slags. SiO_2 and Al_2O_3 are mainly from raw materials such as iron ores, coke and coal. CaO and MgO are mainly added as flux to adjust the slag composition to obtain the required properties.

Phase equilibrium information of the CaO-MgO-Al_2O_3-SiO_2 system has been studied by many scholars [8–23]. Slag Atlas summarized the CaO-MgO-Al_2O_3-SiO_2 system based on the earlier studies [24]. The phase diagrams reported were presented in the form of CaO-MgO-SiO_2 pseudo-ternary sections at fixed Al_2O_3 concentrations [10] or Al_2O_3-CaO-SiO_2 pseudo-ternary sections at fixed MgO concentrations [11–13]. Due to the limitations of the experimental techniques, earlier studies [8–14] only determined the liquidus temperatures without the composition of the solid solutions. Significant differences were also observed between the reported results [9,10,14]. With the development of the ironmaking techniques, previous phase diagrams are not enough to provide accurate operation guidelines. CaO/SiO_2 is usually fixed for a given blast furnace to obtain stable liquidus temperature of the slag and sulfur removal. Phase equilibria in the $(CaO+SiO_2)$-MgO-Al_2O_3 system have been investigated recently at fixed CaO/SiO_2 weight ratios of 0.9, 1.1, 1.3 and 1.5 [15–18]. With this information, MgO addition corresponding to the Al_2O_3 concentrations in the slag can be adjusted to maintain the stable liquidus temperatures. When a blast furnace has a stable supply of the raw materials, the Al_2O_3/SiO_2 weight ratio in the slag is relatively fixed. Addition of CaO and MgO in different ratios and amounts can confer different properties of the slag including liquidus temperature and sulfur capacity. The existing phase diagrams summarized above cannot provide the detailed phase equilibrium information for various CaO and MgO additions. In addition, powerful thermodynamic software such as Fact-Sage [25] and Thermo-Calc [26] have been developed to predict liquidus temperatures of the slags. It was found that there are still differences between the experimental results and the predicted results, indicating that the thermodynamic database needs to be optimized. Experimentally determined liquidus temperatures and corresponding compositions of the solid solutions can provide accurate data to support the development of a reliable thermodynamic database.

In this study, the phase equilibria of the CaO-MgO-Al_2O_3-SiO_2 system at a fixed Al_2O_3/SiO_2 weight ratio of 0.4 were experimentally investigated using a high-temperature equilibration, quenching and electron probe microanalysis (EPMA) method. Effects of the flux ratio MgO/CaO and quaternary basicity $(CaO+MgO)/(Al_2O_3+SiO_2)$ on liquidus temperature will be discussed. The experimental results will be compared with FactSage predictions to evaluate the existing thermodynamic database.

2. Experimental

The experimental procedure includes sample preparation from pure chemicals, high-temperature equilibration, and analyses of the microstructures and compositions of the phases present in the quenched samples. The experiments were first planned based on the available information, including low-order phase diagrams, FactSage predictions and preliminary experiments. High purity powders of Al_2O_3, SiO_2, MgO and $CaCO_3$ were weighed and mixed according to the experimental plan and pelletized. An approximately 0.2 g pellet was placed in a graphite crucible (inner diameter 5 mm and height 5 mm). Equilibration experiments were carried out in a vertical tube furnace (recrystallized alumina as a reaction tube, 30 mm inner diameter). The graphite crucible, which contained the pelletized mixtures was suspended using a platinum wire (0.5 mm diameter). Argon gas (flow rate: 500 mL/min) was passed through the furnace during the experiment to avoid oxidization of the crucible. The sample was pre-melted at a temperature 20–50 °C higher than the equilibration temperature for 30 min. Then the temperature was lowered to the desired temperature and kept for a time sufficient to achieve equilibrium. The equilibration usually took from 2 to 12 h, depending on the viscosity of the slag, which is related to the composition and temperature. For example, a shorter time was employed at higher temperatures or with lower-silica slag. Several steps were taken to ensure the equilibrium was reached in the system. First, preliminary experiments were carried out for different periods at the beginning of the project. The sufficient time was determined by the same phase compositions obtained over the reaction period. The actual reaction time was always much longer than that from the preliminary experiments. Second, uniform microstructure and liquid composition from different areas of the quenched sample confirmed the equilibrium. Third, self-consistence of all experimental data was checked to confirm the equilibrium.

After equilibration, the sample was dropped into water directly to attain rapid cooling. The quenched samples were mounted in epoxy resin, polished and carbon-coated for electron probe microanalysis (EPMA). A JXA 8200 (Japan Electron Optics Ltd., Tokyo, Japan) Electron Probe Microanalyser with Wavelength Dispersive Spectroscopy (WDS,

Japan Electron Optics Ltd., Tokyo, Japan) was used for microstructural and compositional analyses. The EPMA was operated at an accelerating voltage of 15 kV and a probe current of 15 nA. The beam size was set to "0 μm" to accurately measure the composition of an area larger than 1 μm. The ZAF (Z is atomic number correction factor, A is absorption correction factor, and F is fluorescence correction factor) correction procedure was applied for the data analysis. The standards used for EPMA included alumina (Al_2O_3) for Al, magnesia (MgO) for Mg, and wollastonite ($CaSiO_3$) for Ca and Si. These standards were provided by Charles M Taylor Co., Stanford, CA, USA. The average accuracy of the EPMA measurements was within ±1 wt%.

On rapid cooling, the liquid phase in the equilibrated sample was converted to homogeneous glass and the solid phases were retained in their shapes and compositions. The homogeneity of the phases can be confirmed by the EPMA measurements in different areas of the quenched sample. Usually, 8–20 points of the liquid phase and 3–5 points of the solid phase were measured from different areas by EPMA. The samples with the standard deviation of the phase compositions less than 1% were accepted for phase diagram construction.

FactSage 8.2 [25] was used for the thermodynamic calculations that are compared with the experimental results. The databases of "FactPS" and "FToxid" were used in the "Equilib" module. The solution phases selected in the calculations were "FToxide-SLAGA", "FToxide-SPINC", "FToxide-MeO_A", "FToxide-bC2S", "FToxide-aC2S", "FToxide-Mel", "FToxide-WOLLA", "FToxide-OlivA" and "FToxide-Mull".

3. Results and Discussion

3.1. Description of the Pseudo-Ternary Section

Proper presentation of the multi-component phase diagram is essential for easy use by industrial operators and academic researchers. Figure 1 shows the presentation of the pseudo-ternary phase diagram CaO-MgO-(Al_2O_3+SiO_2) at a fixed Al_2O_3/SiO_2 weight ratio of 0.4. The end members of the pseudo-ternary section are CaO, MgO and (Al_2O_3+SiO_2) with a fixed Al_2O_3/SiO_2 weight ratio of 0.4 in the liquid. This Al_2O_3/SiO_2 weight ratio was selected based on the average compositions of the current blast furnace slags from Shougang [27]. The selection of CaO and MgO as the end members enables us to discuss the effect of flux composition on slag phase equilibrium.

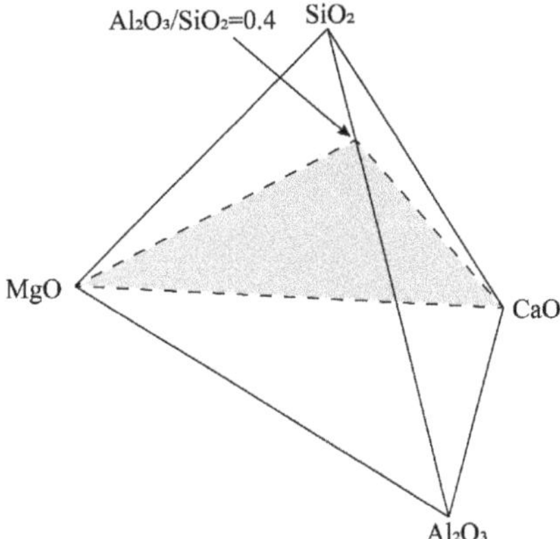

Figure 1. Presentation of the pseudo-ternary phase diagram in the CaO-MgO-(SiO_2+Al_2O_3) system at fixed Al_2O_3/SiO_2 weight ratio of 0.4.

The microstructures and phase compositions of the quenched samples were determined by EPMA. Anorthite (CaO·Al_2O_3·$2SiO_2$), cordierite (2MgO·$2Al_2O_3$·$5SiO_2$), dicalcium silicate (Ca_2SiO_4), melilite (2CaO·MgO·$2SiO_2$-2CaO·Al_2O_3·SiO_2), merwinite

(3CaO·MgO·2SiO$_2$), MgO, mullite (3Al$_2$O$_3$·2SiO$_2$), spinel (MgO·Al$_2$O$_3$) and forsterite (2MgO·SiO$_2$-2CaO·SiO$_2$) were observed in the quenched samples. Typical backscattered SEM micrographs of the quenched samples are shown in Figure 2a–f. Figure 2a–e show the coexistence of liquid with anorthite, cordierite, spinel, Mg$_2$SiO$_4$ and melilite, respectively. Figure 2f–h show the coexistence of liquid with anorthite and Mg$_2$SiO$_4$, melilite and spinel, MgO and merwinite, respectively. It can be seen from the figures that rapid cooling enabled the liquid to be converted to the uniform glass. All solid phases have a sharp edge and large enough size for EPMA measurements.

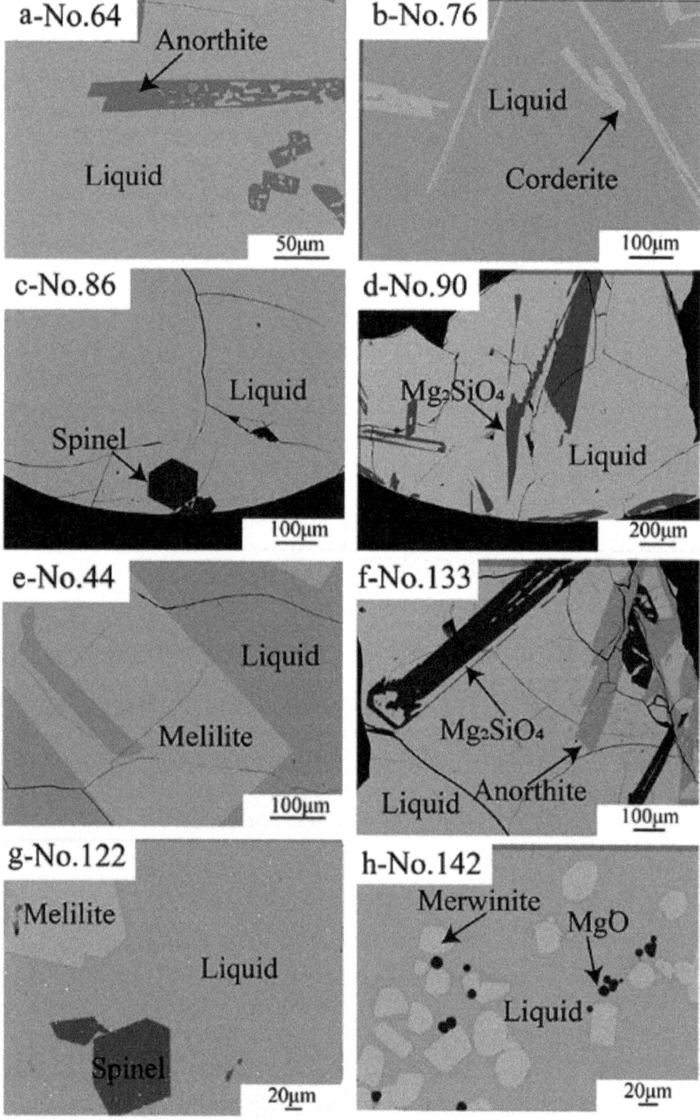

Figure 2. Typical backscattered SEM micrographs of the quenched slags showing the equilibrium of liquid with (**a**) anorthite, (**b**) cordierite, (**c**) spinel, (**d**) Mg$_2$SiO$_4$, (**e**) mellite, (**f**) Mg$_2$SiO$_4$ and anorthite, (**g**) melilite and spinel, (**h**) MgO and merwinite.

Initial compositions of the sample used in the high temperature experiments was presented in Table A1. The compositions of the phases measured by EPMA are given in Table A2 (weight%) and Table A3 (mol%). The experimental data from previous studies [15–18,23] using the same technique are also present in the Table A1. The compositions of the anorthite, dicalcium silicate, merwinite, MgO, mullite, forsterite and spinel are close to their stoichiometry with limited solid solutions. Melilite is the solid solution between akermanite ($2CaO \cdot MgO \cdot 2SiO_2$) and gehlenite ($2CaO \cdot Al_2O_3 \cdot SiO_2$). Cordierite is the solid solution between $2MgO \cdot 2Al_2O_3 \cdot 5SiO_2$ and $2CaO \cdot 2Al_2O_3 \cdot 5SiO_2$. The solubilities of the solid phases are summarized below:

Anorthite ($CaO \cdot Al_2O_3 \cdot 2SiO_2$): contains 24.5–25.3 mol% CaO, 0.3–1.1 mol% MgO, 23.1–24.7 mol% Al_2O_3 and 49.8–51.3 mol% SiO_2. The $SiO_2/(Al_2O_3+CaO+MgO)$ molar ratio is around 1.0 in the anorthite.

Dicalcium silicate (Ca_2SiO_4): up to 3.7 wt% MgO is present in the dicalcium silicate, the $(CaO+MgO)/SiO_2$ molar ratio varies from 1.99 to 2.04.

Merwinite ($3CaO \cdot MgO \cdot 2SiO_2$): the CaO/MgO molar ratios vary in the range of 4.02–4.24 and MgO/SiO_2 molar ratios vary in the range of 0.33–0.38 in the merwinite.

MgO: up to 0.3 wt% CaO and 1.2 wt% Al_2O_3 are present in the MgO. Unexpectedly, more Al_2O_3 than CaO is present in the MgO.

Mullite ($3Al_2O_3 \cdot 2SiO_2$): up to 0.3 wt% CaO and 0.4 wt% MgO are present in the mullite. The molar ratio of Al_2O_3 to SiO_2 is up to 1.68.

Spinel ($MgO \cdot Al_2O_3$): up to 0.3 wt% CaO and 0.4 wt% SiO_2 are present in the spinel. The molar ratio of Al_2O_3 to MgO is up to 1.62.

Forsterite (Mg_2SiO_4): up to 4.2 wt% CaO and 0.5 wt% Al_2O_3 are present in the forsterite. The molar ratio of (CaO+MgO) to SiO_2 is 1.91–2.04.

Melilite: the molar ratio of akermanite ($2CaO \cdot MgO \cdot 2SiO_2$) to gehlenite ($2CaO \cdot Al_2O_3 \cdot SiO_2$) in the melilite phase varies from 0 to 5.59.

Cordierite: $2MgO \cdot 2Al_2O_3 \cdot 5SiO_2$ was initially reported to be a compound in the system MgO-Al_2O_3-SiO_2 [28]. In the later studies of the system CaO-SiO_2-Al_2O_3-MgO, the cordierite primary phase field was reported again in the high-MgO and high-Al_2O_3 sections [10–13]. However, both the experimental studies and FactSage predictions did not mention the solubility of CaO in the $2MgO \cdot 2Al_2O_3 \cdot 5SiO_2$. $2CaO \cdot 2Al_2O_3 \cdot 5SiO_2$ phase was not reported in the literature. In the present study, four experiments (74–77) reported the exitance of the cordierite in equilibrium with liquid at high temperature. Analysis of the EPMA measurements shows that the cordierite was composed of CaO 21.3–22.7, MgO 0.4–0.8, Al_2O_3 21.4–22.8 and SiO_2 53.7–56.8 mol%. The $SiO_2/(Al_2O_3+CaO+MgO)$ molar ratio in the cordierite is around 1.25 which is higher than that in the anorthite. Figure 2b shows a typical microstructure of cordierite in a quenched sample where the solid phase is lighter than the liquid phase. In contrast, it can be seen from Figure 2a that the anorthite is darker than the liquid phase confirming that cordierite is a phase different from the anorthite although they have close compositions. The new information will provide useful information to develop a cordierite solid solution in the thermodynamic database.

The experimental data given in Table A1 have been used to construct the pseudo-ternary section $(Al_2O_3+SiO_2)$-CaO-MgO as shown in Figure 3. The boundaries were drawn by passing through the three phase points where liquid was in equilibrium with two solid phases (Exp No. 121–158). If three phase points were not available, the boundaries were drawn between the experimental points in different primary phase fields. It can be seen from the figure that most of the experimental points are in the primary phase fields of anorthite, melilite, forsterite, MgO, mullite and dicalcium silicate. The solid symbols are the liquid compositions determined in the present study and the open symbols are the liquid compositions reported in the previous studies using the same technique. The thin solid thin lines represent experimentally determined isotherms and the dashed thin lines represent the estimated isotherms. The thick lines represent the boundaries between the primary phase fields.

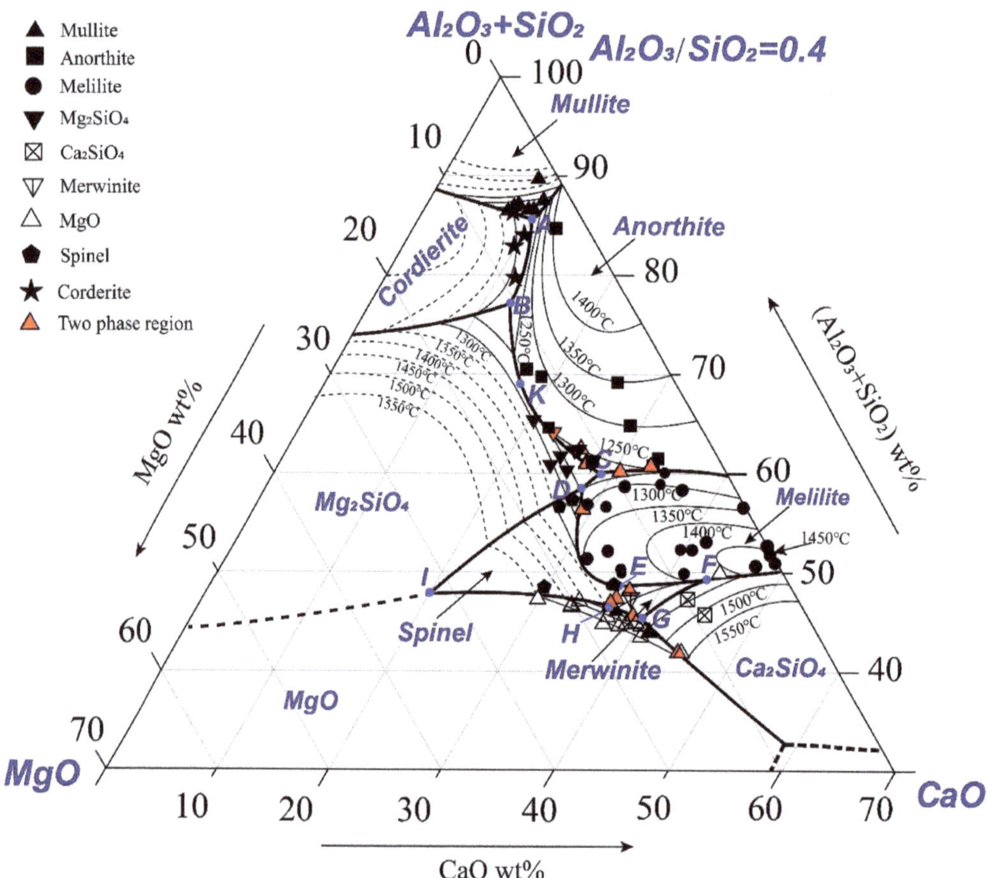

Figure 3. Experimentally determined pseudo-ternary phase diagram (SiO$_2$+Al$_2$O$_3$)-MgO-CaO with Al$_2$O$_3$/SiO$_2$ weight ratio of 0.4.

Nine special points in the system are estimated according to the boundaries and isothermals, as shown in Figure 4. The compositions and temperatures of the special points are given in Table 1. Point K on the boundary line BC joining the primary phase fields of anorthite and forsterite represents a local maximum temperature (1260 ± 5 °C). Experimental results reported by Osborn et al. [10] and Cavalier et al. [12] are shown in Figure 4 for comparison. The number next to the symbol is the liquidus temperature reported. The primary phases reported by Osborn et al. [10] show general agreement with the present results; however, the liquidus temperatures reported by Osborn et al. [10] are lower than the present results in some areas. Two points reported by Cavalier et al. [12] are shown in the figure. The point in the cordierite primary phase field agrees with the present result. However, another point was reported in the anorthite primary phase field with the liquidus temperature of 1272 °C by Cavalier et al. [12] which is in the forsterite primary phase field with the liquidus temperature of 1350 °C according to the present study.

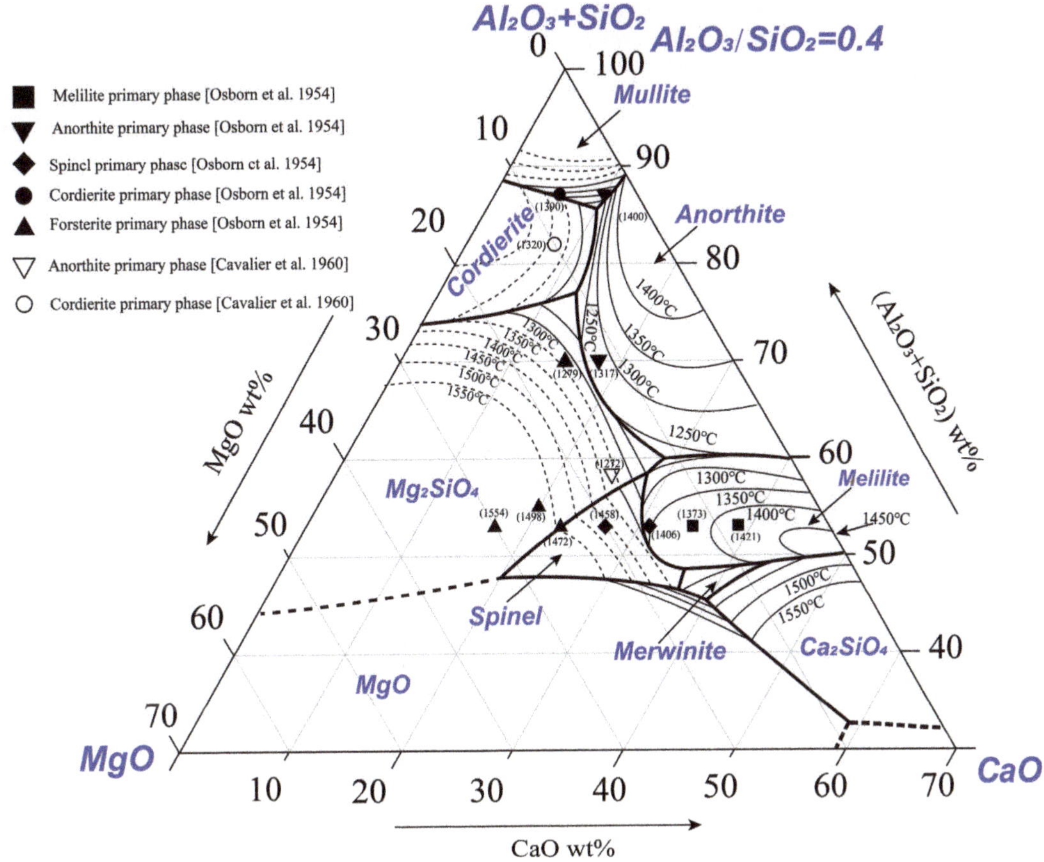

Figure 4. Comparison of the present results with the literature data [10,12].

Table 1. Estimated compositions and temperatures of the special points in the CaO-MgO-Al₂O₃-SiO₂ system with Al₂O₃/SiO₂ weight ratio of 0.4.

Special Point	Description	T (°C)	Composition (wt%)			
			CaO	MgO	Al₂O₃	SiO₂
A	L + mullite ↔ cordierite + anorthite	1250 (±5)	7.6	6.6	24.5	61.3
B	L ↔ cordierite + anorthite + Mg₂SiO₄	1240 (±10)	10.0	14.3	21.6	54.1
C	L ↔ melilite + anorthite + Mg₂SiO₄	1235 (±5)	27.5	11.6	17.4	43.6
D	L + spinel ↔ Mg₂SiO₄ + melilite	1270 (±10)	28.2	13.3	16.7	41.8
E	L + merwinite ↔ melilite + spinel	1380 (±10)	37.2	14.0	14.0	34.9
F	L + melilite + Ca₂SiO₄ ↔ merwinite	1455 (±5)	45.6	4.5	14.3	35.7
G	L + Ca₂SiO₄ ↔ merwinite + MgO	1480 (±10)	36.7	16.6	13.3	33.3
H	L + MgO ↔ merwinite + spinel	1410 (±10)	43.1	13.2	12.5	31.2
I	L + MgO + Mg₂SiO₄ ↔ + spinel	1650 (±20)	20.0	32.1	13.7	34.2

3.2. Application of Phase Diagram in Ironmaking Process

Pseudo-binary phase diagrams are often used by operators and researchers to discuss the effect of slag composition on liquidus temperatures. A typical composition of Shougang BF slag [27] is shown in Figure 5. It can be seen that the BF composition is in the melilite primary phase field with the liquidus temperature 1400–1450 °C. Shougang has a stable supply of raw materials for the BF, resulting a relative stable Al₂O₃/SiO₂ weight ratio of 0.4

in the slag. The additions of MgO and CaO can be adjusted according to the requirements. It is more convenient to use a pseudo-binary phase diagram to discuss the effect of a single parameter on liquidus temperature. The red line, as shown in Figure 5, indicates how the pseudo-binary phase diagrams are interpolated from the pseudo-ternary phase diagram. Based on the pseudo-ternary phase diagram determined, the effects of MgO/CaO, and quaternary basicity $(CaO+MgO)/(Al_2O_3+SiO_2)$ on the liquidus temperature can be considered from different directions, as shown in Figure 5.

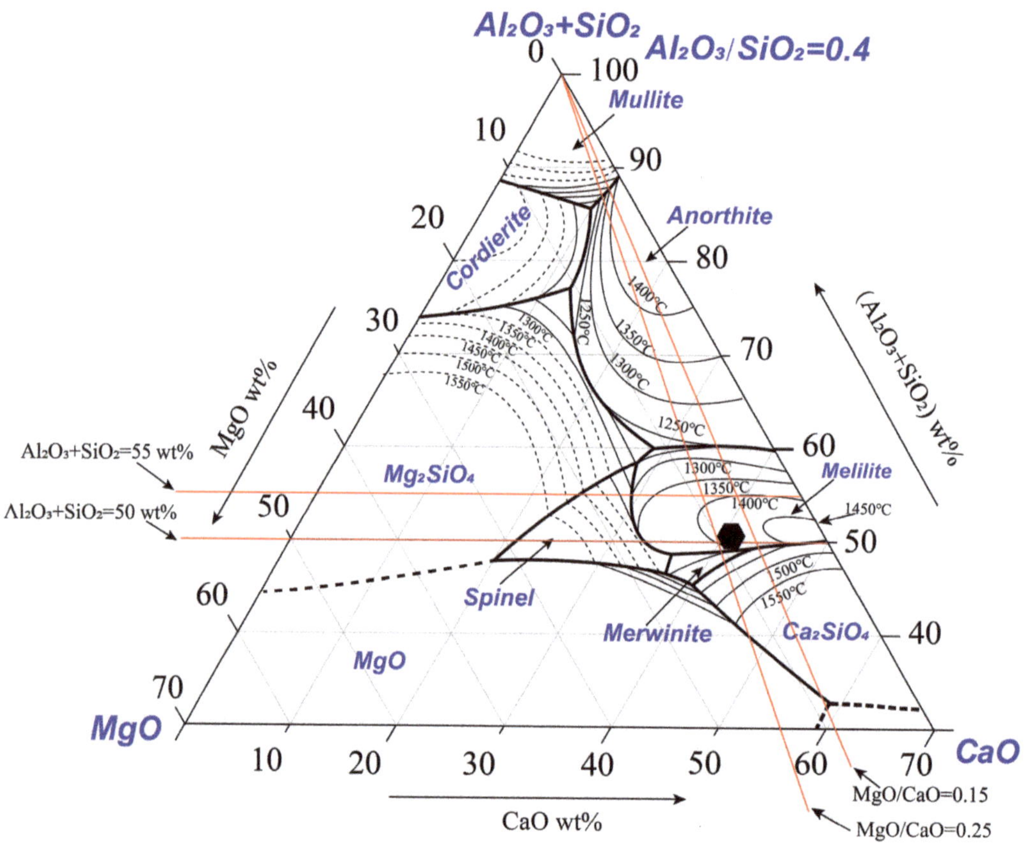

● Average composition of Shougang BF slag [Xu et al. 2016]

Figure 5. A typical composition of Shougang blast furnace slag [27] and effects of slag composition on the liquidus temperature.

3.2.1. Effect of MgO/CaO Weight Ratio on Liquidus Temperature

Figure 6 shows the liquidus temperatures as a function of MgO/CaO weight ratio in liquid at fixed $(Al_2O_3+SiO_2)$ of 50 and 55 wt%, respectively. The typical composition of Shougang BF slag and FactSage predictions are also shown in the figure. It can be seen that melilite is the primary phase at low MgO/CaO weight ratio and spinel is the primary phase at high MgO/CaO weight ratio. The liquidus temperatures decrease in the melilite primary phase field and increase in the spinel primary phase field with increasing MgO/CaO weight ratio. In both melilite and spinel primary phase fields, higher $(Al_2O_3+SiO_2)$ concentration (55 wt%) results in a lower liquidus temperature at a given MgO/CaO weight ratio. It is possible to decrease the liquidus temperature of the BF slag by increasing the MgO/CaO weight ratio up to 0.52. However, there is a limitation for the MgO/CaO ratio where the primary phase changes to the spinel. In comparison between the experimental results and FactSage predictions, it can be seen that the sizes of the melilite and spinel primary phase fields determined in the present study are much larger than the FactSage predictions. For example, FactSage predicts the MgO primary phase field when the MgO/CaO weight ratio is higher than 0.93. However, the MgO primary phase only appears from MgO/CaO weight ratio 1.39 according to the experimental results. Nearly 100 °C difference of the liquidus temperature occurs between the experimental results and FactSage predictions at MgO/CaO ratio 0.52.

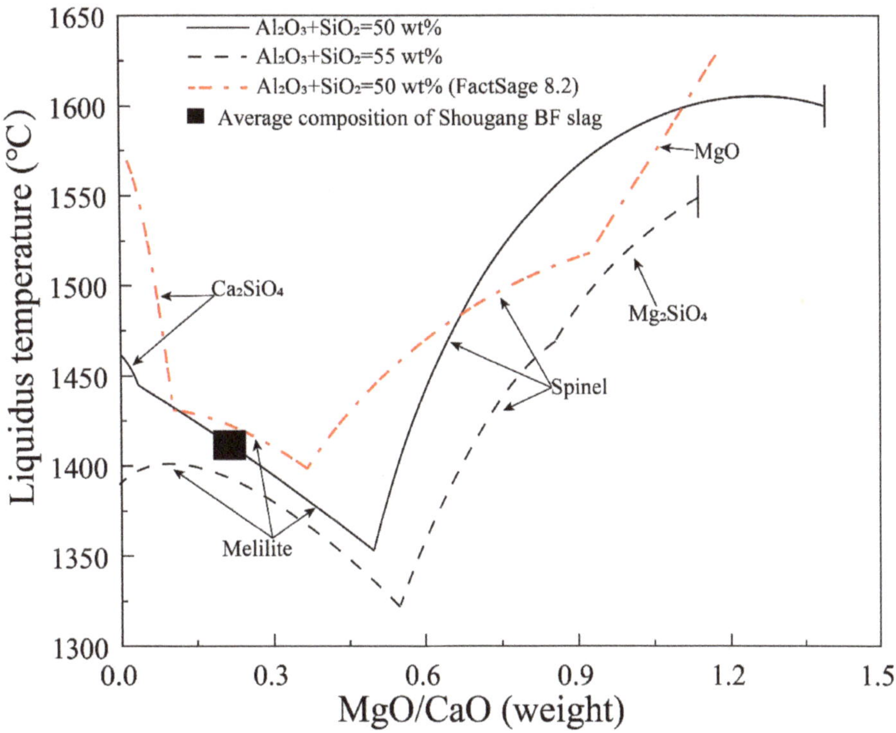

Figure 6. Liquidus temperature as a function of MgO/CaO in liquids at fixed (Al$_2$O$_3$+SiO$_2$) of 50 and 55 wt% in the CaO-MgO-SiO$_2$-Al$_2$O$_3$ system with Al$_2$O$_3$/SiO$_2$ weight ratio of 0.4.

3.2.2. Effect of Quaternary Basicity on Liquidus Temperature

Due to the significant impact of the concentrations of Al$_2$O$_3$ and MgO in slag on the actual basicity and desulfurization capacity of the BF slag, the CaO/SiO$_2$ binary basicity cannot reflect the basicity of high Al$_2$O$_3$ slag objectively [5]. Therefore, the desulfurization coefficient or capacity is more closely related to quaternary basicity (CaO+MgO)/(SiO$_2$+Al$_2$O$_3$). It is worth noting that the quaternary basicity will be forced to increase when the magnesium flux pellet is widely used in the blast furnace, which can greatly affect the softening–melting properties of burden materials in the cohesive zone [7]. Hence, it is necessary to understand the effect of quaternary basicity on the liquidus temperature of the BF slags. Figure 7 shows the liquidus temperature as a function of (CaO+MgO)/(SiO$_2$+Al$_2$O$_3$) in liquids at fixed MgO/CaO weight ratios of 0.15 and 0.25. FactSage predictions are also shown in the figure for comparison. As can be seen, mullite, anorthite, melilite, merwinite and Ca$_2$SiO$_4$ are the primary phases in the composition range investigated. The quaternary basicity has a significant influence on the liquidus temperatures of all primary phase fields.

The typical Shougang BF slag composition is in the melilite primary phase field close to the merwinite (MgO/CaO = 0.25) or Ca_2SiO_4 (MgO/CaO = 0.15) primary phase field. Decrease of the quaternary basicity from the current slag composition can decrease the liquidus temperature significantly. Further increase of the quaternary basicity will bring the slag into the merwinite or Ca_2SiO_4 primary phase fields causing significant increase of the liquidus temperature. It can be seen that increase of the MgO/CaO ratio from 0.15 to 0.25 can slightly decrease the liquidus temperature in the anorthite and melilite primary phase fields but significantly decrease the liquidus temperature in the Ca_2SiO_4 primary phase field. In conclusion, the Shougang BF slag with a quaternary basicity of 1.05 has the optimum composition at the MgO/CaO weight ratio 0.15. If the MgO/CaO weight ratio is increased to 0.25, the quaternary basicity can be further increased, which will decrease the liquidus temperature and increase the desulfurization capacity. The predicted liquidus temperatures in the anorthite and Ca_2SiO_4 primary phase fields are close to the experimental results. In the melilite primary phase field, there is up to 50 °C difference between the FactSage predictions and experimental results. Experimentally determined merwinite phase is not predicted by FactSage.

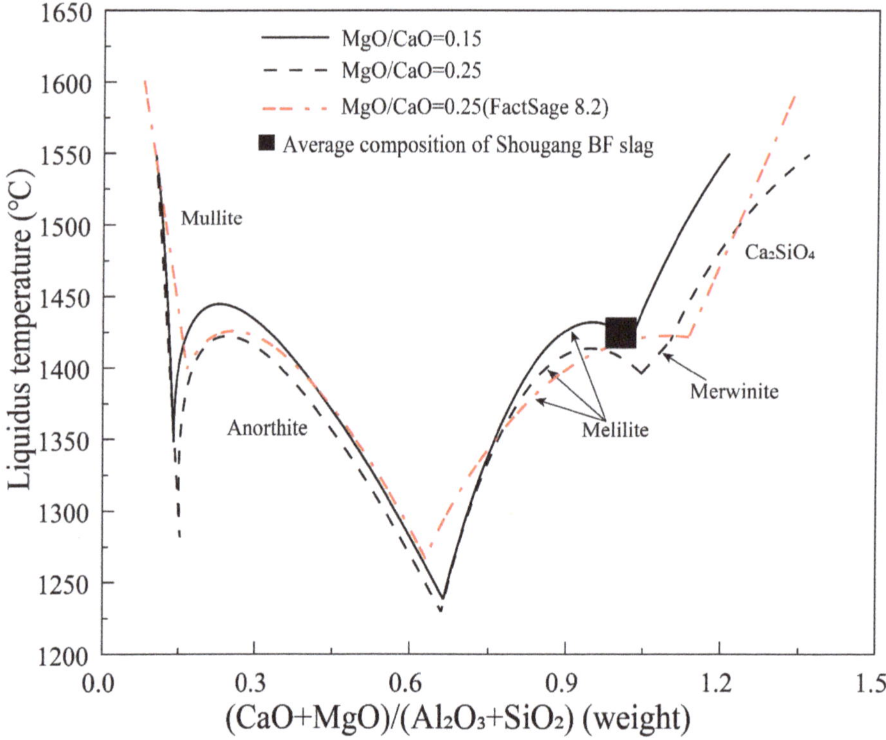

Figure 7. Liquidus temperature as a function of quaternary basicity in liquid at fixed MgO/CaO weight ratios of 0.15 and 0.25 in the CaO-MgO-SiO_2-Al_2O_3 system with Al_2O_3/SiO_2 weight ratio of 0.4.

3.3. Comparison of Experimental Data and FactSage Predictions

FactSage is one of the most successful thermodynamic models in predicting the liquidus temperatures of oxide slags [25]. Accurate experimental data can be used to evaluate the accuracy of the FactSage predictions and support the optimization of the existing database and new database development. Clarifying the difference between the calculated and experimental data is of great significance for optimizing thermodynamic database and avoiding misleading industrial practices. As shown in Figures 6 and 7, a significant difference occurs on the primary phase and the liquidus temperatures between the FactSage predictions and experimental results.

As discussed above, the current composition of the Shougang blast furnace slag is usually in the melilite primary phase field as shown in Figures 6 and 7. Melilite is the solid solution between akermanite ($2CaO \cdot MgO \cdot 2SiO_2$) and gehlenite ($2CaO \cdot Al_2O_3 \cdot SiO_2$). Figure 8 presents the comparison of the experimental data and FactSage predictions. The experiments No. 44–63 listed in Table A1 were used to calculate the primary phases and their liquidus temperatures by FactSage 8.2. Melilite is the primary phase in experiments No. 44–63. However, it can be seen from Figure 8 that anorthite, spinel and Ca_2SiO_4 are also predicted by the FactSage in addition to melilite. Up to 75 °C difference of the liquidus temperature is observed between the predictions and experimental results.

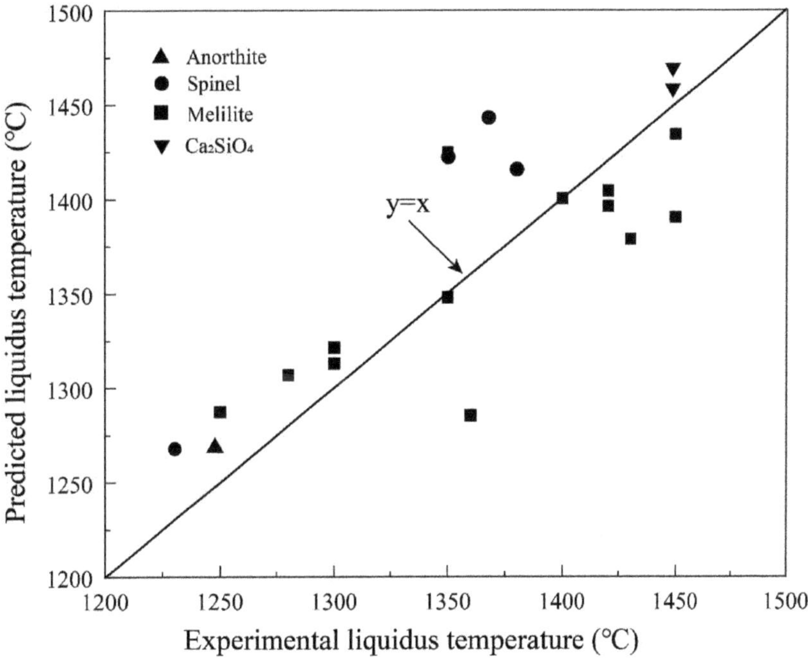

Figure 8. Comparison of the experimental results and FactSage predictions on primary phase and liquidus temperatures. All experimental liquidus are in the melilite primary phase field. The predicted primary phases are shown in the figure.

The experiments No. 99–100 listed in Table A1 show that Mg_2SiO_4 is the primary phase in the quenched samples. The liquid compositions of these samples were used to calculate the primary phase and liquidus temperatures and the results are shown in Table 2. It can be seen that anorthite, spinel, melilite and merwinite are predicted by the FactSage showing that the current database needs to be optimized. The discrepancies between the experimental and predicted liquidus temperatures are up to 112 °C.

Table 3 shows the comparison of the experimental results and FactSage predictions where merwinite is the primary phase of the experiments No. 115–119. FactSage predicted melilite, MgO and spinel as the primary phases but no merwinite. The liquidus temperature predicted by FactSage 8.2 is 189 °C higher than the experimental data (No. 117). Lack of accurate experimental data in the composition range causes inaccuracy in the thermodynamic database.

Table 2. Comparison of the experimental data with the FactSage predictions on primary phase and liquidus temperature, Mg_2SiO_4 is the primary phase in all experiments.

Exp. No.	Experimental Liquidus Temperature (°C)	Liquid Composition (wt%)				Predicted Liquidus Temperature (°C)	Predicted Primary Phase
		CaO	MgO	Al_2O_3	SiO_2		
90	1250	25.8	11.6	18.7	44.0	1287	Anorthite
91	1250	27.6	11.7	17.2	43.5	1258	Spinel
92	1250	26.9	11.8	18.1	42.8	1287	Spinel
93	1300	20.5	14.4	18.5	46.6	1294	Anorthite
94	1300	24.6	13.9	18.2	43.2	1315	Spinel
95	1300	25.9	14.1	17.3	42.7	1307	Spinel
96	1300	40.9	11.6	7.9	39.6	1412	Merwinite
97	1330	24.2	15.3	17.8	42.7	1333	Spinel
98	1350	29.2	16.6	14.0	40.1	1325	Melilite
99	1400	27.7	18.5	13.7	40.0	1342	Mg_2SiO_4
100	1400	28.0	14.7	15.8	41.4	1306	Spinel

Table 3. Comparison of the experimental data with the FactSage prediction on primary phase and liquidus temperature; merwinite is the primary phase in all experiments.

Exp. No.	Experimental Liquidus Temperature (°C)	Liquid Composition (wt%)				Predicted Liquidus Temperature (°C)	Predicted Primary Phase
		CaO	MgO	Al_2O_3	SiO_2		
115	1400	41.1	8.9	14.7	35.3	1427	Melilite
116	1400	37.9	14.2	13.1	34.7	1421	Spinel
117	1430	39.8	15.7	13.4	31.1	1619	MgO
118	1450	40.1	15.1	13.1	31.7	1555	MgO
119	1450	43.1	9.6	13.8	33.4	1425	Melilite

4. Conclusions

The phase equilibria in the CaO-MgO-SiO_2-Al_2O_3 system with an Al_2O_3/SiO_2 weight ratio of 0.4 have been experimentally investigated in the composition range related to blast furnace slags. A pseudo-ternary phase diagram $(CaO+MgO)$-SiO_2-Al_2O_3 has been constructed using the experimentally data. Pseudo-binary phase diagrams are used to discuss the effects of quaternary basicity and MgO/CaO weight ratio on liquidus temperatures of the blast furnace slags. The experimentally determined liquidus temperatures and solid solution compositions are compared with the FactSage predictions to provide useful information for optimization of the thermodynamic database.

Author Contributions: Methodology, B.Z.; Validation, B.Z.; Formal analysis, J.L.; Investigation, J.L. and G.Q.; Resources, G.Q.; Data curation, J.L.; Writing—original draft, J.L. and G.Q.; Writing—review & editing, B.Z.; Supervision, B.Z.; Project administration, B.Z. All authors have read and agreed to the published version of the manuscript.

Funding: This research received no external funding.

Data Availability Statement: Not applicable.

Acknowledgments: The authors would like to acknowledge Shougang Research Institute of Technology for providing the financial support.

Conflicts of Interest: The authors declare that they have no conflict of interest.

Appendix A

Table A1. Initial compositions of the sample used in the high temperature experiments.

No.	Composition (wt%)			
	CaO	MgO	Al_2O_3	SiO_2
M1	31.3	6.3	17.9	44.5
M2	41.7	8.3	14.3	35.7
M3	37.0	7.4	15.9	39.7
M4	45.4	9.1	13.0	32.5
M5	27.8	5.6	19.0	47.6
M6	39.3	7.9	15.1	37.7
M7	34.3	6.9	16.8	42.0
M8	25.6	5.1	19.8	49.5
M9	12.4	2.5	24.3	60.8
M10	10.4	2.1	25.0	62.5
M11	6.3	1.3	26.4	66.0
M12	9.0	1.8	25.5	63.7
M13	32.5	20.5	14.0	33.0
M14	14.7	2.9	30.9	51.5
M15	27.8	5.6	25.0	41.6
M16	11.4	4.6	24.0	60.0
M17	7.9	3.2	25.4	63.5
M18	9.4	3.8	24.8	62.0
M19	22.1	8.9	19.7	49.3
M20	28.5	11.4	17.2	42.9
M21	35.4	14.2	14.4	36.0
M22	36.4	14.6	14.0	35.0
M23	39.4	15.8	12.8	32.0
M24	42.9	17.1	11.4	28.6
M25	32.1	12.9	15.7	39.3
M26	25.3	10.1	18.4	46.2
M27	34.9	25.1	11.4	28.6
M28	29.1	20.9	14.3	35.7
M29	23.3	16.7	17.1	42.9
M30	32.0	23.0	12.8	32.2
M31	26.7	19.2	15.4	38.7
M32	20.4	14.7	18.6	46.3
M33	30.5	21.9	13.6	34.0
M34	16.6	12.0	20.4	51.0
M35	9.7	7.0	23.8	59.5
M36	5.8	4.2	25.7	64.3
M37	35.4	21.2	12.4	31.0
M38	25.3	15.2	17.0	42.5
M39	7.8	4.7	25.0	62.5
M40	10.0	6.0	24.0	60.0
M41	18.8	11.3	20.0	49.9
M42	21.9	13.1	18.6	46.4
M43	23.0	13.8	18.1	45.1
M44	26.9	16.1	16.3	40.7
M45	31.3	18.8	14.3	35.6
M46	37.5	22.5	11.4	28.6
M47	33.6	20.2	13.2	33.0
M48	24.0	14.4	17.6	44.0
M49	12.5	7.5	22.8	57.2
M50	5.6	3.4	26.0	65.0
M51	16.4	6.6	22.0	55.0
M52	7.7	5.6	24.8	61.9

Table A2. Experimental Results in the System CaO-MgO-Al_2O_3-SiO_2 system with Al_2O_3/SiO_2 weight ratio of 0.4.

Mixture				Compositions (wt%)				
No.	No.	T (°C)	Phases	CaO	MgO	Al_2O_3	SiO_2	Al_2O_3/SiO_2
				Liquid only				
M20	1	1250	Liquid	28.9	11.5	17.3	42.3	0.41
M26	2	1250	Liquid	27.1	9.6	19.7	43.6	0.45
M34	3	1250	Liquid	16.5	12.1	20.4	51.0	0.40
M16	4	1280	Liquid	11.6	4.6	23.8	60.0	0.40
M16	5	1300	Liquid	11.6	4.5	24.0	59.9	0.40
M34	6	1300	Liquid	16.4	12.1	20.3	51.2	0.40
M35	7	1300	Liquid	9.6	7.0	23.3	60.1	0.39
M36	8	1300	Liquid	6.8	4.9	21.5	66.8	0.32
M42	9	1300	Liquid	22.6	12.7	18.9	45.8	0.41
M48	10	1300	Liquid	24.8	14.0	18.0	43.2	0.42
M42	11	1310	Liquid	22.2	13.2	18.7	45.9	0.41
M19	12	1320	Liquid	22.7	9.0	19.7	48.6	0.41
M42	13	1320	Liquid	24.3	13.9	17.9	43.9	0.41
M25	14	1330	Liquid	31.6	13.0	15.9	39.5	0.40
M25	15	1330	Liquid	32.6	12.0	16.2	39.2	0.41
M48	16	1340	Liquid	22.9	12.6	19.3	45.2	0.43
M13	17	1350	Liquid	32.9	20.2	13.5	33.4	0.40
M16	18	1350	Liquid	11.6	4.5	23.6	60.3	0.39
M19	19	1350	Liquid	22.1	8.7	19.9	49.3	0.40
M25	20	1350	Liquid	31.9	12.8	15.6	39.7	0.39
M51	21	1350	Liquid	17.9	6.8	23.1	52.2	0.44
M51	22	1350	Liquid	18.5	7.3	23.8	50.4	0.47
M29	23	1350	Liquid	23.2	16.7	17.2	42.9	0.40
M42	24	1350	Liquid	24.4	12.3	20.3	43.0	0.47
M44	25	1350	Liquid	26.7	16.4	16.4	40.5	0.40
M49	26	1350	Liquid	12.9	7.6	23.1	56.4	0.41
M16	27	1350	Liquid	11.5	4.6	23.9	60.0	0.40
M35	28	1350	Liquid	9.9	7.0	23.9	59.2	0.40
M22	29	1390	Liquid	36.5	14.4	14.1	35.0	0.40
M22	30	1400	Liquid	36.5	14.0	13.7	35.8	0.38
M28	31	1400	Liquid	29.8	21.0	14.2	35.0	0.41
M41	32	1400	Liquid	20.9	11.9	22.5	44.7	0.50
M49	33	1400	Liquid	14.7	8.4	25.4	51.5	0.49
M21	34	1400	Liquid	35.6	14.2	14.5	35.7	0.41
M23	35	1450	Liquid	38.7	14.1	12.8	34.4	0.37
M23	36	1450	Liquid	39.4	15.7	12.8	32.1	0.40
M47	37	1450	Liquid	33.8	18.9	13.2	34.1	0.39
M23	38	1450	Liquid	38.8	15.4	13.6	32.2	0.42
M14	39	1480	Liquid	15.0	2.9	30.6	51.5	0.59
M14	40	1500	Liquid	15.0	3.0	30.5	51.5	0.59
M23	41	1500	Liquid	39.5	15.8	12.9	31.8	0.40
M30	42	1500	Liquid	45.0	1.1	17.9	36.0	0.50
M47	43	1500	Liquid	33.9	19.1	13.5	33.5	0.40
				Liquid + Melilite				
M20	44	1230	Liquid	27.7	11.4	17.7	43.2	0.41
			melilite	41.0	12.1	6.4	40.5	
M7	45	1250	Liquid	33.0	6.2	17.6	43.2	0.41
			melilite	41.1	10.4	10.4	38.1	
M43	46	1250	Liquid	27.4	11.3	17.9	43.4	0.41
			melilite	40.5	12.4	5.8	41.3	
	47 *	1280	Liquid	31.7	9.5	16.3	42.5	0.38
			melilite	8.8	40.6	39.1	11.5	

Table A2. *Cont.*

Mixture				Compositions (wt%)				
No.	No.	T (°C)	Phases	CaO	MgO	Al₂O₃	SiO₂	Al₂O₃/SiO₂
M7	48	1300	Liquid	34.5	6.5	16.6	42.4	0.39
			melilite	41.7	8.4	12.4	37.5	
	49 *	1300	Liquid	37.0	4.7	16.6	41.7	0.40
			melilite	32.9	41.0	18.7	7.4	
M21	50	1350	Liquid	31.8	16.8	14.7	36.7	0.40
			melilite	41.1	9.6	13.2	36.1	
M17	51	1350	Liquid	8.6	3.5	23.4	64.5	0.36
			melilite	0.1	0.2	73.1	26.6	
M3	52	1350	Liquid	36.4	7.2	15.4	41.0	0.37
			melilite	41.0	8.3	15.8	34.9	
	53 *	1360	Liquid	43.2	0.0	16.3	40.5	0.40
			melilite	40.6	0.0	36.6	22.8	
M22	54	1370	Liquid	35.3	15.9	13.7	35.1	0.39
			melilite	40.8	8.6	16.1	34.5	
M21	55	1380	Liquid	35.4	14.2	14.7	35.7	0.41
			melilite	41.7	8.0	16.6	33.7	
M6	56	1400	Liquid	39.5	8.0	14.6	37.9	0.39
			melilite	41.6	5.6	22.6	30.2	
	57 *	1420	Liquid	41.6	5.1	15.3	38.0	0.40
			melilite	41.2	3.9	27.0	27.9	
	58 *	1420	Liquid	40.9	6.8	14.7	37.6	0.39
			melilite	41.0	5.1	24.3	29.6	
	59 *	1430	Liquid	47.4	0.0	15.1	37.5	0.40
			melilite	41.4	0.0	36.4	22.2	
	60 *	1450	Liquid	44.6	5.4	14.5	35.5	0.41
			melilite	41.1	2.8	29.8	26.3	
	61 *	1450	Liquid	48.0	0.0	14.9	37.1	0.40
			melilite	41.0	0.0	36.9	22.1	
	62 *	1450	Liquid	47.4	2.0	14.1	36.5	0.39
			melilite	41.0	1.4	33.4	24.2	
	63 *	1450	Liquid	48.9	0.0	14.4	36.7	0.39
			melilite	40.9	0.0	37.2	21.9	
			Liquid + Anorthite					
M26	64	1250	Liquid	27.7	11.3	17.0	44.0	0.39
			Anorthite	20.3	0.2	35.4	44.1	
M35	65	1250	Liquid	32.1	6.7	16.3	44.9	0.36
			Anorthite	20.2	0.3	35.8	43.7	
M43	66	1280	Liquid	22.0	13.3	18.3	46.4	0.39
			Anorthite	20.2	0.4	35.6	43.8	
	67 *	1280	Liquid	33.4	5.2	16.8	44.6	0.38
			Anorthite	36.1	43.4	20.3	0.2	
M34	68	1290	Liquid	17.2	12.4	20.6	49.8	0.41
			Anorthite	20.1	0.6	34.5	44.8	
M41	69	1300	Liquid	18.6	11.5	19.5	50.4	0.39
			Anorthite	20.0	0.6	34.4	45.0	
M19	70	1300	Liquid	22.3	9.7	17.4	50.6	0.35
			Anorthite	20.1	0.5	35.7	43.7	
M5	71	1300	Liquid	29.1	6.3	16.8	47.8	0.35
			Anorthite	20.5	0.3	36.0	43.2	
M8	72	1350	Liquid	26.1	5.3	18.8	49.8	0.38
			anorthite	20.4	0.3	35.6	43.7	
M9	73	1400	Liquid	12.3	2.6	23.5	61.6	0.38
			anorthite	19.8	0.2	36.3	43.7	
			Liquid + Cordierite					
M17	74	1250	Liquid	9.7	7.2	23.7	59.4	0.40
			Cordierite	18.0	0.5	32.6	48.9	

Table A2. *Cont.*

Mixture				Compositions (wt%)				
No.	No.	T (°C)	Phases	CaO	MgO	Al_2O_3	SiO_2	Al_2O_3/SiO_2
M49	75	1300	Liquid	11.7	8.5	21.8	58.0	0.38
			Cordierite	18.6	0.5	33.9	47.0	
M40	76	1300	Liquid	10.0	6.0	23.8	60.2	0.39
			Cordierite	18.3	0.3	33.6	47.8	
M10	77	1300	Liquid	9.9	2.4	24.5	63.2	0.39
			Cordierite	17.5	0.2	32.1	50.2	
			Liquid + Mullite					
M17	78	1250	Liquid	9.0	3.7	23.5	63.8	0.37
			Mullite	0.2	0.4	68.5	30.9	
M18	79	1320	Liquid	9.6	3.7	24.7	62.0	0.40
			Mullite	0.2	0.2	71.1	28.5	
M39	80	1350	Liquid	8.0	4.7	24.7	62.6	0.39
			Mullite	0.1	0.3	70.9	28.7	
M36	81	1400	Liquid	6.2	4.5	23.5	65.8	0.36
			Mullite	0.2	0.3	72.6	26.9	
M12	82	1400	Liquid	9.5	1.8	23.9	64.8	0.37
			Mullite	0.1	0.1	71.9	27.9	
M11	83	1450	Liquid	8.4	1.8	25.1	64.7	0.39
			Mullite	0.1	0.0	74.0	25.9	
M50	84	1500	Liquid	5.9	3.5	24.4	66.2	0.37
			Mullite	0.1	0.1	73.2	26.6	
			Liquid + Spinel					
M29	85	1320	Liquid	25.2	14.9	18.5	41.4	0.45
			Spinel	0.0	28.2	71.7	0.1	
M44	86	1350	Liquid	27.1	16.3	16.3	40.3	0.40
			Spinel	0.1	28.2	71.6	0.1	
M22	87	1370	Liquid	35.6	15.6	13.8	35.0	0.39
			Spinel	0.0	28.1	71.8	0.1	
	89 *	1500	liquid	30.4	21.7	14.1	33.8	0.42
			Spinel	0.3	0.4	70.9	28.4	
			Liquid + Mg_2SiO_4					
M48	90	1250	Liquid	25.8	11.6	18.7	43.9	0.42
			Mg_2SiO_4	1.0	56.9	0.1	42.0	
M38	91	1250	Liquid	27.6	11.7	17.2	43.5	0.40
			Mg_2SiO_4	1.6	55.8	0.1	42.5	
M44	92	1250	Liquid	27.0	11.8	18.2	43.0	0.42
			Mg_2SiO_4	1.4	56.4	0.4	41.8	
M32	93	1300	Liquid	20.5	14.4	18.5	46.6	0.40
			Mg_2SiO_4	0.6	56.7	0.1	42.6	
M29	94	1300	Liquid	24.6	13.9	18.2	43.3	0.42
			Mg_2SiO_4	1.1	56.4	0.3	42.2	
M38	95	1300	Liquid	25.9	14.1	17.3	42.7	0.41
			Mg_2SiO_4	1.3	56.0	0.2	42.5	
M28	96	1300	Liquid	40.9	11.6	7.9	39.6	0.20
			Mg_2SiO_4	4.2	53.8	0.2	41.8	
M29	97	1330	Liquid	24.2	15.3	17.8	42.7	0.42
			Mg_2SiO_4	1.1	56.6	0.2	42.1	
M31	98	1350	Liquid	29.2	16.6	14.0	40.2	0.35
			Mg_2SiO_4	2.4	55.2	0.5	41.9	
M31	99	1400	Liquid	27.7	18.5	13.7	40.1	0.34
			Mg_2SiO_4	2.6	55.1	0.3	42.0	
M44	100	1400	Liquid	28.0	14.7	15.8	41.5	0.38
			Mg_2SiO_4	0.6	56.7	0.1	42.6	
			Liquid + MgO					
M30	101	1450	Liquid	33.4	19.3	13.6	33.7	0.40
			MgO	0.2	98.9	0.9	0.0	

Table A2. Cont.

Mixture				Compositions (wt%)				
No.	No.	T (°C)	Phases	CaO	MgO	Al_2O_3	SiO_2	Al_2O_3/SiO_2
M23	102	1470	Liquid	39.5	15.5	12.8	32.2	0.40
			MgO	0.3	98.9	0.8	0.0	
	103 *	1500	Liquid	30.0	22.7	13.5	33.8	0.40
			MgO	0.1	0.2	1.2	98.5	
	104 *	1500	Liquid	41.4	14.6	12.3	31.7	0.39
			MgO	0.3	99.0	0.7	0.0	
M27	105	1500	Liquid	38.7	16.3	12.6	32.4	0.39
			MgO	0.2	99.0	0.8	0.0	
M30	106	1500	Liquid	32.9	20.3	13.5	33.3	0.40
			MgO	0.1	98.8	1.1	0.0	
M30	107	1500	Liquid	32.9	20.2	13.5	33.4	0.40
			MgO	0.2	98.7	1.1	0.0	
M30	108	1500	Liquid	32.9	20.3	13.5	33.3	0.40
			MgO	0.2	98.8	1.0	0.0	
M37	109	1500	Liquid	36.6	17.6	13.1	32.7	0.40
			MgO	0.3	98.8	0.9	0.0	
M27	110	1550	Liquid	38.1	17.1	12.8	32.0	0.40
			MgO	0.1	99.0	0.9	0.0	
M24	111	1550	Liquid	44.9	13.0	12.1	30.0	0.40
			MgO	0.3	99.1	0.6	0.0	
M46	112	1550	Liquid	41.1	14.8	12.6	31.5	0.40
			MgO	0.3	99.0	0.7	0.0	
M46	113	1550	Liquid	40.6	15.7	12.9	30.8	0.42
			MgO	0.3	98.9	0.8	0.0	
M37	114	1550	Liquid	36.6	18.4	12.8	32.2	0.40
			MgO	0.2	98.8	1.0	0.0	
			Liquid + Merwinite					
	115 *	1400	Liquid	38.0	14.2	13.1	34.7	0.38
			Merwinite	51.2	12.6	0.1	36.1	
M2	116	1400	Liquid	41.1	8.9	14.7	35.3	0.42
			Merwinite	51.1	12.2	0.1	36.6	
M23	117	1430	Liquid	39.8	15.7	13.4	31.1	0.43
			Merwinite	52.4	13.0	0.1	34.5	
	118 *	1450	Liquid	43.2	9.6	13.8	33.4	0.41
			Merwinite	51.3	12.1	0.1	36.5	
M23	119	1450	Liquid	40.1	15.1	13.1	31.7	0.41
			Merwinite	51.8	12.7	0.1	35.4	
			Liquid + Ca_2SiO_4					
M4	120	1500	Liquid	45.1	9.1	13.4	32.4	0.41
			Ca_2SiO_4	61.6	2.9	0.2	35.3	
			Liquid + Melilite + Spinel					
M25	121	1300	Liquid	29.2	13.8	16.2	40.8	0.40
			Melilite	40.9	10.0	11.5	37.6	
			Spinel	0.1	28.2	71.6	0.1	
M25	122	1320	Liquid	31.0	12.2	16.7	40.1	0.42
			Melilite	41.2	10.0	11.9	36.9	
			Spinel	0.1	28.5	71.3	0.1	
			Liquid + Melilite + Anorthite					
M15	123	1250	Liquid	30.4	9.3	17.6	42.7	0.41
			Anorthite	20.5	0.4	36.6	42.5	
			Melilite	41.1	11.0	9.5	38.4	
M1	124	1250	Liquid	32.1	6.8	16.5	44.6	0.37
			Anorthite	20.5	0.1	35.9	43.5	
			Melilite	40.9	11.4	7.3	40.4	

Table A2. *Cont.*

Mixture				Compositions (wt%)				
No.	No.	T (°C)	Phases	CaO	MgO	Al_2O_3	SiO_2	Al_2O_3/SiO_2
			Liquid + Melilite + Mg_2SiO_4					
M20	125	1220	Liquid	27.1	11.0	18.8	43.1	0.44
			Mg_2SiO_4	1.3	56.5	0.3	41.9	
			Melilite	40.8	11.7	7.0	40.5	
M25	126	1250	Liquid	27.2	12.9	17.9	42.0	0.43
			Mg_2SiO_4	1.6	56.3	0.2	41.9	
			Melilite	40.3	9.8	12.2	37.7	
M45	127	1300	Liquid	29.0	13.8	15.5	41.7	0.37
			Mg_2SiO_4	2.4	55.2	0.4	42.0	
			Melilite	40.7	9.9	12.1	37.3	
M44	128	1300	Liquid	25.4	14.2	12.6	47.8	0.56
			Mg_2SiO_4	40.7	11.0	8.6	39.7	
			Melilite	2.1	54.8	1.7	41.4	
M28	129	1350	Liquid	31.0	16.4	12.8	39.8	
			Mg_2SiO_4	40.9	11.9	6.8	40.4	
			Melilite	3.9	53.8	0.3	42.0	
M45	130	1350	Liquid	30.1	16.4	13.5	40.0	0.34
			Mg_2SiO_4	4.1	54.0	0.2	41.7	
			Melilite	40.3	12.5	7.1	40.1	
M45	131	1450	Liquid	34.0	15.9	10.4	39.7	0.34
			Mg_2SiO_4	4.1	53.5	0.4	42.0	
			Melilite	41.2	11.7	7.8	39.3	
			Liquid + Mg_2SiO_4 + Anorthite					
M42	132	1200	Liquid	23.2	15.3	13.9	47.6	0.29
			Anorthite	19.7	0.8	34.6	44.9	
			Mg_2SiO_4	1.2	56.5	0.2	42.1	
M42	133	1230	Liquid	22.8	13.4	17.5	46.3	0.38
			Anorthite	20.2	0.5	35.7	43.6	
			Mg_2SiO_4	0.8	57.0	0.1	42.1	
M20	134	1230	Liquid	25.9	11.9	17.2	45.0	0.38
			Anorthite	20.1	0.2	35.6	44.1	
			Mg_2SiO_4	0.9	56.2	0.3	42.6	
M34	135	1240	Liquid	16.2	12.1	15.6	56.1	0.28
			Anorthite	19.9	1.0	33.1	46.0	
			Mg_2SiO_4	0.3	57.2	0.1	42.4	
M34	136	1250	Liquid	17.4	13.0	15.8	53.8	0.29
			Anorthite	19.8	0.9	33.7	45.6	
			Mg_2SiO_4	0.3	57.0	0.1	42.6	
M41	137	1250	Liquid	16.5	13.0	15.8	54.7	0.29
			Anorthite	19.7	0.9	34.1	45.3	
			Mg_2SiO_4	0.5	57.2	0.1	42.2	
M42	138	1250	Liquid	26.9	11.8	17.4	43.9	0.40
			Anorthite	20.3	0.5	34.8	44.4	
			Mg_2SiO_4	0.8	56.6	0.2	42.4	
			Liquid + Mg_2SiO_4 + Spinel					
M28	139	1400	Liquid	31.5	20.0	11.3	37.2	0.30
			Spinel	0.0	28.1	71.5	0.4	
			Mg_2SiO_4	4.4	53.4	0.3	41.9	
M28	140	1400	Liquid	30.7	19.7	11.4	38.2	0.30
			Spinel	0.0	28.5	71.3	0.2	
			Mg_2SiO_4	4.7	53.5	0.3	41.5	
M28	141	1400	Liquid	30.8	19.7	11.8	37.7	0.31
			Spinel	0.0	28.1	71.5	0.4	
			Mg_2SiO_4	4.4	53.4	0.3	41.9	

Table A2. *Cont.*

Mixture				Compositions (wt%)				
No.	No.	T (°C)	Phases	CaO	MgO	Al$_2$O$_3$	SiO$_2$	Al$_2$O$_3$/SiO$_2$
			Liquid + Merwinite + MgO					
M23	142	1450	Liquid	38.8	15.4	13.6	32.2	0.42
			MgO	0.3	98.9	0.8	0.0	
			Merwinite	50.8	12.6	0.1	36.5	
			Liquid + MgO + Ca$_2$SiO$_4$					
M24	143	1550	Liquid	44.6	13.3	12.0	30.1	0.40
			Ca$_2$SiO$_4$	60.9	3.7	0.2	35.2	
			MgO	0.3	99.0	0.7	0.0	
M24	144	1550	Liquid	42.3	12.8	17.1	27.8	0.30
			Ca$_2$SiO$_4$	51.5	12.2	0.2	36.1	
			MgO	0.3	98.9	0.8	0.0	
			Liquid + MgO + Spinel					
M47	145	1430	Liquid	34.9	18.5	12.5	34.1	0.37
			Spinel	0.2	28.1	71.6	0.1	
			MgO	0.2	99.0	0.8	0.0	
M33	146	1450	Liquid	30.5	21.7	12.6	35.2	0.36
			Spinel	0.0	28.4	71.4	0.2	
			MgO	0.1	99.0	0.9	0.0	
			Liquid + Merwinite + Spinel					
	147 *	1400	Liquid	37.6	14.2	14.0	34.2	0.41
			Spinel	0.1	28.0	71.8	0.1	
			Merwinite	51.1	12.4	0.1	36.4	
	148 *	1420	Liquid	36.8	15.8	13.3	34.1	0.39
			Spinel	0.6	27.9	71.1	0.4	
			Merwinite	50.7	12.7	0.1	36.5	
	149 *	1420	Liquid	36.5	16.5	13.3	33.7	0.40
			Spinel	0.1	28.5	71.3	0.1	
			Merwinite	50.8	12.6	0.1	36.5	
			Liquid + Mullite + Anorthite					
M52	151	1250	Liquid	7.4	6.0	24.4	62.2	0.39
			Mullite	0.1	0.6	70.2	29.1	
			Anorthite	17.7	0.4	33.3	48.6	
M39	152	1250	Liquid	8.2	5.1	23.0	63.7	0.36
			Mullite	0.1	0.5	70.7	28.7	
			Anorthite	17.1	0.4	31.8	50.7	
M18	153	1270	Liquid	6.9	5.4	22.8	64.9	0.37
			Mullite	0.1	0.4	70.9	28.6	
			Anorthite	17.4	0.5	32.3	49.8	
M18	154	1270	Liquid	7.7	5.2	23.8	63.3	0.38
			Mullite	0.1	0.3	71.2	28.4	
			Anorthite	17.7	0.4	32.4	49.5	
M39	155	1300	Liquid	8.0	4.7	23.2	64.1	0.36
			Mullite	0.1	0.4	71.2	28.3	
			Anorthite	18.0	0.4	33.4	48.2	
M52	156	1300	Liquid	8.0	5.6	24.2	62.2	0.39
			Mullite	0.1	0.4	71.4	28.1	
			Anorthite	18.7	0.5	34.6	46.2	
M10	157	1300	Liquid	8.7	3.0	23.4	64.9	0.36
			Mullite	0.2	0.1	72.2	27.5	
			Anorthite	17.4	0.2	31.8	50.6	
M18	158	1310	Liquid	9.1	4.2	24.8	61.9	0.40
			Mullite	0.2	0.2	71.7	27.9	
			Anorthite	18.3	0.3	33.7	47.7	

* Experimental data from previous work [15–18,23].

Table A3. Experimental Results in the System CaO-MgO-Al_2O_3-SiO_2 system with Al_2O_3/SiO_2 weight ratio of 0.4.

Mixture				Compositions (mol%)				
No.	No.	T (°C)	Phases	CaO	MgO	Al_2O_3	SiO_2	Al_2O_3/SiO_2
				Liquid only				
M20	1	1250	Liquid	30.7	17.2	10.1	42.0	0.41
M26	2	1250	Liquid	29.4	14.6	11.8	44.2	0.45
M34	3	1250	Liquid	17.9	18.4	12.2	51.5	0.40
M16	4	1280	Liquid	13.3	7.4	15.0	64.3	0.40
M16	5	1300	Liquid	13.3	7.3	15.2	64.2	0.40
M34	6	1300	Liquid	17.8	18.4	12.1	51.7	0.40
M35	7	1300	Liquid	10.9	11.1	14.5	63.5	0.39
M36	8	1300	Liquid	7.7	7.8	13.5	71.0	0.32
M42	9	1300	Liquid	24.2	19.0	11.1	45.7	0.41
M48	10	1300	Liquid	26.2	20.7	10.4	42.7	0.42
M42	11	1310	Liquid	23.6	19.7	11.0	45.7	0.41
M19	12	1320	Liquid	24.8	13.8	11.8	49.6	0.41
M42	13	1320	Liquid	25.7	20.6	10.4	43.3	0.41
M25	14	1330	Liquid	33.1	19.1	9.2	38.6	0.40
M25	15	1330	Liquid	34.3	17.7	9.4	38.6	0.41
M48	16	1340	Liquid	24.5	18.9	11.4	45.2	0.43
M13	17	1350	Liquid	33.0	28.4	7.4	31.2	0.40
M16	18	1350	Liquid	13.3	7.2	14.9	64.6	0.39
M19	19	1350	Liquid	24.2	13.4	12.0	50.4	0.40
M25	20	1350	Liquid	33.4	18.8	9.0	38.8	0.39
M51	21	1350	Liquid	20.1	10.7	14.3	54.9	0.44
M51	22	1350	Liquid	20.8	11.5	14.7	53.0	0.47
M29	23	1350	Liquid	24.1	24.3	9.8	41.8	0.40
M42	24	1350	Liquid	26.2	18.5	12.0	43.3	0.47
M44	25	1350	Liquid	27.7	23.8	9.3	39.2	0.40
M49	26	1350	Liquid	14.5	12.0	14.3	59.2	0.41
M16	27	1350	Liquid	13.2	7.4	15.1	64.3	0.40
M35	28	1350	Liquid	11.2	11.1	14.9	62.8	0.40
M22	29	1390	Liquid	37.6	20.8	8.0	33.6	0.40
M22	30	1400	Liquid	37.6	20.2	7.8	34.4	0.38
M28	31	1400	Liquid	29.8	29.6	7.8	32.8	0.41
M41	32	1400	Liquid	22.8	18.2	13.5	45.5	0.50
M49	33	1400	Liquid	16.6	13.3	15.8	54.3	0.49
M21	34	1400	Liquid	36.8	20.6	8.2	34.4	0.41
M23	35	1450	Liquid	39.7	20.3	7.2	32.8	0.37
M23	36	1450	Liquid	40.0	22.4	7.2	30.4	0.40
M47	37	1450	Liquid	34.0	26.7	7.3	32.0	0.39
M23	38	1450	Liquid	39.7	22.0	7.6	30.7	0.42
M14	39	1480	Liquid	17.9	4.8	20.0	57.3	0.59
M14	40	1500	Liquid	17.9	5.0	20.0	57.1	0.59
M23	41	1500	Liquid	40.1	22.5	7.2	30.2	0.40
M30	42	1500	Liquid	50.0	1.7	10.9	37.4	0.50
M47	43	1500	Liquid	34.1	26.9	7.5	31.5	0.40
				Liquid + Melilite				
M20	44	1230	Liquid	29.6	17.0	10.4	43.0	0.41
			melilite	41.3	17.1	3.5	38.1	
M7	45	1250	Liquid	36.0	9.5	10.6	43.9	0.41
			melilite	42.4	15.0	5.9	36.7	
M43	46	1250	Liquid	29.2	16.9	10.5	43.4	0.41
			melilite	40.6	17.4	3.2	38.8	
	47 *	1280	Liquid	33.8	14.2	9.6	42.4	0.38
			melilite	9.0	58.2	21.9	10.9	

Table A3. *Cont.*

Mixture				Compositions (mol%)				
No.	**No.**	**T (°C)**	**Phases**	**CaO**	**MgO**	**Al₂O₃**	**SiO₂**	**Al₂O₃/SiO₂**
M7	48	1300	Liquid	37.4	9.9	9.9	42.8	0.39
			melilite	43.7	12.4	7.2	36.7	
	49 *	1300	Liquid	40.4	7.2	10.0	42.4	0.40
			melilite	30.6	53.4	9.6	6.4	
M21	50	1350	Liquid	32.6	24.1	8.3	35.0	0.40
			melilite	43.0	14.1	7.6	35.3	
M17	51	1350	Liquid	9.9	5.7	14.8	69.6	0.36
			melilite	0.2	0.4	61.4	38.0	
M3	52	1350	Liquid	39.0	10.8	9.1	41.1	0.37
			melilite	43.6	12.4	9.3	34.7	
	53 *	1360	Liquid	48.0	0.0	10.0	42.0	0.40
			melilite	49.6	0.0	24.5	25.9	
M22	54	1370	Liquid	36.1	22.8	7.7	33.4	0.39
			melilite	43.5	12.8	9.4	34.3	
M21	55	1380	Liquid	36.5	20.6	8.4	34.5	0.41
			melilite	44.6	12.0	9.8	33.6	
M6	56	1400	Liquid	42.1	11.9	8.5	37.5	0.39
			melilite	46.2	8.7	13.8	31.3	
	57 *	1420	Liquid	44.9	7.7	9.1	38.3	0.40
			melilite	47.2	6.2	16.9	29.7	
	58 *	1420	Liquid	43.7	10.2	8.6	37.5	0.39
			melilite	46.0	8.0	15.0	31.0	
	59 *	1430	Liquid	52.2	0.0	9.2	38.6	0.40
			melilite	50.5	0.0	24.3	25.2	
	60 *	1450	Liquid	47.9	8.1	8.5	35.5	0.41
			melilite	47.7	4.6	19.1	28.6	
	61 *	1450	Liquid	52.9	0.0	9.0	38.1	0.40
			melilite	50.0	0.0	24.8	25.2	
	62 *	1450	Liquid	51.6	3.0	8.4	37.0	0.39
			melilite	48.9	2.3	21.9	26.9	
	63 *	1450	Liquid	53.6	0.0	8.7	37.7	0.39
			melilite	50.0	0.0	25.0	25.0	
			Liquid + Anorthite					
M26	64	1250	Liquid	29.5	16.9	10.0	43.6	0.39
			Anorthite	25.0	0.3	24.0	50.7	
M35	65	1250	Liquid	34.8	10.2	9.7	45.3	0.36
			Anorthite	24.9	0.5	24.3	50.3	
M43	66	1280	Liquid	23.4	19.8	10.7	46.1	0.39
			Anorthite	24.9	0.7	24.1	50.3	
	67 *	1280	Liquid	36.5	7.9	10.1	45.5	0.38
			Anorthite	33.3	56.2	10.3	0.2	
M34	68	1290	Liquid	18.6	18.8	12.3	50.3	0.41
			Anorthite	24.6	1.0	23.2	51.2	
M41	69	1300	Liquid	20.1	17.4	11.6	50.9	0.39
			Anorthite	24.5	1.0	23.1	51.4	
M19	70	1300	Liquid	24.1	14.7	10.3	50.9	0.35
			Anorthite	24.7	0.9	24.2	50.2	
M5	71	1300	Liquid	31.7	9.6	10.1	48.6	0.35
			Anorthite	25.3	0.5	24.4	49.8	
M8	72	1350	Liquid	28.9	8.2	11.4	51.5	0.38
			anorthite	25.1	0.5	24.1	50.3	
M9	73	1400	Liquid	14.2	4.2	15.0	66.6	0.38
			anorthite	24.5	0.3	24.7	50.5	
			Liquid + Cordierite					
M17	74	1250	Liquid	11.0	11.4	14.8	62.8	0.40
			Cordierite	21.9	0.9	21.8	55.4	

Table A3. *Cont.*

Mixture				Compositions (mol%)				
No.	No.	T (°C)	Phases	CaO	MgO	Al_2O_3	SiO_2	Al_2O_3/SiO_2
M49	75	1300	Liquid	13.1	13.3	13.4	60.2	0.38
			Cordierite	22.7	0.9	22.8	53.6	
M40	76	1300	Liquid	11.4	9.6	14.9	64.1	0.39
			Cordierite	22.4	0.5	22.6	54.5	
M10	77	1300	Liquid	11.6	3.9	15.7	68.8	0.39
			Cordierite	21.3	0.3	21.5	56.9	
			Liquid + Mullite					
M17	78	1250	Liquid	10.4	6.0	14.9	68.7	0.37
			Mullite	0.3	0.8	56.0	42.9	
M18	79	1320	Liquid	11.1	6.0	15.8	67.1	0.40
			Mullite	0.3	0.4	59.0	40.3	
M39	80	1350	Liquid	9.2	7.6	15.7	67.5	0.39
			Mullite	0.2	0.6	58.8	40.4	
M36	81	1400	Liquid	7.1	7.3	14.9	70.7	0.36
			Mullite	0.3	0.6	60.8	38.2	
M12	82	1400	Liquid	11.1	3.0	15.4	70.5	0.37
			Mullite	0.2	0.2	60.0	39.6	
M11	83	1450	Liquid	9.9	3.0	16.2	70.9	0.39
			Mullite	0.2	0.0	62.6	37.2	
M50	84	1500	Liquid	6.9	5.7	15.6	71.8	0.37
			Mullite	0.2	0.2	61.6	38.0	
			Liquid + Spinel					
M29	85	1320	Liquid	26.6	22.0	10.7	40.7	0.45
			Spinel	0.0	50.0	49.9	0.1	
M44	86	1350	Liquid	28.0	23.7	9.3	39.0	0.40
			Spinel	0.1	50.0	49.8	0.1	
M22	87	1370	Liquid	36.4	22.4	7.8	33.4	0.39
			Spinel	0.0	49.9	50.0	0.1	
	89 *	1500	liquid	30.4	30.4	7.8	31.4	0.42
			Spinel	0.5	0.8	58.7	40.0	
			Liquid + Mg_2SiO_4					
M48	90	1250	Liquid	27.6	17.4	11.0	44.0	0.42
			Mg_2SiO_4	0.8	66.5	0.0	32.7	
M38	91	1250	Liquid	29.3	17.4	10.1	43.2	0.40
			Mg_2SiO_4	1.3	65.5	0.0	33.2	
M44	92	1250	Liquid	28.8	17.7	10.7	42.8	0.42
			Mg_2SiO_4	1.2	66.0	0.2	32.6	
M32	93	1300	Liquid	21.7	21.4	10.8	46.1	0.40
			Mg_2SiO_4	0.5	66.3	0.0	33.2	
M29	94	1300	Liquid	26.1	20.6	10.6	42.7	0.42
			Mg_2SiO_4	0.9	66.1	0.1	32.9	
M38	95	1300	Liquid	27.3	20.8	10.0	41.9	0.41
			Mg_2SiO_4	1.1	65.6	0.1	33.2	
M28	96	1300	Liquid	41.6	16.5	4.4	37.5	0.20
			Mg_2SiO_4	3.5	63.5	0.1	32.9	
M29	97	1330	Liquid	25.4	22.5	10.3	41.8	0.42
			Mg_2SiO_4	0.9	66.2	0.1	32.8	
M31	98	1350	Liquid	29.9	23.8	7.9	38.4	0.35
			Mg_2SiO_4	2.0	65.0	0.2	32.8	
M31	99	1400	Liquid	28.1	26.3	7.6	38.0	0.34
			Mg_2SiO_4	2.2	64.8	0.1	32.9	
M44	100	1400	Liquid	29.2	21.5	9.1	40.2	0.38
			Mg_2SiO_4	0.5	66.3	0.0	33.2	
			Liquid + MgO					
M30	101	1450	Liquid	33.7	27.2	7.5	31.6	0.40
			MgO	0.1	99.5	0.4	0.0	

Table A3. *Cont.*

Mixture				Compositions (mol%)				
No.	No.	T (°C)	Phases	CaO	MgO	Al$_2$O$_3$	SiO$_2$	Al$_2$O$_3$/SiO$_2$
M23	102	1470	Liquid	40.1	22.1	7.2	30.6	0.40
			MgO	0.2	99.5	0.3	0.0	
	103 *	1500	Liquid	29.8	31.6	7.4	31.2	0.40
			MgO	0.1	0.3	0.7	98.9	
	104 *	1500	Liquid	42.2	20.8	6.9	30.1	0.39
			MgO	0.2	99.5	0.3	0.0	
M27	105	1500	Liquid	39.3	23.1	7.0	30.6	0.39
			MgO	0.1	99.6	0.3	0.0	
M30	106	1500	Liquid	33.0	28.5	7.4	31.1	0.40
			MgO	0.1	99.5	0.4	0.0	
M30	107	1500	Liquid	33.0	28.4	7.4	31.2	0.40
			MgO	0.1	99.5	0.4	0.0	
M30	108	1500	Liquid	33.0	28.5	7.4	31.1	0.40
			MgO	0.1	99.5	0.4	0.0	
M37	109	1500	Liquid	37.0	24.9	7.3	30.8	0.40
			MgO	0.2	99.4	0.4	0.0	
M27	110	1550	Liquid	38.5	24.2	7.1	30.2	0.40
			MgO	0.1	99.5	0.4	0.0	
M24	111	1550	Liquid	46.0	18.6	6.8	28.6	0.40
			MgO	0.2	99.6	0.2	0.0	
M46	112	1550	Liquid	41.9	21.1	7.1	29.9	0.40
			MgO	0.2	99.5	0.3	0.0	
M46	113	1550	Liquid	41.2	22.4	7.2	29.2	0.42
			MgO	0.2	99.5	0.3	0.0	
M37	114	1550	Liquid	36.8	25.9	7.1	30.2	0.40
			MgO	0.1	99.5	0.4	0.0	
			Liquid + Merwinite					
	115*	1400	Liquid	38.9	20.4	7.4	33.3	0.38
			Merwinite	49.8	17.2	0.1	32.9	
M2	116	1400	Liquid	43.5	13.2	8.5	34.8	0.42
			Merwinite	49.8	16.7	0.1	33.4	
M23	117	1430	Liquid	40.5	22.4	7.5	29.6	0.43
			Merwinite	50.9	17.7	0.1	31.3	
	118 *	1450	Liquid	45.2	14.1	8.0	32.7	0.41
			Merwinite	50.0	16.6	0.1	33.3	
M23	119	1450	Liquid	40.9	21.6	7.3	30.2	0.41
			Merwinite	50.3	17.4	0.1	32.2	
			Liquid + Ca$_2$SiO$_4$					
M4	120	1500	Liquid	47.2	13.4	7.7	31.7	0.41
			Ca$_2$SiO$_4$	62.4	4.1	0.1	33.4	
			Liquid + Melilite + Spinel					
M25	121	1300	Liquid	30.6	20.3	9.3	39.8	0.40
			Melilite	42.4	14.6	6.6	36.4	
			Spinel	0.1	49.9	49.9	0.1	
M25	122	1320	Liquid	32.8	18.1	9.7	39.4	0.42
			Melilite	42.8	14.6	6.8	35.8	
			Spinel	0.1	50.3	49.5	0.1	
			Liquid + Melilite + Anorthite					
M15	123	1250	Liquid	32.7	14.0	10.4	42.9	0.41
			Anorthite	25.4	0.7	24.9	49.0	
			Melilite	42.1	15.8	5.4	36.7	
M1	124	1250	Liquid	34.8	10.3	9.8	45.1	0.37
			Anorthite	25.3	0.2	24.4	50.1	
			Melilite	41.5	16.2	4.1	38.2	

Table A3. *Cont.*

Mixture				Compositions (mol%)				
No.	No.	T (°C)	Phases	CaO	MgO	Al_2O_3	SiO_2	Al_2O_3/SiO_2
Liquid + Melilite + Mg_2SiO_4								
M20	125	1220	Liquid	29.1	16.6	11.1	43.2	0.44
			Mg_2SiO_4	1.1	66.2	0.1	32.6	
			Melilite	41.3	16.6	3.9	38.2	
M25	126	1250	Liquid	28.8	19.2	10.4	41.6	0.43
			Mg_2SiO_4	1.3	65.9	0.1	32.7	
			Melilite	42.0	14.3	7.0	36.7	
M45	127	1300	Liquid	30.3	20.2	8.9	40.6	0.37
			Mg_2SiO_4	2.0	64.9	0.2	32.9	
			Melilite	42.3	14.5	6.9	36.3	
M44	128	1300	Liquid	26.2	20.6	7.2	46.0	0.56
			Mg_2SiO_4	41.5	15.8	4.8	37.9	
			Melilite	1.8	64.8	0.8	32.6	
M28	129	1350	Liquid	31.5	23.4	7.2	37.9	
			Mg_2SiO_4	41.3	16.8	3.8	38.1	
			Melilite	3.3	63.6	0.1	33.0	
M45	130	1350	Liquid	30.8	23.5	7.6	38.1	0.34
			Mg_2SiO_4	3.5	63.7	0.1	32.7	
			Melilite	40.7	17.7	3.9	37.7	
M45	131	1450	Liquid	34.4	22.5	5.8	37.3	0.34
			Mg_2SiO_4	3.5	63.3	0.2	33.0	
			Melilite	41.8	16.6	4.4	37.2	
Liquid + Mg_2SiO_4 + Anorthite								
M42	132	1200	Liquid	24.0	22.2	7.9	45.9	0.29
			Anorthite	24.1	1.4	23.3	51.2	
			Mg_2SiO_4	1.0	66.1	0.1	32.8	
M42	133	1230	Liquid	24.1	19.9	10.2	45.8	0.38
			Anorthite	24.9	0.9	24.2	50.0	
			Mg_2SiO_4	0.7	66.5	0.0	32.8	
M20	134	1230	Liquid	27.5	17.8	10.0	44.7	0.38
			Anorthite	24.8	0.3	24.1	50.8	
			Mg_2SiO_4	0.8	65.9	0.1	33.2	
M34	135	1240	Liquid	17.2	18.0	9.1	55.7	0.28
			Anorthite	24.1	1.7	22.1	52.1	
			Mg_2SiO_4	0.2	66.9	0.0	32.9	
M34	136	1250	Liquid	18.4	19.3	9.2	53.1	0.29
			Anorthite	24.1	1.5	22.6	51.8	
			Mg_2SiO_4	0.2	66.7	0.0	33.1	
M41	137	1250	Liquid	17.5	19.3	9.2	54.0	0.29
			Anorthite	24.0	1.5	22.9	51.6	
			Mg_2SiO_4	0.4	66.9	0.0	32.7	
M42	138	1250	Liquid	28.6	17.6	10.2	43.6	0.40
			Anorthite	24.9	0.9	23.5	50.7	
			Mg_2SiO_4	0.7	66.2	0.1	33.0	
Liquid + Mg_2SiO_4 + Spinel								
M28	139	1400	Liquid	31.4	27.9	6.2	34.5	0.30
			Spinel	0.0	49.8	49.7	0.5	
			Mg_2SiO_4	3.7	63.2	0.1	33.0	
M28	140	1400	Liquid	30.7	27.6	6.3	35.4	0.30
			Spinel	0.0	50.4	49.4	0.2	
			Mg_2SiO_4	4.0	63.2	0.1	32.7	
M28	141	1400	Liquid	30.8	27.6	6.5	35.1	0.31
			Spinel	0.0	49.8	49.7	0.5	
			Mg_2SiO_4	3.7	63.2	0.1	33.0	

Table A3. *Cont.*

Mixture				Compositions (mol%)				
No.	No.	T (°C)	Phases	CaO	MgO	Al$_2$O$_3$	SiO$_2$	Al$_2$O$_3$/SiO$_2$
			Liquid + Merwinite + MgO					
M23	142	1450	Liquid	39.7	22.0	7.6	30.7	0.42
			MgO	0.2	99.5	0.3	0.0	
			Merwinite	49.5	17.2	0.1	33.2	
			Liquid + MgO + Ca$_2$SiO$_4$					
M24	143	1550	Liquid	45.6	19.0	6.7	28.7	0.40
			Ca$_2$SiO$_4$	61.5	5.2	0.1	33.2	
			MgO	0.2	99.5	0.3	0.0	
M24	144	1550	Liquid	44.3	18.8	9.8	27.1	0.30
			Ca$_2$SiO$_4$	50.3	16.7	0.1	32.9	
			MgO	0.2	99.5	0.3	0.0	
			Liquid + MgO + Spinel					
M47	145	1430	Liquid	35.0	26.1	6.9	32.0	0.37
			Spinel	0.3	49.8	49.8	0.1	
			MgO	0.1	99.6	0.3	0.0	
M33	146	1450	Liquid	30.3	30.2	6.9	32.6	0.36
			Spinel	0.0	50.3	49.5	0.2	
			MgO	0.1	99.5	0.4	0.0	
			Liquid + Merwinite + Spinel					
	147 *	1400	Liquid	38.7	20.5	7.9	32.9	0.41
			Spinel	0.1	49.8	50.0	0.1	
			Merwinite	49.8	17.0	0.1	33.1	
	148 *	1420	Liquid	37.5	22.6	7.5	32.4	0.39
			Spinel	0.8	49.3	49.4	0.5	
			Merwinite	49.4	17.3	0.1	33.2	
	149 *	1420	Liquid	37.2	23.5	7.4	31.9	0.40
			Spinel	0.1	50.4	49.4	0.1	
			Merwinite	49.5	17.2	0.1	33.2	
			Liquid + Mullite + Anorthite					
M52	151	1250	Liquid	8.5	9.6	15.4	66.5	0.39
			Mullite	0.1	1.3	57.9	40.7	
			Anorthite	21.6	0.7	22.4	55.3	
M39	152	1250	Liquid	9.4	8.2	14.5	67.9	0.36
			Mullite	0.2	1.1	58.4	40.3	
			Anorthite	20.8	0.7	21.2	57.3	
M18	153	1270	Liquid	7.9	8.6	14.3	69.2	0.37
			Mullite	0.2	0.8	58.8	40.2	
			Anorthite	21.1	0.9	21.6	56.4	
M18	154	1270	Liquid	8.8	8.4	15.0	67.8	0.38
			Mullite	0.2	0.6	59.2	40.0	
			Anorthite	21.5	0.7	21.7	56.1	
M39	155	1300	Liquid	9.2	7.6	14.7	68.5	0.36
			Mullite	0.2	0.8	59.1	39.9	
			Anorthite	22.0	0.7	22.4	54.9	
M52	156	1300	Liquid	9.2	9.0	15.3	66.5	0.39
			Mullite	0.2	0.8	59.4	39.6	
			Anorthite	22.9	0.9	23.3	52.9	
M10	157	1300	Liquid	10.1	4.9	14.9	70.1	0.36
			Mullite	0.3	0.2	60.4	39.1	
			Anorthite	21.1	0.3	21.2	57.4	
M18	158	1310	Liquid	10.5	6.8	15.8	66.9	0.40
			Mullite	0.3	0.4	59.8	39.5	
			Anorthite	22.4	0.5	22.7	54.4	

* Experimental data from previous work [15–18,23].

References

1. Geerdes, M.; Chaigneau, R.; Lingiardi, O.; Molenaar, R.; van Opbergen, R.; Sha, Y.; Warren, P. Chapter XI—Operation Challenges. In *Modern Blast Furnace Ironmaking: An Introduction*, 4th ed.; Ios Press: Amsterdam, The Netherlands, 2020; pp. 197–226, ISBN 978-1-64368-122-8.
2. Sunahara, K.; Nakano, K.; Hoshi, M.; Inada, T.; Komatsu, S.; Yamamoto, T. Effect of high Al_2O_3 slag on the blast furnace operations. *ISIJ Int.* **2008**, *48*, 420–429. [CrossRef]
3. Yi, Z.; Liu, Q.; Shao, H. Effect of MgO on highly basic sinters with high Al_2O_3. *Min. Metall. Explor.* **2021**, *38*, 2175–2183. [CrossRef]
4. Talapaneni, T.; Yedla, N.; Sarkar, S.; Pal, S. Effect of basicity, Al_2O_3 and MgO content on the softening and melting properties of the CaO-MgO-SiO_2-Al_2O_3 high alumina quaternary slag system. *Metall. Res. Technol.* **2016**, *113*, 501.
5. Hu, X.; Lin, L.; Li, L. Laboratory research on property of higher Al_2O_3 content slag. *China Metall.* **2006**, *16*, 36–41. (In Chinese)
6. Wang, P.; Meng, Q.; Long, H.; Li, J. Influence of basicity and MgO on fluidity and desulfurization ability of high aluminum slag. *High Tem. Mat. Process.* **2016**, *35*, 669–675. [CrossRef]
7. Bai, K.; Zuo, H.; Liu, W.; Wang, J.; Chen, J.; Xue, Q. Effect of quaternary basicity on softening–melting behavior of primary slag based on magnesium flux pellet. *J. Iron Steel Res. Int.* **2022**, *29*, 1185–1193. [CrossRef]
8. Prince, A.T. Phase Equilibrium relationships in a portion of the system MgO-Al_2O_3-$2CaO·SiO_2$. *J. Am. Ceram. Soc.* **1951**, *34*, 44–51. [CrossRef]
9. DeVries, R.C.; Osborn, E.F. Phase equilibria in High-alumina part of the system CaO-MgO-Al_2O_3-SiO_2. *J. Am. Ceram. Soc.* **1957**, *40*, 6–15. [CrossRef]
10. DeVries, R.C.; Osborn, E.F.; Gee, K.H.; Kraner, H.M. Optimum composition of blast furnace slag as deduced for the quaternary system CaO-MgO-Al_2O_3-SiO_2. *Trans. AIME* **1954**, *6*, 33–45.
11. Prince, A.T. Liquidus relationships on 10% MgO plane of the system lime-magnesia-alumina-silica. *J. Am. Ceram. Soc.* **1954**, *37*, 402–408. [CrossRef]
12. Cavalier, G.; Sandrea-Deudon, M. Quaternary slags CaO-MgO-Al_2O_3-SiO_2: Liquidus surfaces and crystallisation paths for constant magnesia concentrations. *Rev. Metall.* **1960**, *5*, 1143–1157. [CrossRef]
13. Gutt, W.; Russel, A.D. Studies of the system CaO-SiO_2-Al_2O_3-MgO in relation to the stability of blast furnace slag. *J. Mater. Sci.* **1977**, *12*, 1869–1878. [CrossRef]
14. Dahl, F.; Brandberg, J.; Du, S. Characterization of melting of some slags in the Al_2O_3-CaO-MgO-SiO_2 quaternary system. *ISIJ Int.* **2006**, *46*, 614–616.
15. Ma, X.; Wang, G.; Wu, S.; Zhu, J.; Zhao, B. Phase equilibria in the CaO-SiO_2-Al_2O_3-MgO System with CaO/SiO_2 ratio of 1.3 relevant to iron blast furnace slags. *ISIJ Int.* **2015**, *55*, 2310–2317.
16. Ma, X.; Zhang, D.; Zhao, Z.; Evans, T.; Zhao, B. Phase equilibria studies in the CaO-SiO_2-Al_2O_3-MgO system with CaO/SiO_2 ratio of 1.10. *ISIJ Int.* **2016**, *56*, 513–519. [CrossRef]
17. Kou, M.; Wu, S.; Ma, X.; Wang, L.; Chen, M.; Cai, Q.; Zhao, B. Phase equilibrium studies of CaO-SiO_2-Al_2O_3-MgO system with binary basicity of 1.5 related to blast furnace slag. *Metall. Mater. Trans. B* **2016**, *47*, 1093–1102. [CrossRef]
18. Wang, D.; Chen, M.; Jiang, Y.; Wang, S.; Zhao, Z.; Evans, T.; Zhao, B. Phase equilibria studies in the CaO-SiO_2-Al_2O_3-MgO system with CaO/SiO_2 ratio of 0.9. *J. Am. Ceram. Soc.* **2020**, *103*, 7299–7309.
19. Gran, J.; Wang, Y.; Du, S. Experimental determination of the liquidus in the high basicity region in the Al_2O_3 (30 mass%)-CaO-MgO-SiO_2 system. *Calphad* **2011**, *35*, 249–254. [CrossRef]
20. Gran, J.; Yan, B.; Du, S. Experimental determination of the liquidus in the high-basicity region in the Al_2O_3 (25 mass pct)-CaO-MgO-SiO_2 and Al_2O_3 (35 mass pct)-CaO-MgO-SiO_2 systems. *Metall. Mater. Trans. B* **2011**, *42*, 1008–1016. [CrossRef]
21. Lyu, S.; Ma, X.; Chen, M.; Huang, Z.; Wang, G. Application of phase equilibrium studies of CaO–SiO_2–Al_2O_3–MgO system for oxide inclusions in Si-deoxidized steels. *Calphad* **2020**, *68*, 101721. [CrossRef]
22. Lyu, S.; Ma, X.; Huang, Z.; Yao, Z.; Lee, H.; Jiang, Z.; Wang, G.; Zou, J.; Zhao, B. Inclusion characterization and formation mechanisms in spring steel deoxidized by silicon. *Metall. Mater. Trans. B* **2019**, *5*, 732–747. [CrossRef]
23. Yao, Z.; Ma, X.; Lyu, S. Phase equilibria of the Al_2O_3–CaO–SiO_2-(0%, 5%, 10%) MgO slag system for non-metallic inclusions control. *Calphad* **2021**, *72*, 102227. [CrossRef]
24. *Verein Duetscher Eisenhuttenleute: Slag Atlas*; Verlag Sthaleisen GmbH: Düsseldorf, Germany, 1995; p. 156.
25. Bale, C.W.; Bélisle, E.; Chartrand, P.; Decterov, S.; Eriksson, G.; Gheribi, A.; Hack, K.; Jung, I.H.; Kang, Y.B.; Melançon, J.; et al. Reprint of: FactSage thermochemical software and databases, 2010–2016. *Calphad* **2016**, *55*, 1–19. [CrossRef]
26. Sundman, B.; Jansson, B.; Andersson, J.O. The thermo-calc databank system. *Calphad* **1985**, *9*, 153–190.
27. Xu, R.Z.; Zhang, H.S.; Zhang, J.L.; Liu, C.J.; Wang, Z.Y. Evaluation of Jingtang blast furnace slag properties and analysis of low MgO slag performance. *Chin. Metall.* **2016**, *26*, 16–20. (In Chinese)
28. Muan, A.; Osborn, E.F. *Phase Equilibria Among Oxides in Steelmaking*; Addison-Wesley Publishing Company: Boston, MA, USA, 1965; pp. 148–157.

 *metals*

Article

Comparative Analysis on the Corrosion Resistance to Molten Iron of Four Kinds of Carbon Bricks Used in Blast Furnace Hearth

Cui Wang [1], Jianliang Zhang [2,*], Wen Chen [3], Xiaolei Li [3], Kexin Jiao [2], Zhenping Pang [3], Zhongyi Wang [2], Tongsheng Wang [3] and Zhengjian Liu [2]

[1] State Key Laboratory of Advanced Metallurgy, University of Science and Technology Beijing, Beijing 100083, China; cui_wang1988@163.com
[2] School of Metallurgical and Ecological Engineering, University of Science and Technology Beijing, Beijing 100083, China; jiaokexin_ustb@126.com (K.J.); 18235198438@163.com (Z.W.); liuzhengjian@126.com (Z.L.)
[3] WISDRI Handan Wupeng New Lining Material Co., Ltd., Handan 056200, China; whchenwen@163.com (W.C.); 13627216298@163.com (X.L.); pangzhp@163.com (Z.P.); 15527050612@163.com (T.W.)
* Correspondence: zhang.jianliang@hotmail.com; Tel.: +86-13910019986

Abstract: The corrosion resistance to molten iron of four kinds of carbon bricks used in blast furnace hearth were investigated to elaborate the corrosion mechanism through the macroscopic and microscopic analysis of carbon bricks before and after reaction and thermodynamic analysis. The macroscopic analysis showed that brick A had the lowest degree of corrosion and highest uniformity at different heights, attributing to its moderate carbon content of 76.15%, main phases of C, Al_2O_3, SiC, and $Al_6Si_2O_{13}$ (mullite), and lower resistance to molten iron infiltration, etc. The microscopic analysis showed that all the carbon bricks had more and larger pores than the original carbon bricks. The phenomena of the iron beads adhering to carbon brick and iron infiltration were observed between the interface of carbon brick and molten iron. In addition, the obvious corrosion process was presented that the carbon matrix was broken and peeled off during the iron infiltration process. For the carbon brick being corroded, the dissolution of carbon was the predominant reaction. The higher the carbon solubility of the molten iron, the easier the corrosion on the carbon brick. Al_2O_3 and SiC enhanced the corrosion resistance to molten iron of carbon bricks, and SiO_2 could react with carbon to form pores as channels for the penetration of molten iron and increase the corrosion on carbon bricks. A higher graphitization degree of carbon bricks was beneficial to lessen their corrosion degree. The corrosion on carbon bricks by molten iron could be attributed to three aspects: carburization, infiltration, and scouring of molten iron. The carburization process of molten iron was the main reaction process. The molten iron infiltration into the carbon bricks facilitated the dissolution of carbon and destroyed the structure and accelerated the corrosion of the carbon bricks. The scouring of molten iron subjected the iron–carbon interface to interaction forces, promoting the separation of the exfoliated fragmented carbon brick from the iron–carbon interface to facilitate a new round of corrosion process.

Keywords: carbon brick; corrosion resistance to molten iron; carburization; iron infiltration

Citation: Wang, C.; Zhang, J.; Chen, W.; Li, X.; Jiao, K.; Pang, Z.; Wang, Z.; Wang, T.; Liu, Z. Comparative Analysis on the Corrosion Resistance to Molten Iron of Four Kinds of Carbon Bricks Used in Blast Furnace Hearth. *Metals* **2022**, *12*, 871. https://doi.org/10.3390/met12050871

Academic Editor: Mark E. Schlesinger

Received: 27 April 2022
Accepted: 17 May 2022
Published: 20 May 2022

Publisher's Note: MDPI stays neutral with regard to jurisdictional claims in published maps and institutional affiliations.

1. Introduction

With the continuous development of blast furnaces towards large-scale and high smelting strength, their safety and longevity have been unprecedentedly threatened [1–3]. To meet demand, the refractories for blast furnace hearth are constantly updated to adapt to high smelting strength and maintain the safety and longevity of blast furnaces [4–7]. Among the refractories such as carbon brick, carbon composite brick, corundum brick, and castable, carbon brick is still a dominant refractory, and its corrosion resistance to molten iron and slag, oxidation resistance, and thermal conductivity affect the service life of the hearth. Therein, the corrosion resistance to molten iron is a key link in determining the longevity of the blast furnace. To improve the corrosion resistance, especially the corrosion resistance to molten iron, some carbon brick manufacturers have been making continuous efforts, including adding different additives, such as aluminum powder; improving the microporous properties of carbon brick, such as adjusting the production process; in situ generation of whiskers, such as silicon powder addition; and adding titanium-containing substances to reduce the contact of molten iron and carbon and decrease the update rate of molten iron at the interface, which results in higher mechanical strength, lower porosity sizes, etc., greatly improving the corrosion resistance to molten iron of carbon brick [8,9] and enhancing the compressive strength and thermal shock performance [10–12].

Clarifying the influencing factors and mechanism of the corrosion resistance to molten iron of carbon brick can provide theoretical guidance for the carbon brick quality improvement, the carbon brick selection for the blast furnace hearth, and even the adjustment of blast furnace operation so as to prolong the service life of blast furnace hearth. The previous studies on the dissolution of carbonaceous materials in molten iron mainly focus on iron–coke, iron–graphite, etc. [13–17], aiming at enhancing their carburizing ability and optimizing the smelting process. In contrast, few studies on the corrosion of carbon brick in molten iron were reported, especially on the corrosion of the new generation of super-microporous carbon bricks. Deng et al. [18–20] investigated the dissolution mechanism of carbon bricks into molten iron; the dissolution reaction of carbon was considered as the dominant reaction and controlled by interfacial reaction and mass transfer of carbon when the phosphorus content was up to 0.2% in molten iron. Stec et al. [21] undertook the molten metal infiltration into micropore carbon refractory materials using X-ray computed tomography; changes were observed in the micropore carbon refractory material's microstructure, and the elements of the open pore structure that were crucial in molten metal infiltration were identified. Jiao et al. [22] investigated the corrosion behavior of alumina–carbon composite brick in blast furnace slag and iron; the corrosion caused by iron decreased with increasing slag basicity, the dissolution of carbon was one of the corrosion reasons, and the reaction between carbon and silica in the brick could also be promoted by the appearance of iron. With respect to the fact there is scarce and unsystematic research on the corrosion resistance to molten iron of carbon bricks, it is urgent to illustrate the influence mechanism of the chemical composition, physical and chemical properties, and microstructure of carbon bricks on the corrosion resistance performance of different types of carbon bricks in molten iron, allowing the safety and longevity of blast furnace as well as high-efficiency and low-consumption smelting.

2. Experimental

2.1. Sample Preparation

The experimental samples were taken from four types of carbon bricks used extensively in blast furnace hearths worldwide. The chemical composition is presented in Table 1. Among them, A, B, and C are super-microporous carbon bricks, and D is a microporous carbon brick. The carbon bricks were cut into cylinders (φ30 mm × 50 mm) with a concentric and through-hole cylinder using a special drill as shown in Figure 1. To make the surface of the cylinders smooth, they were coarsely ground and finely ground with sandpaper of different particle sizes and then immersed in absolute ethanol for ultrasonic treatment to remove impurities. After cleaning, they were dried in a drying oven at 105 °C for 4 h to avoid the influence of moisture on the experimental results, then stored hermetically for later use. Before the test, the cylinders were inserted into a special ceramic rod to stir the molten iron. The matching ceramic rod is 8 mm in diameter, and its length is determined by the distance between the motor and the molten iron surface. One end of the ceramic rod has a 60 mm long thread and a matching nut as also shown in Figure 1.

Table 1. Chemical composition of the four types of carbon bricks.

Carbon Brick	Chemical Composition/%					
	C	Al_2O_3	SiO_2	SiC	TiO_2	Others
A	76.15	8.74	7.41	6.91	0.22	0.57
B	84.13	4.43	6.76	3.23	0.33	1.12
C	71.68	0.91	14.34	10.78	0.42	1.87
D	79.54	1.33	11.80	4.43	0.21	2.69

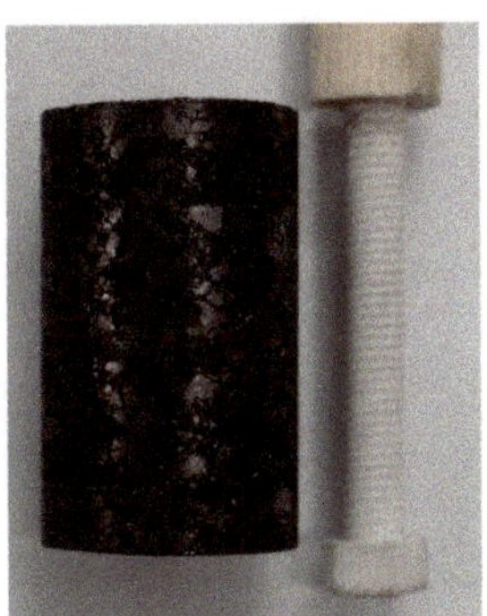

Figure 1. Carbon brick sample and matching ceramic rod.

Molten iron samples were prepared using reduced iron powder (AR, purity greater than 98%) (Macklin, Shanghai, China), graphite powder (CP, purity greater than 99.85%) (Macklin, Shanghai, China), silicon metal powder (purity greater than 99%) (Sinopharm, Shanghai, China), manganese powder (purity greater than 99%) (Aladdin, Shanghai, China), phosphorus powder (purity greater than 99%) (Alfa Aesar, Shanghai, China), and FeS_2 powder (AR, purity greater than 99%) (Rhawn, Shanghai, China). The composition of molten iron is preset as shown in Table 2. The mixture of 770.77 g molten iron sample was ground in the mortar to increase the contact between the particles and ensure that the mixture was thoroughly blended.

Table 2. Chemical composition of the molten iron sample.

Molten iron	Chemical Composition/%					
	Fe	C	Si	Mn	P	S
	95.87	3.50	0.30	0.15	0.15	0.03

2.2. Experimental Apparatus and Procedure

The experimental apparatus is described in Figure 2. It mainly includes a BLMT–1700 °C high-temperature tube furnace (BLMT, Luoyang, Henan, China), argon cylinder, JJ-1B constant-speed electric stirrer, SRS13A precision temperature controller, etc. The Si-Mo heating rods are adopted as the heating element to control the tube furnace temperature from room temperature to 1700 °C at a maximum rate of 5 K/min. The constant temperature zone in the tube furnace is about 100 mm in length. The constant-speed electric stirrer is placed above the tube furnace, the non-threaded end of the ceramic rod is connected to the motor with an electric stirrer to simulate the circulation velocity of molten iron in the hearth by controlling the rotation speed, and the carbon brick sample is immersed in the molten iron to carry out the experiment.

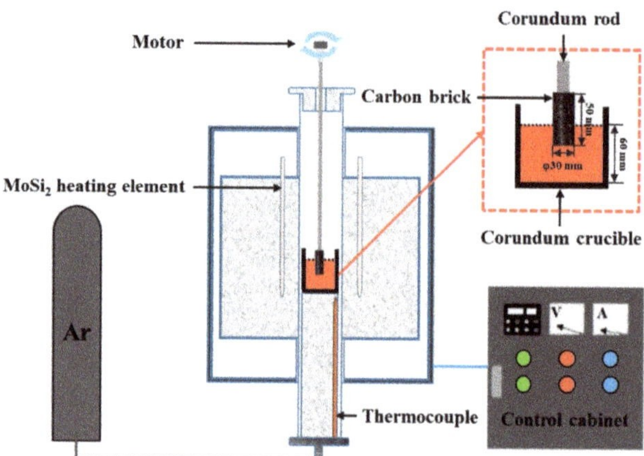

Figure 2. Schematic diagram of the experimental apparatus.

The molten iron sample was compacted into a ceramic crucible with an outer diameter of 62 mm, a height of 300 mm, and a wall thickness of 3 mm, and then placed in the middle of the constant temperature zone to prevent adhesion to the sidewall of the ceramic furnace tube. The high-purity argon gas (99.9%) was purged into the furnace tube throughout the experiment at a flow rate of 3 L/min. When the tube furnace was heated to 1500 °C, the temperature was maintained for 60 min and the molten iron stirred with a quartz tube to ensure the complete melting and uniformity of the sample. After 40 min of constant temperature, the carbon brick sample was put into the furnace tube above the molten iron to preheat for 20 min, then was slowly submerged to 30 mm below the molten iron surface. The experiment was started after setting the stirring speed to 60 r/min and the time to 120 min. During the experimental process, the molten iron was extracted 3~5 g using a quartz tube with 4 mm in inner diameter every 30 min, and then water quenched quickly for carbon content detection. When extracting molten iron samples, the rotation of the stirring motor was suspended, and the extraction time was controlled within 1 min. After the experiment, the carbon brick sample was lifted 50 mm to separate from the molten iron, and the motor was turned on for 5 min to shake off the molten iron attached to the surface of the carbon brick sample. When the furnace temperature dropped to room temperature, the carbon brick sample was taken out for subsequent characterization.

3. Results and Discussion

3.1. Analysis of the Original Carbon Bricks

It can be seen from Table 1 that B has the highest proportion of carbon component, which is much larger than the other three types of carbon bricks. Ceramic additives such as Al$_2$O$_3$, SiO$_2$, and SiC in A and B are added in larger amounts; the content of SiO$_2$ in B is relatively high, and the proportion of Al$_2$O$_3$ in A is relatively high. The components of C and D are obviously different from A and B. C and D have less Al$_2$O$_3$, and SiO$_2$ and SiC are the main ceramic additives. From the XRD patterns (Rigku SmartLab, Japan) in Figure 3, it can be determined that the main phases in A are C, Al$_2$O$_3$, SiC, and Al$_6$Si$_2$O$_{13}$ (mullite), the ones in B are C, Al$_2$O$_3$, and SiC, while the main phases of C and D are C, SiO$_2$, and SiC. Due to the non-uniformity of carbon brick composition and the relatively low content of SiO$_2$ in A and B, the SiO$_2$ phase in A and B is not detected, while Al$_6$Si$_2$O$_{13}$ phase is detected in A. The physical and chemical properties of these carbon bricks are listed in Table 3. The apparent porosity of A and B is higher than that of C and D, while the pore volume (<1 μm) of A is superior to C and D, which helps to prevent the penetration of molten iron into the carbon bricks. Therefore, the resistance to molten iron infiltration of A is lower than that of the other three carbon bricks. The thermal conductivity of A and B is also superior to C and D.

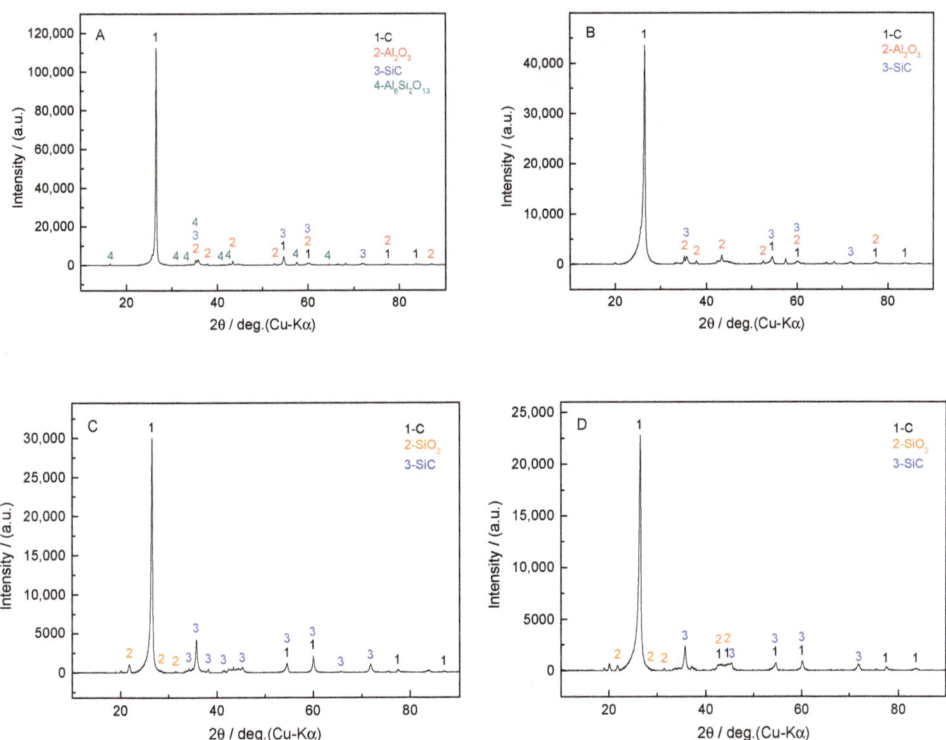

Figure 3. XRD patterns of the four types of carbon bricks (**A–D** are the types of carbon bricks).

Table 3. Physical and chemical properties of the four types of carbon bricks.

Item		Unit	A	B	C	D
Apparent porosity		%	13.4	18.4	11	12
Bulk density		g/cm^3	1.81	1.69	1.76	1.72
Compressive strength (room temperature)		MPa	48.1	42.8	50	48
Average pore diameter		μm	0.034	0.08	0.05	0.1
Pore volume (<1 μm)		%	88.6	88.2	86	78
Resistance to molten iron infiltration		%	18.7	21.4	20	22
Gas permeability		mDa	0.39	0.71	0.8	3
Oxidation rate		%	0.82	1.67	6	8
Thermal conductivity	300 °C	W/(m·K)	29.1	≥22	≥10	11
	600 °C		26.2	15.4	≥14	15

3.2. Experimental Results

The macroscopic morphology of carbon bricks before and after corrosion by molten iron is shown in Figure 4. Obvious appearance differences were observed on the carbon bricks after corrosion under laboratory conditions.

Figure 4. Macroscopic morphology of carbon bricks before and after corrosion (**A**–**D** are the types of carbon bricks).

The surface of the original brick A is the smoothest and densest, with larger carbon aggregates in a compact distribution. The surface of the original brick B is relatively rough with many tiny holes of a relatively shallow depth, and the carbon aggregate particles are relatively small. The surface of the original bricks C and D is relatively smooth and dense, but a few micro-cracks are exhibited. The micro-cracks on the surface of D are more prominent than those on C. A has the lowest corrosion degree, with small variation in the diameter of the partial brick immersed into molten iron and uniformity in the corrosion degree at different heights, while a large number of holes are formed on the surface due to the shedding of carbon aggregates. The surface of B is smoother than A after corrosion, and no obvious pores are evident on the surface, but the corrosion is severe at the gas–refractory–iron three-phase interface, showing curved recesses. After being eroded, irregular surfaces appear on C and D, the degree of corrosion is between A and B, and there exists lighter corrosion characteristics of three-phase interface. It also should be noted that C adheres to more iron beads.

The weight and average diameter of carbon bricks before and after corrosion were measured. The mass change rate, diameter change rate, and corrosion rate are defined as follows:

$$\eta = \frac{m_0 - m_{\mathrm{f}}}{m_0} \times 100\% \tag{1}$$

$$\Delta d = \frac{d_0 - d_{\mathrm{f}}}{d_0} \tag{2}$$

$$v = \frac{\pi \times l \times \left(d_0^2 - d_{\mathrm{f}}^2\right) \times \rho \times w_{\mathrm{C}}}{400 \times t \times S} \tag{3}$$

where η is the mass change rate, %; m_0 is the weight of carbon brick before the experiment, g; m_{f} is the weight of carbon brick after corrosion, g; Δd is the diameter change rate, %; d_0 is the diameter of the carbon brick before experiment, mm; d_{f} is the diameter of the carbon brick after corrosion, mm; v is the corrosion rate of carbon brick, g/(h·cm^2); l is the depth of carbon brick immersed in molten iron, cm; ρ is the density of carbon brick, g/cm^3; w_C is the carbon content of carbon brick, %; t is the reaction time, h; and S is the reaction area, cm^2.

It can be seen from Figure 5 that the mass change rate, diameter change rate, and corrosion rate of A are far lower than those of the other three carbon bricks, indicating that A possesses better corrosion resistance to molten iron and has a good positive effect on prolonging the life of the blast furnace.

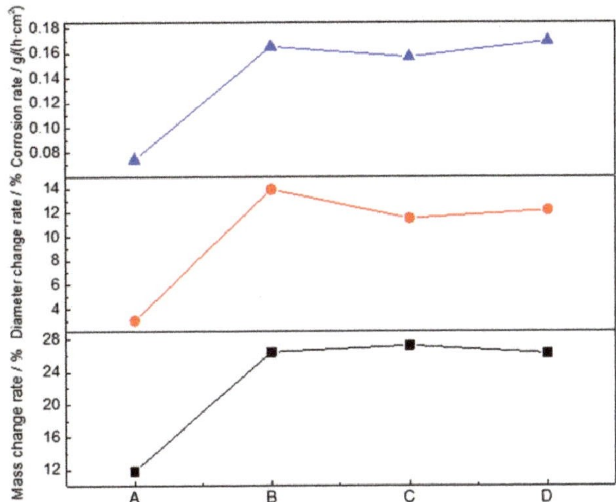

Figure 5. Mass change rate, diameter change rate, and corrosion rate of the four carbon bricks after corrosion (A, B, C and D are the types of carbon bricks).

3.3. Microscopic Analysis of Reaction Interface

The original carbon bricks were randomly sampled from the whole carbon brick for microscopic observation by scanning electron microscope (SEM) (FEI Quanta 250, Eindhoven, The Netherlands) on the original morphology, structure characteristics, and phase distribution, etc. The samples after corrosion were cut along the axis of the concentric cylinder, and the outer surfaces, including three-phase interface and refractory–iron interface, were also subjected to microscopic characterization as Figure 6 shows.

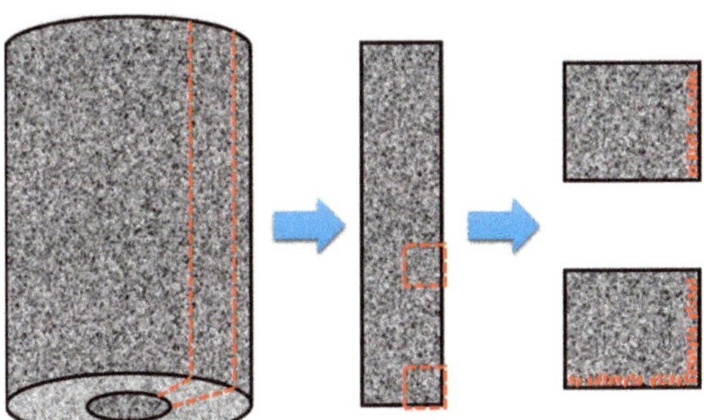

Figure 6. Schematic diagram of sample cutting before and after corrosion.

3.3.1. Analysis of the Microscopic Appearance of the Original Carbon Brick

The microscopic morphology of the original brick A is taken under different scaleplates, as shown in Figure 7. It can be seen from Figure 7a that the boundaries between the carbon matrix, white ceramic, and black carbon aggregates are obvious. The internal surface is rough, with more micro-pores in size of 20~70 µm as presented from Figure 7b. From Figure 7c,d, it indicates the dispersed distribution of ceramic in the carbon matrix, thereby avoiding the adverse effect of the single ceramic assemblage on the thermal conductivity of the carbon brick. The local carbon matrix and mineral layer are in a strip shape, which isolate the carbonaceous aggregates at different locations. This phase distribution can effectively block the contact between the carbon aggregate and the molten iron at the iron–carbon interface during the dissolution process and slow down the dissolution process of carbon brick. Some whiskers are found in the pores of A, as shown in Figure 7e.

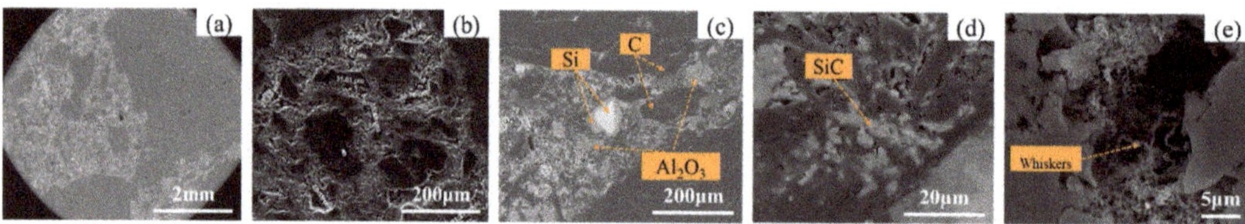

Figure 7. (**a–e**) Microscopic morphology of the original brick A under different scaleplates.

As a large amount of flake graphite is added to B to boost the thermal conductivity, the carbon content is relatively high, which meets the chemical composition in Table 1. B contains less ceramic seen from Figure 8a compared with A. The boundary of the black carbon aggregate layer is not obvious, so the overall interface is mainly carbon matrix with the ceramic dispersed and discontinuous distribution, as shown in Figure 8c. From the secondary electron image in Figure 8b, it can be observed that the surface of B is smooth with fewer micro-pores; those it does have are mostly in the size range of 20~50 µm. On the other hand, there are many pores on the surface of the original brick under the scaleplate of 2 mm with larger depth, as shown in Figure 8a. It shows another distribution of the ceramic additives, Al_2O_3 in this region performs lamellar. Similarly, higher silicide is located in the interface between Al_2O_3 and the carbon matrix.

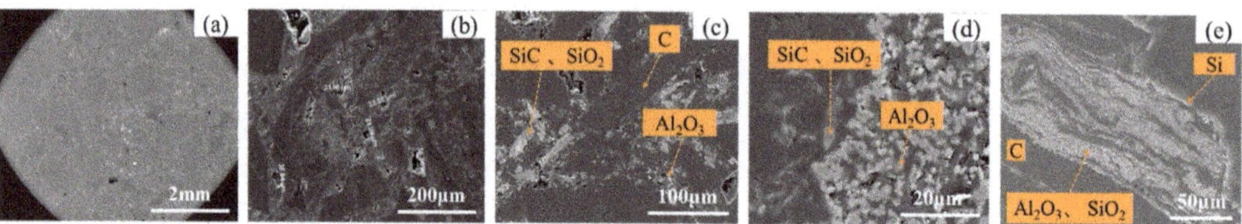

Figure 8. (**a–e**) Microscopic morphology of the original brick B under different scaleplates.

The microscopic morphology of the original brick C is shown in Figure 9. As seen from Figure 9a, it exhibits more dispersed distribution of ceramic additive and a greater number of small-sized carbon aggregates compared with that in A and B, and an obvious boundary between the carbon aggregate and the non-aggregate is observed. There are cracks on the surface of the original brick C, while the number of micropores is relatively few with thin and long feature. It is discovered from Figure 9c that the ceramic additives are mainly silicon compounds, which meet the chemical composition in Table 1. The SiO_2 and SiC ceramic phases are uniformly distributed in the carbon matrix, and there is no Al_2O_3 aggregation on the surface.

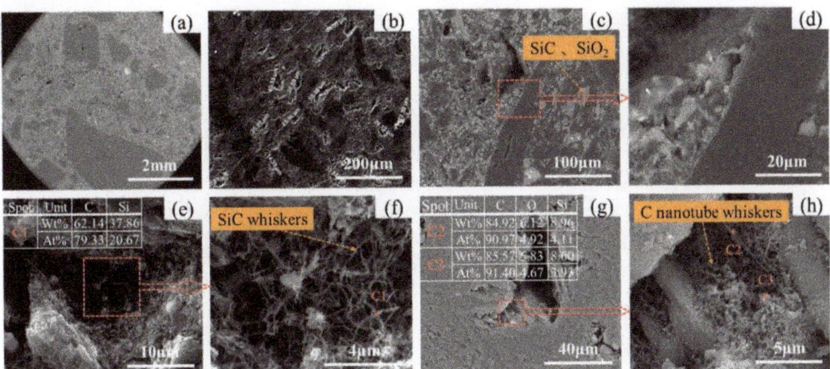

Figure 9. (**a–h**) Microscopic morphology of the original brick C under different scaleplates.

There exists a mass of flocculent whiskers in the pores of the original brick C as presented in Figure 9c,d, covering the inner walls of the pores. Further magnified in Figure 9e,f, we may see the whiskers connect to each other and form a network structure, filling the pores of the carbon brick. By spot scanning, we can ascertain the whiskers mainly contain two elements: C and Si, indicating that this area is dominated by SiC whiskers. The whiskers in the pores of the carbon aggregates are shown in Figure 9g,h. By spot scanning C2 and C3, we find the C element accounts for >90%. Combined with the brick composition and production process design, it can be inferred that the whiskers in this area are mainly carbon nanotubes.

The microscopic morphology of the original brick D is shown in Figure 10; in comparison with C, denser and larger pores on the surface are confirmed in Figure 10a,b with the size mainly ranging from 30 μm to 50 μm. Through the comparative analysis of Figures 9c and 10c, it can be found that the internal density of D is lower than that of C. A large number of whiskers are found inside of the crack at the junction of the carbon aggregate and the carbon matrix, shown in Figure 10e. By EDS analysis of spots D1 and D2, we find the main component of the whiskers is carbon, and they contain a small amount of Si element. Combined with the brick composition and production design, the whiskers can be deduced to be mainly carbon nanotubes, and part of the SiC whiskers are mixed.

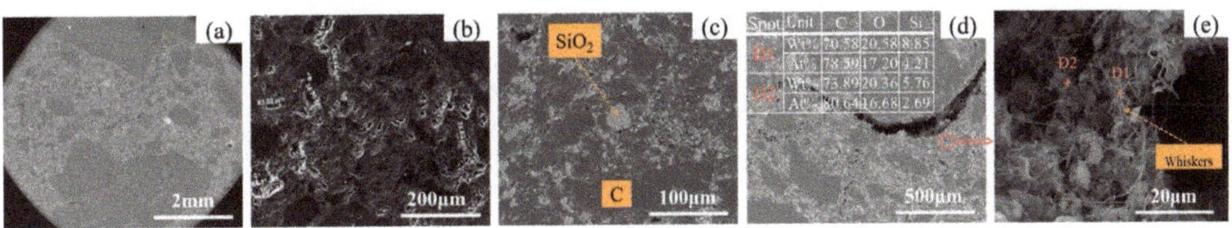

Figure 10. (**a–e**) Microscopic morphology of the original brick D under different scaleplates.

3.3.2. Analysis of the Microscopic Appearance of the Carbon Brick after Corrosion

Figure 11a–d show the microscopic morphology of the side surface of A after corrosion. The uneven morphology is still maintained on the reaction interface as shown in Figure 11a,b. There exist a few cracks and pores on the internal interface, and the pores are mainly concentrated on the surface of the carbon aggregates with the diameter more than 100 μm. The phenomenon of the iron beads adhering to the carbon brick appears on the surface in Figure 11c. The small iron beads mainly gather at the carbon aggregates, and the mixed part of the carbon matrix and the ceramic phase is mainly iron infiltration. As shown in Figure 11d, only strip-shaped carbon aggregates remain in the iron-infiltrated part.

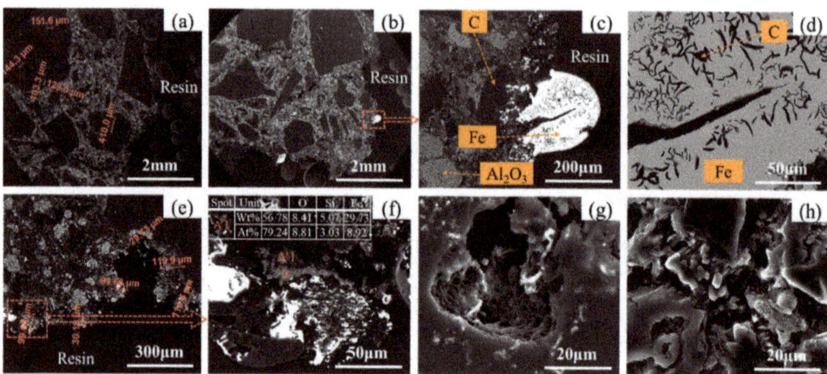

Figure 11. (a–h) Microscopic morphology of A after corrosion.

The bottom surface of A immersed in molten iron after corrosion is presented in Figure 11e,f. The reaction interface retains a good iron infiltration phenomenon. This not only adheres to small iron beads but also an iron infiltration layer in the carbon aggregates. The whole iron infiltration layer is between 30 μm and 120 μm, and the thinner thickness of the carbon matrix and ceramic is discovered due to the ceramic presence. It is strong proof that the ceramic material enhances the carbon brick's resistance to molten iron corrosion and penetration. From Figure 11f, it is found that the small pieces of carbon aggregates have been almost separated by iron infiltration from the carbon matrix; a nebula-like, exfoliated, and dissolved iron–carbon mixture emerges, with 8.92% iron and 56.78% carbon in atomic ratio.

The internal whiskers in the pores close to the reaction interface have been dissolved by molten iron, and the inner surface of the pores is covered by small, dense pits as analyzed from Figure 11g, while a large number of whiskers still exist in the pores away from the reaction interface as shown in Figure 11h. This illustrates that the whiskers in the pores are firstly dissolved by the molten iron, thereby retarding the process of carbon brick being corroded by the molten iron.

The microscopic morphology of B after corrosion is presented in Figure 12. From Figure 12a, an obvious arc is observed at the three-phase interface. Compared with the A corrosion morphology, the reaction interface of B is relatively flat and there are fewer iron beads adhering on the surface, most of which are iron infiltration with thickness nearly 50 μm. In addition, a lot of holes are distributed on the surface of B, the maximum diameter of which can reach 370.4 μm. However, there are very few cracks on the surface of graphite carbon aggregates. From the mapping of the local carbon aggregate in Figure 12d, we can see a small amount of Al_2O_3 and SiO_2 ceramic additives dispersed within. The large-scale cracks in the carbon aggregate itself have iron accumulation; it can be inferred that they will act as a channel for iron penetration when the molten iron is corroded to a certain extent and accelerate the corrosion rate of the carbon brick in this part. Obviously, the number and size of internal pores and cracks in carbon brick have a significant negative impact on its ability to resist molten iron corrosion of carbon brick.

Figure 13a–d show the microscopic morphology of C after corrosion; it can be observed that C has a smaller amount of large-particle carbon aggregates and very few pores in the carbon brick, but many large cracks exist, and the crack length can reach 763.7 μm. What is more, the reaction interface on side surface is relatively flat, gaps appear at the junction of the bottom surface and the side surface, and a large number of iron beads are bonded on the reaction interface in the gap. As shown in Figure 13d, both iron infiltration and iron bead adhesion arise in the carbon aggregates, and the thickness of the iron infiltration layer varies from 50 μm to 100 μm. An iron infiltration zone inside the aggregates can destroy the structure of the carbon brick near the reaction interface and accelerate the subsequent dissolution of the carbon brick in this area.

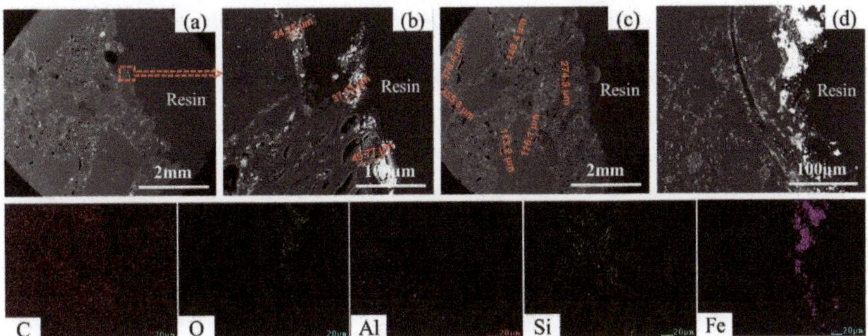

Figure 12. (**a–d**) Microscopic morphology of B after corrosion.

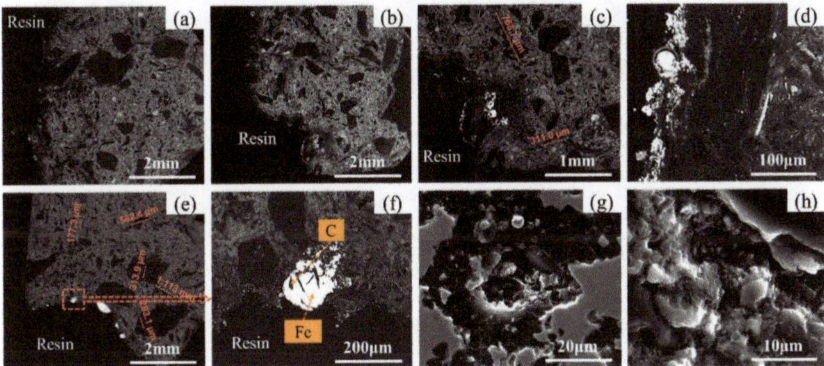

Figure 13. (**a–h**) Microscopic morphology of C after corrosion.

The microscopic morphology of the bottom surface of C after corrosion is presented in Figure 13e,f, cracks of up to 1.113 mm appear in this area, which will reduce the stability performance of carbon brick. In addition, there is a phenomenon of iron beads adhering to the reaction interface in the area of non-carbon aggregates. The strip-shaped black phases inside the iron beads are deduced to be graphite phase saturated with carbon and graphite aggregate wrapped and isolated during the process of carbon aggregate dissolution by molten iron through EDS analysis. The mixing and contact area of iron bead and carbon brick indicates that, as the penetration of molten iron deepens, the main boundary structure of C is severely damaged, a large number of small particles in carbon brick matrix and carbon aggregate are stripped and accumulate to form a layer of stripped small particles at the reaction interface. Figure 13g,h show the microscopic morphology of the pores near and away from the reaction interface, respectively. It can be observed that the SiC whiskers and carbon nanotube whiskers near the reaction interface have been almost completely decomposed and disappeared, which is similar to the corrosion seen in A.

The microscopic morphology of the side and bottom surface of D after corrosion is shown in Figure 14. Under 2 mm scaleplate, D has a small quantity of pores, while the number of internal cracks is significantly greater compared with C after corrosion, and the crack length can reach more than 2 mm. The phenomenon of iron bead adhering to the reaction interface is not obvious. Figure 14c shows that the iron beads in this area have been embedded in the carbon aggregate, and a certain degree of iron permeation layer at the reaction interface exists. It is foreseeable that, as the molten iron further dissolves and penetrates, the carbon particles bordered by the yellow dashed line will be stripped off. Further enlargement of Figure 14c allows us to observe that the iron beads at this location do not exist in the form of being embedded inside the carbon particles. There is still an iron penetration layer and carbon matrix separation between the iron beads and the carbon particles. Therefore, iron penetration will cause the deterioration of the carbon brick, and then the carbon matrix separates from the structure. Figure 14e shows the microscopic morphology of the carbon aggregate after corrosion; the corrosion of carbon aggregate is not entirely caused by molten iron corrosion. The penetration of molten iron in carbon aggregate will also cause local cracking and stripping of the carbon brick.

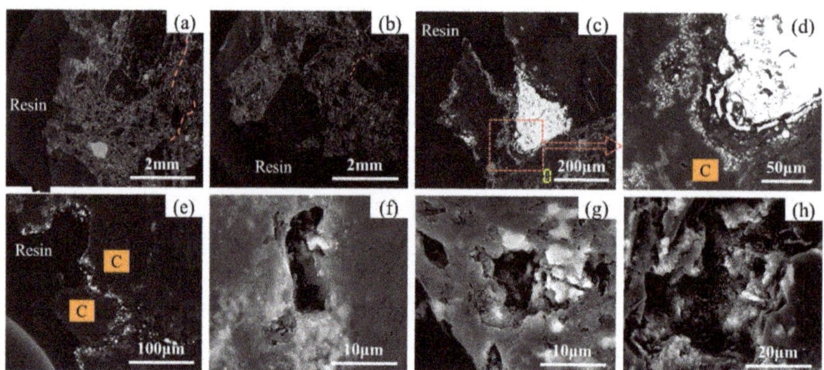

Figure 14. (**a–h**) Microscopic morphology of D after corrosion.

The characterization results of the internal pores in the radial direction of D sample are presented in Figure 14f–h. Figure 14f shows the internal pores in the yellow dashed frame in Figure 14c, which are closest to the reaction interface. SiC whiskers, or carbon nanotubes, are not found inside. From Figure 14g,h close to the pores inside D in sequence, only some whiskers exist in the pores in Figure 14g, while a large number of whiskers retain in Figure 14h. Therefore, the presence of whiskers helps to delay partial dissolution of pores when the carbon brick contacts with molten iron.

3.4. Analysis on the Mechanism of Corrosion Resistance to Molten Iron of Carbon Brick

3.4.1. Thermodynamic Analysis

There are two reactions that may occur in the carburization process of molten iron; one is the dissolution of solid carbon into molten iron, and the other is the formation of Fe_3C from the reaction of solid carbon and iron atom within the temperature range of 1809 K to 2000 K. According to thermodynamic analysis in Figure 15, the dissolution of carbon is intended to be the predominant reaction.

$$C_{(s)} = [C] \quad \Delta G^\theta = 22{,}590 - 42.26T, \quad J/mol \tag{4}$$

$$C_{(s)} + 3Fe_{(l)} = Fe_3C_{(s)} \quad \Delta G^\theta = 10{,}530 - 10.2T, \quad J/mol \tag{5}$$

The carbon content in molten iron is always unsaturated in the production process of blast furnaces. Therefore, the carbon bricks with carbon as the main component will have continuous direct contact with the molten iron to cause carbon dissolution, resulting in carbon brick corrosion. Higher carbon content in carbon bricks can give rise to easier corrosion due to the larger contact area between the molten iron and the carbon in the carbon brick. The molten iron is also in constant motion. The renewal of molten iron will also cause continuous corrosion on carbon bricks. Therefore, the method to reduce the corrosion of carbon bricks by molten iron is centered on the following factors.

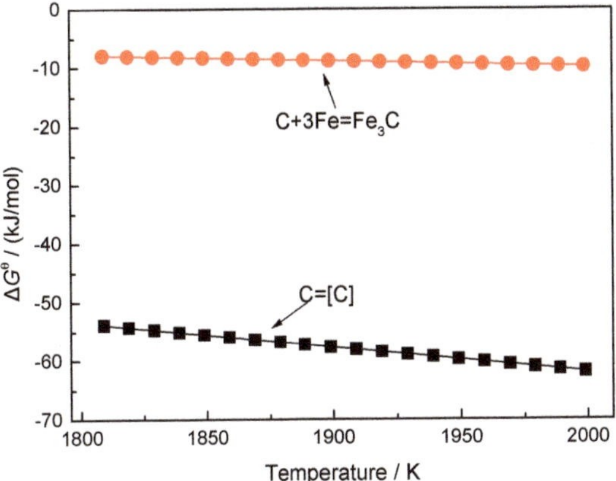

Figure 15. Variation of the standard Gibbs free energy of carburizing reactions with temperature.

(1)　Carbon solubility in molten iron

When the reaction of $C_{(s)}$ = [C] reaches equilibrium, the standard Gibbs free energy of the reaction is:

$$\Delta G^{\theta} = -RT\ln K^{\theta} = -2.303RT\lg K^{\theta} = -2.303RT\lg a_{[C]} = -2.303RT(\lg[C] + \lg f_{C}) \quad (6)$$

$$\lg[C] = -\frac{1179.81}{T} + 2.21 - \lg f_{C} \quad (7)$$

According to Wagner's model, the activity coefficient of carbon in multi-system molten iron can be expressed as follows:

$$\lg f_{C} = e_{C}^{C}[C] + e_{C}^{Si}[Si] + e_{C}^{Mn}[Mn] + e_{C}^{P}[P] + e_{C}^{S}[S] \quad (8)$$

Combining Equations (7) and (8), we can gain the carbon content in molten iron.

$$\lg[C] = -\frac{1179.81}{T} + 2.21 - e_{C}^{C}[C] - e_{C}^{Si}[Si] - e_{C}^{Mn}[Mn] - e_{C}^{P}[P] - e_{C}^{S}[S] \quad (9)$$

where e_{C}^{j} is the activity interaction coefficient of the j element and carbon element in the molten iron.

There are two factors affecting the carbon solubility in molten iron. One is the temperature of the molten iron. The carbon solubility will increase with the increase in temperature under the constant content of each component in molten iron. The other is the product of the content of each component and the activity interaction coefficient of this component and carbon component. When $e_{C}^{j} > 0$, the activity coefficient of carbon increases with the content of j element increasing, resulting in a decrease in the carbon solubility. When $e_{C}^{j} < 0$, the activity coefficient of carbon will decrease with the increase in j element, leading to an increase in the solubility of carbon in the molten iron. The carbon solubility in molten iron can reflect the corrosion ability of molten iron on carbon brick, that is, the higher the carbon solubility, the easier the carbon brick is to corrode.

(2)　Ceramic phases and whiskers

Al$_2$O$_3$, SiO$_2$, and SiC are the main ceramic phases contained in the present carbon bricks. Al$_2$O$_3$ cannot react at all at the experimental temperature, and it is not wet with molten iron [23]. Therefore, the addition of Al$_2$O$_3$ enhances the corrosion resistance to molten iron of carbon bricks. SiC is another factor enhancing the corrosion resistance to molten iron, because the special preparation process of carbon bricks can promote the formation of SiC whiskers in situ. The presence of whiskers in the pores will withstand the molten iron, delaying carbon brick corrosion. Carbon nanotube whiskers can also delay the corrosion of carbon bricks to a certain extent, because the channels where molten iron enters the pores will preferentially erode the carbon nanotube whiskers in the pores.

SiO_2 can react with carbon to produce FeSi and CO under the catalysis of iron, as Equation (10) shows [24,25]. The pores formed after the reaction will become channels for the penetration of molten iron and increase the corrosion of carbon bricks.

$$SiO_{2(s)} + 2C_{(s)} + Fe_{(l)} = FeSi + 2CO \quad \Delta G^{\theta} = 649,404 - 383.53T, \quad J/mol \tag{10}$$

Ceramic additives such as Al_2O_3, SiO_2, and SiC in A and B are added in larger amounts, and the proportion of Al_2O_3 in A is relatively high. The components of C and D are obviously different from A and B. C and D have less Al_2O_3, and SiO_2 and SiC are the main ceramic additives. Additionally, A also contains some whiskers, so its performance is better.

(3)   Graphitization degree of carbon brick

From Figure 3, the sharp (002) carbon peaks of the four carbon bricks can be clearly observed. The sharper the carbon peak, the higher in the degree of the ordering of carbon structure or graphitization [26]. Generally, the average stacking height (L_C) of the layered structure is used to characterize the lattice dimensions of crystalline carbon. L_C can be obtained by using the classical Scherrer's Equation [27–29] with crystallites in the absence of lattice strain or distortion.

$$L_C = \frac{0.89\lambda}{B_{002}\cos(\theta_{002})} \tag{11}$$

where λ is the wavelength of the X-ray radiation, B_{002} is the full-width at half-maximum intensity of the (002) carbon peak, and θ_{002} is the diffraction angle of the (002) band. Accordingly, a sharper (002) carbon peak indicates a greater crystalline order or graphitization degree of carbon materials.

Based on Bragg's law [30], the interlayer spacing (d_{002}) and average layer number (N_C) can be calculated as follows:

$$d_{002} = \frac{\lambda}{2\sin(\theta_{002})} \tag{12}$$

$$N_C = \frac{L_C}{d_{002}} \tag{13}$$

The calculation results of L_C, d_{002}, and N_C are shown in Table 4.

Table 4. Structural parameters of carbon bricks measured from the XRD spectra.

Carbon Brick	$2\theta_{002}/°$	$d_{002}/(nm)$	$B_{002}/°$	$L_C/(nm)$	N_C
SGL	26.48	0.34	0.287	28.44	85
NDK	26.41	0.34	0.532	15.34	45
SM	26.54	0.34	0.431	18.94	55
CM	26.52	0.34	0.601	13.58	40

The L_C and N_C values of carbon bricks present in the same order of $L_C(A) > L_C(C) > L_C(B) > L_C(D)$ and $N_C(A) > N_C(C) > N_C(B) > N_C(D)$, demonstrating a decreasing graphitization degree and crystalline structure of carbon. Moreover, the graphitization degree is almost inversely proportional to the experimental results of the corrosion resistance to molten iron of four carbon bricks, that is, the higher the graphitization degree, the smaller the mass change rate, diameter change rate, and corrosion rate. This demonstrates that the crystalline structure of carbon is closely related to its dissolution behavior in molten iron, and the higher the graphitization degree, the less easily the carbon brick will be corroded. The findings are different from those of previous studies [13,31,32], but are consistent with Guo's investigation [33].

3.4.2. Corrosion Mechanism of Carbon Brick in Molten Iron

According to macroscopic and microscopic morphologies of carbon bricks before and after corrosion by molten iron, and thermodynamic analysis, the process of molten iron corroding carbon brick can be attributed to three aspects: the carburization, infiltration, and scouring of molten iron. The carburization process of molten iron has been demonstrated using the thermodynamic reaction analysis; its dissolution into molten iron can be further divided into three steps. The first step is the separation of carbon atoms from the solid carbon structure; the second step is the accumulation of the separated carbon atoms at the reaction interface; the third step is that the carbon atoms are adsorbed to the interstitial spaces of iron atoms from the reaction interface. In addition to the dissolution of the carbon into the molten iron, the infiltration of molten iron into carbon brick is another important cause of carbon brick corrosion. When molten iron penetrates the interior of the carbon brick, it will interact with the materials inside the carbon brick to form an iron infiltration layer, destroying the structure of the carbon brick at the iron–carbon interface, and then producing structural separation. The molten iron diffuses into the interior of the carbon brick through the pores and micro-cracks, and further deepens in combination with the carbon dissolution process. During the whole infiltration and carburization process, the channels that the molten iron diffused into from different paths may connect, resulting in local carbon brick to peel off from the main body to form a separation layer. Furthermore, the exfoliated particles will leave the separation layer for subsequent reaction behaviors. The scouring of molten iron, simulated by the stirring speed in the experiment, subjects the iron–carbon interface to interaction forces, which promotes the accelerated separation of the exfoliated fragmented carbon brick from the iron–carbon interface, thereby facilitating a new round of corrosion process as shown in Figure 16.

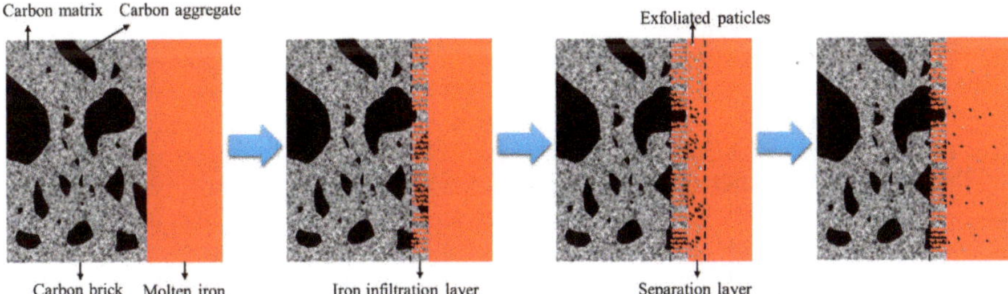

Figure 16. Schematic diagram of the corrosion mechanism of carbon brick in molten iron.

In the practical carbon brick production process, attention should be paid to the combination of carbon, Al_2O_3, SiO_2, and SiC contents in carbon brick. Higher carbon and SiO_2 contents will increase the corrosion by molten iron due to the reaction. Higher Al_2O_3 and SiC contents will hinder the contact between the carbon brick and the molten iron, delaying the corrosion of the carbon brick. The graphitization degree of carbon brick will also affect its corrosion; the higher of the graphitization degree, the lower of the corrosion degree. In addition, the microporous properties will also influence the corrosion of molten iron on carbon brick. Smaller average pore diameter and pore volume (<1 μm) contribute to a lower corrosion degree. Therefore, the formation of carbon nanotubes and SiC whiskers can also delay the corrosion of carbon brick while reducing the average pore diameter and retarding reaction. In terms of blast furnace operation, the temperature and composition of molten iron will affect the degree of dissolution of carbon in carbon brick into molten iron, that is, the greater the solubility of carbon in molten iron, the greater the degree of corrosion to carbon brick. Thus, the adjustment of temperature, content of each component, and activity interaction coefficient of this component and the carbon component for the reduction in carbon solubility in molten iron can weaken the dissolution of carbon in carbon brick into molten iron. In general, reducing the corrosion of molten iron on carbon brick is to reduce the carburization process of molten iron, the molten iron infiltration into carbon brick, and the scouring of molten iron to carbon brick.

4. Conclusions

The corrosion resistance to molten iron of four kinds of carbon bricks used in a blast furnace hearth were investigated to illustrate the corrosion mechanism through the macroscopic and microscopic analysis of carbon bricks before and after reaction and thermodynamic analysis. It was found that the corrosion resistance to molten iron of carbon brick could be improved by the combination of carbon, Al_2O_3, SiO_2, and SiC contents in the carbon bricks, a high graphitization degree of carbon brick, a small average pore diameter and pore volume (<1 μm) of carbon brick, the formation of carbon nanotubes and SiC whiskers in carbon brick, a low solubility of carbon in molten iron, and the weakened scouring of molten iron to carbon brick, etc. The summarized specific consequences are as follows:

The chemical composition and phase analyses show that B has the highest carbon content, up to 84.13%, which is much greater than the other three types of carbon bricks. The main phases in A are C, Al_2O_3, SiC, and $Al_6Si_2O_{13}$, the ones in B are C, Al_2O_3, and SiC, while the main phases of C and D are C, SiO_2, and SiC. The analysis of physical and chemical properties shows that the resistance to molten iron infiltration of A is lower than that of the other three carbon bricks. The thermal conductivity of A and B is superior to C and D.

The macroscopic morphology analysis of carbon bricks after corrosion by molten iron shows that A has the lowest corrosion degree and highest uniformity at different heights, but a large number of holes are formed on the surface. The surface of B is smoother than A after corrosion, but the corrosion is severe at the gas–refractory–iron three-phase interface. C and D have irregular surfaces, their corrosion degree is between A and B, and there exists lighter corrosion characteristics of three-phase interface.

The microscopic morphology analysis of the original carbon bricks shows that A has obvious boundaries between the carbon matrix, white ceramic, and black carbon aggregates. Its internal surface is rough, with more micro-pores in the range of 20~70 μm. The ceramic in A carbon matrix is in dispersed distribution. There are some whiskers in the A pores. The boundary of the black carbon aggregate layer in B is not obvious, and the carbon matrix and ceramic are in dispersed and discontinuous distribution. Its surface is smooth, with fewer micro-pores; those that do appear are mostly in the size range of 20~50 μm. An obvious boundary is observed between the carbon aggregate and the non-aggregate in C, and cracks on its surface, but the micropores are relatively few, with long and thin features. The SiO_2 and SiC ceramic phases are uniformly distributed in the carbon matrix, and no Al_2O_3 aggregation on the surface. A mass of flocculent whiskers exists in the pores of the original brick C, which are mainly determined to be SiC whiskers, and some carbon nanotubes exist in the pores of the carbon aggregates. D has denser and larger pores on the surface, with the size mainly ranging from 30 μm to 50 μm. A large number of whiskers are also found inside of the cracks, which are mainly inferred to carbon nanotubes, and part of SiC whiskers are mixed.

The microscopic morphology analysis of the carbon bricks after corrosion shows that all the carbon bricks have more and larger pores than their original carbon bricks. There exist pores on the internal interface of A with the diameter more than 100 μm, and the phenomenon of the iron beads adhering to the carbon brick mainly happens in the carbon aggregates. The mixed part of the carbon matrix and the ceramic phase is mainly iron infiltration. The surface of B is relatively flat after corrosion compared with that of A. There exist fewer iron beads adhering on the surface, most of which are iron infiltration with a thickness of nearly 50 μm. Many large cracks emerge on the surfaces of C and D with the length exceeding 1 mm in C and 2 mm in D. A large number of iron beads are bonded on the reaction interface of C, and the iron bead adhering to the reaction interface of D is not obvious. The iron penetration degree of C, D, and A is similar. C and D present an obvious corrosion process phenomenon, that is, the carbon matrix is broken and peeled off during the iron infiltration process. In addition, the whiskers inside the pores also tend to decrease as they approach the reaction interface.

The dissolution of carbon is the predominant reaction of the carbon brick being corroded. The higher the carbon solubility in molten iron, the easier it is to corrode the carbon brick. Al_2O_3 and SiC can enhance the corrosion resistance of carbon bricks to molten iron, and SiO_2 can react with carbon to form pores, which will become channels for the penetration of molten iron and increase the corrosion of carbon bricks. A higher graphitization degree of carbon bricks is beneficial to lessen their corrosion degree.

The corrosion of carbon bricks by molten iron can be attributed to three aspects: the carburization, infiltration, and scouring of molten iron. The carburization process of molten iron is the main reaction process of molten iron corroding carbon brick. The infiltration of molten iron into carbon brick will facilitate the dissolution of carbon, destroying the structure of carbon bricks and accelerating the corrosion of carbon bricks. The scouring of molten iron subjects the iron–carbon interface to interaction forces, promoting the separation of the exfoliated fragmented carbon brick from the iron–carbon interface to facilitate a new round of corrosion process.

Author Contributions: Conceptualization, J.Z.; methodology, C.W. and K.J.; validation, W.C. and Z.L.; formal analysis, C.W.; investigation, Z.W. and T.W.; resources, X.L. and Z.P.; writing—original draft preparation, C.W. and Z.W.; and writing—review and editing, J.Z. and C.W. All authors have read and agreed to the published version of the manuscript.

Funding: This research was funded by the National Natural Science Foundation of China, grant number 51874025, and the Project of SKLAM, grant number 41622017 and K22-05.

Institutional Review Board Statement: Not applicable.

Informed Consent Statement: Not applicable.

Data Availability Statement: The data presented in this study are available on request from the corresponding author.

Acknowledgments: WISDRI Handan Wupeng New Lining Material Co., Ltd. is acknowledged for providing materials and collaborative research.

Conflicts of Interest: The authors declare no conflict of interest.

References

1. Zhang, J.L.; Liu, Z.J.; Jiao, K.X.; Xu, R.S.; Li, K.J.; Wang, Z.Y.; Wang, C.; Wang, Y.Z.; Zhang, L. Progress of new technologies and fundamental theory about ironmaking. *Chin. J. Eng.* **2021**, *43*, 1630–1646.
2. Liu, Z.J.; Zhang, J.L.; Zuo, H.B.; Yang, T.J. Recent progress on long service life design of Chinese blast furnace hearth. *ISIJ Int.* **2012**, *52*, 1713–1723. [CrossRef]
3. Guo, Z.Y.; Zhang, J.L.; Jiao, K.X.; Gao, T.L.; Zhang, J. Research on low-carbon smelting technology of blast furnace-optimized design of blast furnace. *Ironmak. Steelmak.* **2021**, *48*, 685–692. [CrossRef]
4. Zagaria, M.; Dimastromatteo, V.; Colla, V. Monitoring erosion and skull profile in blast furnace hearth. *Ironmak. Steelmak.* **2010**, *37*, 229–234. [CrossRef]
5. Niu, Q.; Cheng, S.S.; Xu, W.X.; Niu, W.J. Microstructure and phase of carbon brick and protective layer of a 2800 m^3 industrial blast furnace hearth. *ISIJ Int.* **2019**, *59*, 1776–1785. [CrossRef]
6. Stec, J.; Smulski, R.; Nagy, S.; Szyszkiewicz-Warzecha, K.; Tomala, J.; Filipek, R. Permeability of carbon refractory materials used in a blast furnace hearth. *Ceram. Int.* **2021**, *47*, 16538–16546. [CrossRef]
7. Ibrahim, S.H.; Skibinski, J.; Oliver, G.; Wejrzanowski, T. Microstructure effect on the permeability of the tape-cast open-porous materials. *Mater. Des.* **2019**, *167*, 107639. [CrossRef]
8. Li, Y.W.; Chen, X.L.; Sang, S.B.; Li, Y.B.; Jin, S.L.; Zhao, L.; Ge, S. Microstructures and properties of carbon refractories for blast furnaces with SiO_2 and Al additions. *Metall. Mater. Trans. A* **2010**, *41*, 2085–2092. [CrossRef]
9. Ma, H.X.; Zhang, J.L.; Jiao, K.X.; Chang, Z.Y.; Wang, Y.J.; Zheng, P.C. Analysis of erosion characteristics and causes of blast furnace hearth. *Iron Steel* **2018**, *53*, 14–19.
10. Liao, N.; Li, Y.W.; Jin, S.L.; Sang, S.B.; Harmuth, H. Enhanced mechanical performance of Al_2O_3-C refractories with nano carbon black and in-situ formed multi-walled carbon nanotubes (MWCNTs). *J. Eur. Ceram. Soc.* **2016**, *36*, 867–874. [CrossRef]

11. Luo, M.; Li, Y.W.; Jin, S.L.; Sang, S.B.; Zhao, L.; Wang, Q.H.; Li, Y.B. Microstructure and mechanical properties of multi-walled carbon nanotubes containing Al_2O_3-C refractories with addition of polycarbosilane. *Ceram. Int.* **2013**, *39*, 4831–4838. [CrossRef]

12. Mertke, A.; Aneziris, C.G. The influence of nanoparticles and functional metallic additions on the thermal shock resistance of carbon bonded alumina refractories. *Ceram. Int.* **2015**, *41*, 1541–1552. [CrossRef]

13. Zhang, W.; Hua, F.B.; Dai, J.; Xue, Z.L.; Ma, G.J.; Li, C.Z. Isothermal kinetic mechanism of coke dissolving in hot metal. *Metals* **2019**, *9*, 470. [CrossRef]

14. Jang, D.; Kim, Y.; Shin, M.; Lee, J. Kinetics of carbon dissolution of coke in molten iron. *Metall. Mater. Trans. B* **2012**, *43B*, 1308–1314. [CrossRef]

15. Cham, S.T.; Sakurovs, R.; Sun, H.P.; Sahajwalla, V. Influence of temperature on carbon dissolution of cokes in molten iron. *ISIJ Int.* **2006**, *46*, 652–659. [CrossRef]

16. Wright, J.K.; Baldock, B.R. Dissolution kinetics of particulate graphite injected into iron/carbon melts. *Metall. Trans. B* **1988**, *19*, 375–382. [CrossRef]

17. Mourao, M.B.; Murthy Krishna, G.G.; Elliott, J.F. Experimental investigation of dissolution rates of carbonaceous materials in liquid iron-carbon melts. *Metall. Trans. B* **1993**, *24B*, 629–637. [CrossRef]

18. Deng, Y.; Zhang, J.L.; Jiao, K.X. Dissolution mechanism of carbon brick into molten iron. *ISIJ Int.* **2018**, *58*, 815–822. [CrossRef]

19. Deng, Y.; Liu, R.; Jiao, K.X.; Chen, Y.B. Wetting behavior and mechanism between hot metal and carbon brick. *J. Eur. Ceram. Soc.* **2021**, *41*, 5740–5749. [CrossRef]

20. Deng, Y.; Liu, R.; Jiao, K.X.; Chen, L.D.; Chen, Y.B. Evolution and Mechanism of Dissolutive Wetting between Hot Metal and Carbon Brick. Available online: https://www.sciencedirect.com/science/article/pii/S0955221922002898 (accessed on 12 April 2022).

21. Stec, J.; Tarasiuk, J.; Wro´nski, S.; Kubica, P.; Tomala, J.; Filipek, R. Investigation of molten metal infiltration into micropore carbon refractory materials using X-ray computed tomography. *Metals* **2021**, *14*, 3148. [CrossRef]

22. Jiao, K.X.; Fan, X.Y.; Zhang, J.L.; Wang, K.D.; Zhao, Y.A. Corrosion behavior of alumina-carbon composite brick in typical blast furnace slag and iron. *Ceram. Int.* **2018**, *44*, 19981–19988. [CrossRef]

23. Wei, Y.W.; Shao, Y.; Chen, J.F.; Li, N. Influence of carbonaceous materials on the interactions among (Al_2O_3-C)/Fe system with temperature and soaking time. *J. Eur. Ceram. Soc.* **2017**, *38*, 313–322. [CrossRef]

24. Wu, S.L.; Wang, X.L. *Iron and Steel Metallurgy*, 4th ed.; Metallurgical Industry Press: Beijing, China, 2019; pp. 146–216.

25. Fan, X.Y.; Jiao, K.X.; Zhang, J.L.; Wang, K.D.; Chang, Z.Y. Phase transformation of cohesive zone in a water-quenched blast furnace. *ISIJ Int.* **2018**, *58*, 1775–1780. [CrossRef]

26. Gupta, S.; Sahajwalla, V.; Chaubal, P.; Youmans, T. Carbon structure of coke at high temperatures and its influence on coke fines in blast furnace dust. *Metall. Mater. Trans. B* **2005**, *36B*, 385–394. [CrossRef]

27. Warren, B.E. X-Ray diffraction in random layer lattices. *Phys. Rev.* **1941**, *59*, 693–698. [CrossRef]

28. Arnold, T.; Chanaa, S.; Clarke, S.M.; Cook, R.E.; Larese, J.Z. Structure of an *n*-butane monolayer adsorbed on magnesium oxide (100). *Phys. Rev. B Condens. Matter Mater. Phys.* **2006**, *74*, 085421. [CrossRef]

29. Lu, L.; Sahajwalla, V.; Kong, C.; Harris, D. Quantitative X-ray diffraction analysis and its application to various coals. *Carbon* **2001**, *39*, 1821–1833. [CrossRef]

30. Klug, H.P.; Alexander, L.E. *X-ray Diffraction Procedures: For Polycrystalline and Amorphous Materials*, 2nd ed.; Wiley: New York, NY, USA, 1974; pp. 130–135.

31. Xu, R.S.; Zhang, J.L.; Wang, W.; Zuo, H.B.; Xue, Z.L.; Song, M.M. Dissolution kinetics of solid fuels used in COREX gasifier and its influence factors. *J. Iron Steel Res. Int.* **2018**, *25*, 1–12. [CrossRef]

32. Sun, M.M.; Zhang, J.L.; Li, K.J.; Li, H.T.; Wang, Z.M.; Jiang, C.H.; Ren, S.; Wang, L.; Zhang, H. The interfacial behavior between coke and liquid iron: A comparative study on the influence of coke pore, carbon structure and ash. *JOM* **2020**, *72*, 2174–2183. [CrossRef]

33. Guo, Z.Y.; Jiao, K.X.; Zhang, J.L.; Ma, H.B.; Meng, S.; Wang, Z.Y.; Zhang, J.; Zong, Y.B. Graphitization and performance of deadman coke in a large dissected blast furnace. *ACS Omega* **2021**, *39*, 25430–25439. [CrossRef]

Article

Experimental Study on Desulfurization and Removal of Alkali Behavior of BF Slag System in Low-Slag Ironmaking

Lei Xu [1], Gele Qing [2,*], Xiangfeng Cheng [2], Meng Xu [2], Baojun Zhao [1,3] and Jinfa Liao [1,*]

[1] International Institute for Innovation, Jiangxi University of Science and Technology, Ganzhou 341000, China
[2] Shougang Research Institute of Technology, Beijing 100043, China
[3] Sustainable Minerals Institute, University of Queensland, Brisbane 4072, Australia
*   Correspondence: qinggele_68@163.com (G.Q.); liaojinfa678@outlook.com (J.L.)

Abstract: The increased utilization of pellets in blast furnaces is one of the directions for low-carbon ironmaking. As a result, the low slag rate may affect the desulfurization of the hot metal and the removal of alkali in the blast furnace. Effective desulfurization and the removal of alkali in the low slag ironmaking process have become the focus of the steel industry. In this paper, the effects of slag quantity, temperature, reaction time and slag composition on the desulfurization and removal of alkali were studied using the slag-metal reaction method. It was found that the slag quantity had the same influence trend on the desulfurization and the removal of alkali. The greater the slag quantity, the more effective the desulfurization and the removal of alkali. The slag composition, temperature and reaction time had the opposite effect on the desulfurization and the removal of alkali. High temperature, long reaction time, high MgO concentration, high CaO/SiO_2 ratio and low Al_2O_3 concentration increased the desulfurization of hot metal but reduced the removal rate of alkali from the blast furnace. Applications of the experimental results on high-proportion pellet blast furnace operation are discussed.

Keywords: blast furnace; high-proportion pellet; desulphurization; removal of alkali; ironmaking

Citation: Xu, L.; Qing, G.; Cheng, X.; Xu, M.; Zhao, B.; Liao, J. Experimental Study on Desulfurization and Removal of Alkali Behavior of BF Slag System in Low-Slag Ironmaking. *Metals* **2023**, *13*, 414. https://doi.org/10.3390/met13020414

Academic Editors: Chenguang Bai and Alexios P. Douvalis

Received: 28 December 2022
Revised: 2 February 2023
Accepted: 13 February 2023
Published: 16 February 2023

1. Introduction

The steel industry accounts for approximately 11% of the global carbon dioxide (CO_2) emissions in the world [1]. Under the current policy and technology regime, the energy use and global greenhouse gas emissions of the steel industry are likely to continue increasing because of the increased demand for steel, particularly in developing countries [2]. A blast furnace (BF) is consistently the major technology for ironmaking in which sinter, pellet and lump are the iron-containing feeds. Increasing the proportion of high-grade pellets in the blast furnace is an effective method for carbon emission reduction [3,4]. Compared with sinter production, pellets can be produced with lower energy consumption and lower environmental pollution. In addition, a pellet contains a high concentration of iron which reduces the carbon consumption and CO_2 emission in a blast furnace due to the low slag rate [5–8]. Iron-containing and carbon-containing raw materials are composed of sulfur and alkalis that can affect the hot metal quality and the blast furnace operation. The hot metal from a blast furnace is mainly used for steelmaking where sulfur is a harmful element that affects the quality of the steel products. Alkali oxides are reduced to the metals by the hot carbon and vaporized to the upper of the blast furnace where they are oxidized and move down with the fresh feeds. Accumulation of the alkalis in a blast furnace can damage the coke and the refractories. Most of the sulfur and alkalis in a blast furnace can only be removed by the slag. The desulfurization of hot metal and the removal of alkali from the blast furnace are important roles of the slag. Reduced slag volume could affect the effective desulfurization and the removal of alkali in the blast furnace operation.

The desulfurization and the removal of alkali in the blast furnace have been investigated extensively [9–16]. However, the previous studies have certain limitations: (1) the desulfurization and the removal of alkali experiments were conducted separately which is different from a real blast furnace operation; (2) the slag system studied was focused on high-Al_2O_3 [9–13] or low-MgO [14,15]; (3) none of these studies considered the effect of the slag volume on the desulfurization and the removal of alkali [16,17]. As such, there is no direct theoretical guidance for high pellet and low slag BF operation.

In this study, the effects of the slag quantity, the temperature, the reaction time, Al_2O_3, MgO and CaO/SiO_2 on the desulfurization and the removal of alkali were studied using the slag-metal reaction method. This study aims to provide systematic guidance to support high pellet ratio blast furnace operations.

2. Experimental Section

2.1. Preparation of Sulfur-Containing Iron

The sulfur-containing iron was prepared in a vacuum induction melting furnace. Five Kg pig iron from a blast furnace and a certain amount of FeS (analytically pure) were used as starting materials and placed in a magnesia-alumina spinel crucible. The furnace was flashed three times with high-purity argon (99.999%). The furnace was vacuumized to 10^{-2} Pa before each flashing. The argon was filled to make the furnace slightly negative pressure and subsequently the power was turned on to start heating. The sample was cast after melting for 10 min. The S and C contents in the sulfur-containing iron from different locations were analyzed by a carbon-sulfur analyzer.

2.2. Preparation of Alkali-Containing Slag

The preparation of alkali-containing slag included three setups: sample preparation, high-temperature melting and sample examination.

High-purity reagent powders of Al_2O_3, SiO_2, MgO, $CaCO_3$, Na_2CO_3 and K_2CO_3 were used as starting materials. The oxide powders were mixed at the required ratios in an agate mortar to prepare the samples. Each mixture, approximately 15 g, was pelletized and placed in a graphite crucible.

The mixture in a graphite crucible was heated in a vertical tube furnace as shown in Figure 1. The detailed experiment procedure is given below. Firstly, the sample was introduced and kept at the bottom of the reaction tube and the furnace was properly sealed. Secondly, the reaction tube was flashed with argon gas for 30 min to remove the air. Thirdly, the sample was raised and kept in the hot zone of the furnace. The distance between the graphite crucible and the thermocouple was controlled within 5 mm to ensure the accurate temperature measurement. Fourthly, the furnace was heated to 1500 °C to melt the sample. Fifthly, after 30 min at 1500 °C, the Mo wire was pulled up and the sample was quenched directly into the cooling water. The rapid quenching of the silicate slags to room temperature converted the liquid phase to homogenous glass.

The quenched slag sample was dried on a hot plate and ground in an agate mortar to prepare the samples. The composition of the slag was analyzed by inductively coupled plasma spectrometry (ICP).

2.3. Experiments for Desulfurization and Removal of Alkali

The slag-metal reaction technique was employed to study the desulfurization and removal of alkali by the slag. The schematic diagram of the experimental set-up is shown in Figure 1. The prepared sulfur-containing iron and alkali-containing slag were put into a graphite crucible according to the required weight ratios. The procedure for slag-metal experiments was the same as the preparation of alkali-containing slag. After quenching, the sample was dried on a hot plate, and the slag and the hot metal were separated and analyzed by ICP and carbon-sulfur analysis, respectively. Following the discussions with the blast furnace operators, the slag to metal ratio, the temperature, the reaction time and the slag composition were the variables in the BF operation. The conditions of all slag-metal reaction experiments to cover the range of the industrial variables are summarized in Table 1.

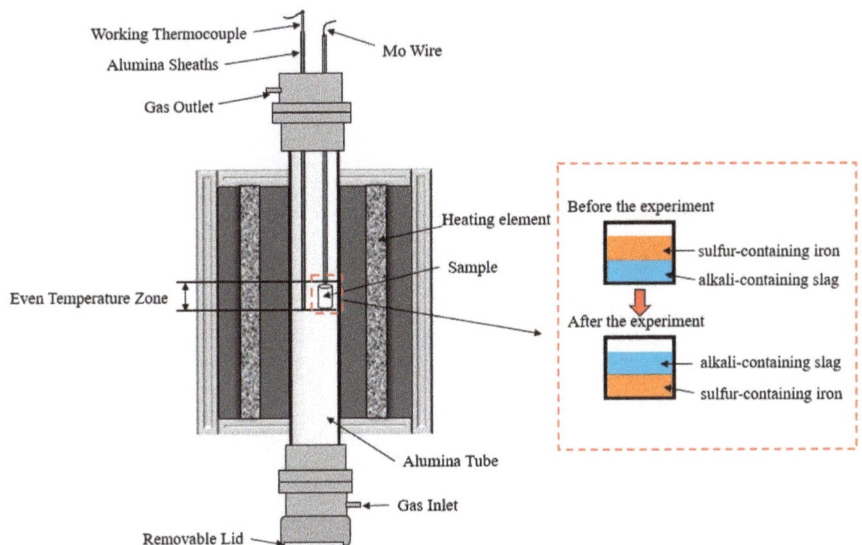

Figure 1. The schematic diagram of the experimental method desulfurization and removal of alkali.

Table 1. Experimental conditions for slag-metal reaction experiments.

Exp No	Metal (g)	Slag (g)	Al_2O_3 (%)	MgO (%)	CaO/SiO_2	Temp (°C)	Time (min)
1	20	4.4	15.6	8.3	1.2	1500	30
2	20	4.0	15.6	8.3	1.2	1500	30
3	20	3.8	15.6	8.3	1.2	1500	30
4	20	3.6	15.6	8.3	1.2	1500	30
5	20	3.2	15.6	8.3	1.2	1500	30
6	20	3.0	15.6	8.3	1.2	1500	30
7	20	4.0	15.6	8.3	1.2	1475	30
8	20	4.0	15.6	8.3	1.2	1525	30
9	20	4.0	15.6	8.3	1.2	1500	15
10	20	4.0	15.6	8.3	1.2	1500	60
11	20	4.0	13.0	8.3	1.2	1500	30
12	20	4.0	18.0	8.3	1.2	1500	30
13	20	4.0	15.6	6.0	1.2	1500	30
14	20	4.0	15.6	11.0	1.2	1500	30
15	20	4.0	15.6	8.3	1.0	1500	30
16	20	4.0	15.6	8.3	1.1	1500	30
17	20	4.0	15.6	8.3	1.3	1500	30

In this study, the desulfurization is directly expressed as sulfur in hot metal. The removal rate of alkali is defined as $R = M_1/M_2$, where M_1 is the weight of Na_2O or K_2O remaining in the slag after high temperature experiments, M_2 is the weight of Na_2O or K_2O in original alkali-containing slag.

2.4. Thermodynamic Calculation

The CALPHAD (CALculation of PHAse Diagram) approach has been developed to simulate high-temperature processes [18,19]. FactSage is one of the most successful thermodynamic models in predicting [20] the equilibria of slag and metal systems. Experimental data are compared with the predictions of FactSage 8.2. "FactPS", "FToxid" and "FTmisc" databases were selected under the "Equilib" module. The solution species selected in the calculations include "FToxid-SLAGA", and "FTmisc-FeLQ".

3. Results and Discussion

High-sulfur iron and high-alkali slag were used as the starting materials for high-temperature experiments of desulfurization and the removal of alkali. The compositions of the sulfur-containing iron and alkali-containing slag are shown in Tables 2 and 3, respectively. It can be seen from the tables that the prepared iron had uniform composition, with a C content of 4.25 wt% and an S content of 0.4 wt%. The slag contained 2.33 wt% alkali (K_2O + Na_2O).

Table 2. C and S contents in the sulfur-containing iron analyzed by a carbon-sulfur analyzer.

Positions	C (wt%)	S (wt%)
Upper part of ingot	4.20	0.39
Lower part of ingot	4.27	0.40

Table 3. Composition of the alkali-containing slag analyzed by ICP.

Na_2O	K_2O	SiO_2	CaO	MgO	Al_2O_3
1.14 wt%	1.19 wt%	34.3 wt%	40.28 wt%	8.16 wt%	15.76 wt%

The metal and slag shown in Tables 1 and 3 were used as a base to evaluate the effects of the slag quantity, the temperature and the reaction time on the desulfurization and the removal of alkali. When the effect of slag composition was discussed, the alkali-containing slag was prepared separately to vary Al_2O_3, MgO and CaO/SiO_2 in the slag. The advantage of the research technique used in the present study is that better simulation conditions occurred in a BF hearth where the hot metal reacted with the slag for the desulfurization and the removal of alkali. The reactions can be expressed as

$$[Fe + S] + (slag) \rightarrow [Fe] + (slag + S) \tag{1}$$

$$[Fe + C] + (slag + Na_2O + K_2O) \rightarrow [Fe] + (slag + Na_2O + K_2O) + Na\uparrow + K\uparrow \tag{2}$$

As a result of the metal-slag reaction, sulfur was transferred into the slag and part of the alkali in the slag was reduced by the carbon in the hot metal and vaporized before the metal and slag were tapped out of the furnace. The effects of different parameters on the desulfurization and the removal of alkali are discussed in the following sections.

3.1. Effect of Slag Quantity

The slag quantity in a blast furnace decreased with an increased proportion of high-grade pellets. Figure 2 shows the effect of the slag quantity on the desulfurization and the removal of alkali at 1500 °C. It can be seen from Figure 2a that sulfur in the hot metal decreased with increasing slag quantity. The sulfur in the hot metal mainly reacted with Ca^{2+} and Mg^{2+} to enter the slag. With the increase of the slag-to-metal ratio, a greater amount of Ca^{2+} and Mg^{2+} in the slag were available to react with S in the iron. Under the experimental conditions, when the slag-to-metal ratio was more than 170 kg/ton, the content of the S in the iron was less than 0.05 wt% which is the limit of sulfur for steelmaking.

It can be seen from Figure 2b that with the decrease in slag-to-metal ratio, the contents of alkali in the slag also decreased. The effect of slag quantity seemed more sensitive to Na removal from the slag. On the other hand, it was easier to remove K from the slag because it is easier to reduce K_2O by carbon to form gaseous K. It was therefore more difficult to remove potassium from the blast furnace through slag. The recycling of potassium in the blast furnace was more severe than sodium. Both the desulfurization and the removal of alkali relied on slag. It was expected that low slag volume would influence the desulfurization of hot metal and the removal of alkali from a BF. Figure 2 gives a quantitative estimation of the effect for the industry to adjust the operating parameters accordingly.

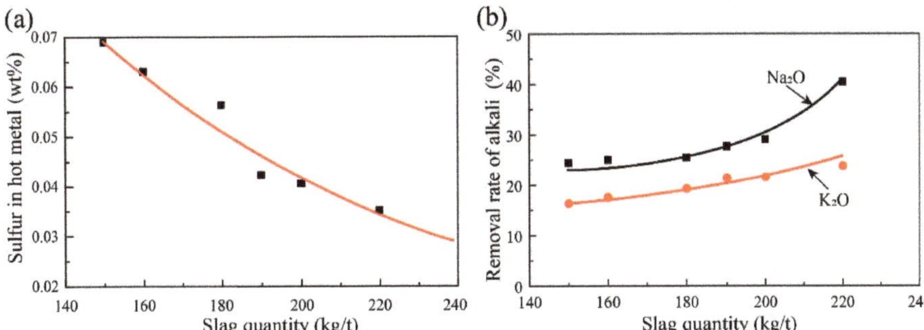

Figure 2. The effect of the slag quantity on the desulfurization and the removal of alkali behavior. (**a**) Desulfurization; (**b**) the removal of alkali.

3.2. Effect of Temperature

Figure 3 shows the effect of temperature on the desulfurization and the removal of alkali under a fixed slag-to-metal ratio of 200 kg/t. It can be seen from Figure 3a that the content of S in the hot metal decreased slightly when the temperature was increased from 1475 to 1525 °C which means that the desulfurization was not sensitive to temperature within the operating range. However, it can be seen from Figure 3b that the removal rate of alkali decreased significantly with increasing temperature. This can be explained by the fact that reaction (2) was much faster at a higher temperature. For a given BF, the slag and the metal temperature in the hearth was relatively stable which provided a hot metal product with stable sulfur. Removal of the alkali from the BF could be operated periodically. Figure 3 shows that a low temperature is beneficial to remove the alkali without a significant increase of sulfur in the hot metal.

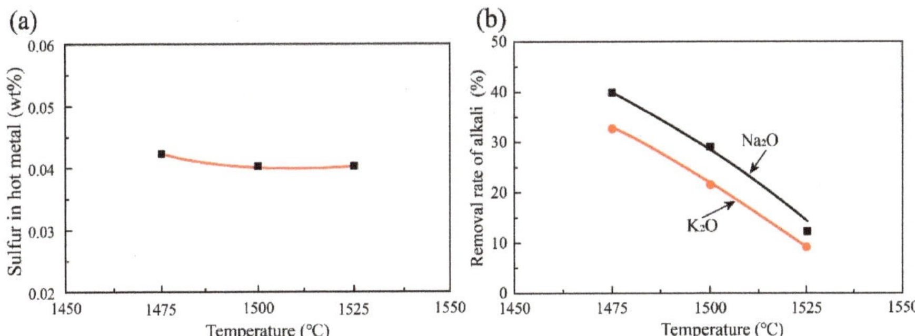

Figure 3. Effect of temperature on the desulfurization and the removal of alkali at slag-to-metal ratio of 200 kg/t (**a**) Desulfurization; (**b**) the removal of alkali.

3.3. Effect of Reaction Time

The effect of the reaction time on the desulfurization and the removal rate of alkali at a fixed slag quantity of 200 kg/t is shown in Figure 4. It can be seen that a longer reaction time between the slag and metal decreased the sulfur content in the hot metal and the alkali in the slag. Specifically, a better desulfurization was achieved by a longer reaction time. However, with the continuous reaction, the alkali in the slag was reduced and volatilized. It can be seen from the figure that the reaction between the slag and metal did not reach equilibrium in 60 min. The reaction time could be adjusted by the tapping period in a BF operation. A short tapping period of the slag and the metal is favorable for keeping the alkali in the slag.

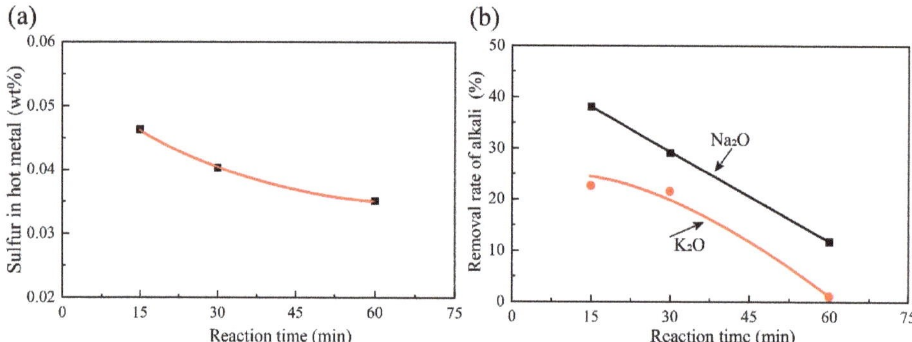

Figure 4. Effect of reaction time on the desulfurization and the removal of alkali at slag-to-metal ratio of 200 kg/t. (**a**) Desulfurization; (**b**) the removal of alkali.

3.4. Effect of MgO

Figure 5 shows the effect of the MgO content on the desulfurization and the removal of alkali at a fixed slag quantity of 200 kg/t. The experimental results showed that high MgO can decrease the sulfur content in the hot metal. When the MgO content was higher than 7 wt%, S in hot metal was lower than 0.05%. As a basic ion, Mg^{2+} tends to react with the acidic sulfur in the hot metal. Conversely, high Mg^{2+} concentration in the slag increased the activities of the alkali such that they could be easily reduced and removed from the slag. Figure 5b shows that alkali in the slag decreased with increasing MgO concentration in the slag. MgO is usually added as a flux in the BF operation. A lower addition of MgO can decrease the slag volume and also increase the removal of alkali from the BF.

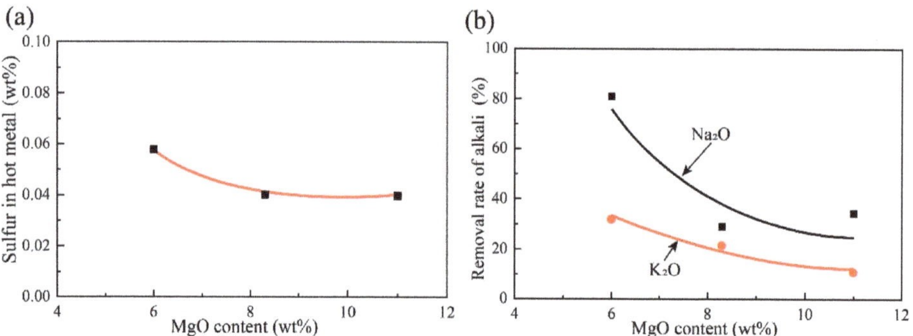

Figure 5. Effect of MgO content on the desulfurization and the removal of alkali at slag-to-metal ratio of 200 kg/t (**a**) Desulfurization; (**b**) the removal of alkali.

3.5. Effect of Al_2O_3

Figure 6 shows the effect of the Al_2O_3 content in the slag on the desulfurization and the removal of alkali at a fixed slag quantity of 200 kg/t. Al_2O_3 had an opposite influence compared to MgO on the desulfurization of the hot metal and the removal of alkali from the BF. High Al_2O_3 in the slag increased the content of S in the hot metal and the alkali contents in the slag. This indicated that Al_2O_3 behaved as a weak acid oxide in the BF slag which tended to keep K_2O and Na_2O in the slag. Al_2O_3 usually comes from burden materials which increases the liquidus temperature and viscosity of the BF slag. The removal of alkali from the BF appeared to be the advantage of using high- Al_2O_3 feeds.

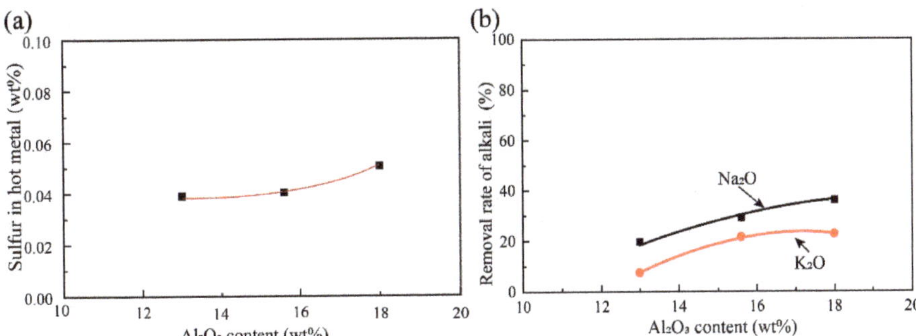

Figure 6. Effect of Al$_2$O$_3$ content on the desulfurization and the removal of alkali at slag-to-metal ratio of 200 kg/t (**a**) Desulfurization; (**b**) the removal of alkali.

3.6. Effect of Binary Basicity

Figure 7 shows the effect of the CaO/SiO$_2$ ratio on the desulfurization of the hot metal at a fixed slag quantity of 200 kg/t. Thermodynamic prediction by FactSage 8.2 is shown in Figure 7a for comparison. It can be seen that both the prediction and the experimental results showed the same trend that the sulfur in hot metal decreased with increasing CaO/SiO$_2$ in the slag. However, the experimental results show a more significant effect of binary basicity on the desulfurization than the predictions. Figure 7b indicates that with the CaO/SiO$_2$ ratio increased from 1.0 to 1.3, and the sulfur in the hot metal decreased from 0.1 to 0.038 wt%. However, the predicted sulfur in the hot metal only decreased from 0.06 to 0.025 wt% when the CaO/SiO$_2$ ratio increased from 1.0 to 1.3.

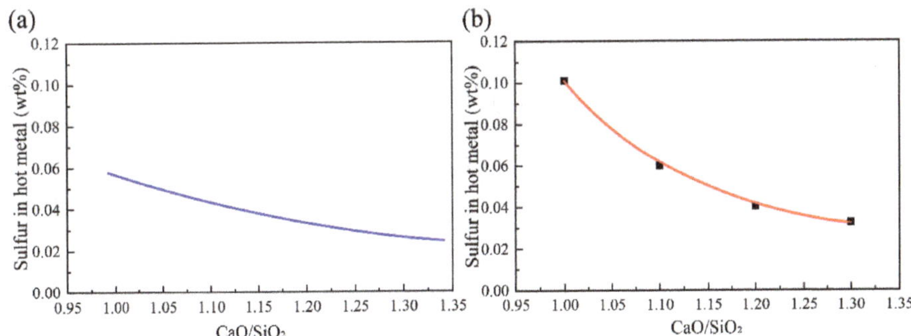

Figure 7. The effect of the CaO/SiO$_2$ on the desulfurization at slag-to-metal ratio of 200 kg/t, (**a**) FactSage calculation, (**b**) experimental results.

Figure 8 presents the effect of the CaO/SiO$_2$ ratio on the removal rate of alkali from a blast furnace. The experimental results are compared with the predictions by FactSage 8.2. As can be seen from Figure 8b, the removal rate of alkali decreased significantly with increasing CaO/SiO$_2$ ratio in the slag. This is understandable because Ca^{2+} is a strong basic ion which increased the activities of the alkali in the slag. As a result, the alkali could be easily reduced and removed from the slag. However, it can be seen from Figure 8a that the removal rate of alkali increased significantly with increasing CaO/SiO$_2$ ratio in the slag according to the FactSage predictions. Although the FactSage software has been developed rapidly and successfully applied to the high-temperature process, there was still a difference between the experimental results and the predictions, indicating that the thermodynamic database needed to be further optimised.

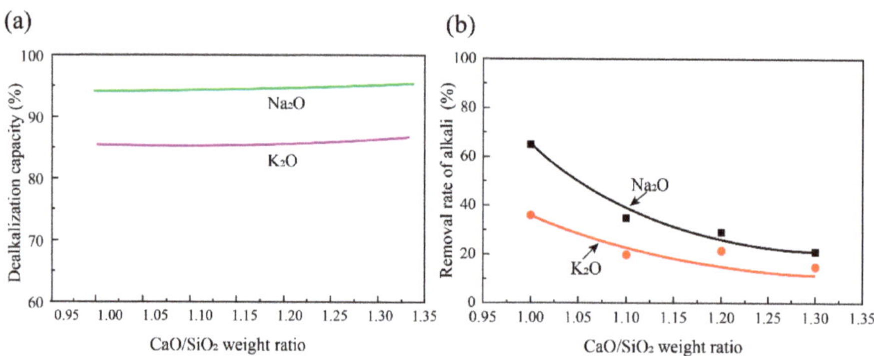

Figure 8. Effect of the CaO/SiO$_2$ ratio on the removal of alkali at slag-to-metal ratio of 200 kg/t, (**a**) FactSage calculation, (**b**) experimental results.

3.7. Applications

The desulfurization of the hot metal and the removal of alkali from the blast furnace were related not only to the metal-slag reaction but also to the sulfur and alkali loads to the blast furnace. The experimental technique used in the present study closely simulated the blast furnace conditions by considering the reaction between the sulfur-containing metal and the alkali-containing slag. However, it was shown in the present study that the metal-slag reaction under the laboratory conditions did not reach equilibrium. The sulphide capacity or desulfurization ability was usually reported to evaluate the ability of a BF slag to absorb sulphur which was an equilibrium property [9–15]. Desulfurization ability is the ratio of the sulfur-in-slag to the sulfur-in-metal after the equilibrium reaction. It was reported that desulfurization ability of the synthetic slag was in the range of 15–30 at 1500 °C [9]. The desulfurization ability increased with increasing CaO/SiO$_2$ and MgO/Al$_2$O$_3$. Under the same conditions, the ratios of the sulfur-in-slag to the sulfur-in-metal were calculated to be 5–11 which was lower than that reported in the literature [9]. However, it can be seen from the above discussion that the same trends were observed in the present study for the effects of CaO/SiO$_2$ and MgO/Al$_2$O$_3$ on desulfurization. Yang et al. [16] reported that Na$_2$O or K$_2$O concentration in the slag decreased with increasing temperature and CaO/SiO$_2$. The same trends were reported in the present study but the rates of the decrement were different. These studies did not use the sulfur-containing iron and the alkali-containing slag simultaneously in the high-temperature experiments which is different from the blast furnace conditions. A large-scale reaction in the blast furnace is significantly different from that in the laboratory condition. It is more appropriate to refer to the trends reported in the present study in regard to blast furnace operation.

The effects of the variables on the desulfurization and the removal of alkali are summarized in Table 4. Only the slag quantity had the same effect on the desulfurization and the removal of alkali. Low slag quantity decreased both the desulfurization and the removal rate of alkali. Al$_2$O$_3$ in the slag had a negative effect on the desulfurization and a positive effect on the removal of alkali. The other factors, including the temperature, the reaction time, MgO and CaO/SiO$_2$, had a positive effect on the desulfurization and a negative effect on the removing of alkalis.

The results from the present study showed that it was impossible to increase the desulfurization of the hot metal and the removal of alkali with a simple variation of the operating parameters. Under the low slag operation of a blast furnace, good desulfurization could not be attained simultaneously with the high removal of alkali. The quality of the hot metal was a continuous control index. Desulfurization therefore required daily operation to keep the quality of the hot metal for steelmaking. A high temperature, long reaction time, and high MgO and CaO/SiO$_2$ were favorable for desulfurization. The damage of alkali to the blast furnace resulted from the accumulation by recirculation. The removal of alkali from the blast furnace could be carried out at regular intervals. Within a certain period, low-temperature operation, short reaction time, and low MgO and CaO/SiO_2 slag could remove the alkali from the blast furnace. During the period of alkali removal, the sulfur content in the raw materials should be controlled to a low level to obtain satisfactory hot metal.

Table 4. Summary of the desulfurization and the removal of alkali by different parameters, + indicates positive effect, − indicates negative effect.

Variables	Desulfurization	Removal of Alkali
Slag quantity	+	+
Temperature	+	−
Reaction time	+	−
CaO/SiO_2	+	−
Al_2O_3	−	+
MgO	+	−

4. Conclusions

An improved slag-metal reaction technique was developed to investigate the desulfurization and the removal of alkali under iron blast furnace conditions. The experimental results showed that the desulfurization and the removal rate of alkali decreased with decreasing slag quantity. Al_2O_3 in the slag was helpful for the removal of alkali but reduced desulfurization. The temperature, the reaction time, MgO and CaO/SiO_2 in the slag increased the desulfurization but influenced the removal of alkali. The change of a single operating parameter did not achieve high desulfurization and the removal rate of alkali simultaneously. It is proposed that the alkali should be removed from the blast furnace periodically with low loads of sulfur and alkali.

Author Contributions: Methodology, B.Z.; Software, L.X. and J.L.; Formal analysis, G.Q., B.Z. and J.L.; Investigation, B.Z., L.X. and J.L.; Resources, G.Q. and M.X.; Data curation, L.X., G.Q., X.C. and M.X.; Writing—original draft, L.X. and J.L.; Writing—review & editing, B.Z. and J.L.; Supervision, B.Z.; Project administration, B.Z.; Funding acquisition, G.Q., X.C. and M.X. All authors have read and agreed to the published version of the manuscript.

Funding: This research received no external funding.

Data Availability Statement: Not applicable.

Acknowledgments: The authors would like to acknowledge Shougang Research Institute of Technology for providing the financial support and samples for this research.

Conflicts of Interest: The authors declare no conflict of interest.

References

1. Hasanbeigi, A. Steel climate impact: An international benchmarking of energy and CO_2 Intensities. *Rep. Glob. Effic. Intell.* **2022**, *624*, c53a07. Available online: https://www.globalefficiencyintel.com/steel-climate-impact-international-benchmarking-energy-co2-intensities (accessed on 7 April 2022).
2. Geerdes, M.; Chaigneau, R.; Lingiardi, O. *Modern Blast Furnace Ironmaking: An Introduction, Chapter XI-Operation Challenges*, 4th ed.; Ios Press: Amsterdam, The Netherland, 2020; pp. 197–226. ISBN 978-1-64368-122-8.
3. Agrawal, A. Blast furnace performance under varying pellet proportion. *Trans. Indian Inst. Met.* **2019**, *72*, 777–787. [CrossRef]
4. Bai, K.; Liu, L.; Pan, Y. A review: Research progress of flux pellets and their application in China. *Ironmak. Steelmak.* **2021**, *48*, 1048–1063. [CrossRef]
5. Zhao, Z.; Saxén, H.; Liu, Y.; She, X.; Xue, Q. Numerical study on the influence of pellet proportion on burden distribution in blast furnace. *Ironmak. Steelmak.* **2022**, *49*, 1–8. [CrossRef]
6. Yang, X.; Zhang, D.; Liu, K.; Jing, F. Substituted pelleting for sintering —An important approach of energy-saving, low-carbon and emission reduction before ironmaking. *J. Eng. Stud.* **2017**, *9*, 44–52. (In Chinese) [CrossRef]
7. Zhang, F.; Wang, Q.; Han, Z. Innovation and application on pelletizing technology of large traveling grate induration machine. In Proceedings of the AISTech 2015 Iron & Steel Technology Conference, Cleveland, OH, USA, 4–7 May 2015.

8. Zhang, F.; Zhang, W.; Qing, G.; Wang, Q.; Han, Z. Design and application on pelletizing technologies and equipments in large-scale belt type roasting machine. *Sinter. Pelletizing* **2021**, *46*, 1–10. (In Chinese)

9. Guo, Y.; Shen, F.; Zheng, H.; Wang, S.; Jiang, X.; Gao, Q. Desulfurization ability of blast furnace slag containing high Al_2O_3 at 1773 K. *Crystals* **2021**, *11*, 910. [CrossRef]

10. Sunahara, K.; Nakano, K.; Hoshi, M.; Inada, T.; Komatsu, S.; Yamamoto, T. Effect of high Al_2O_3 slag on the blast furnace operations. *ISIJ Int.* **2008**, *48*, 420–429. [CrossRef]

11. Hu, X.; Lin, L.; Li, L. Laboratory research on property of higher Al_2O_3 content slag. *China Metall.* **2006**, *16*, 36–41. (In Chinese)

12. Wang, P.; Meng, Q.; Long, H.; Li, J. Influence of basicity and MgO on fluidity and desulfurization ability of high aluminum slag. *High Temp. Mater. Process.* **2016**, *35*, 669–675. [CrossRef]

13. Zhang, J.; Lv, X.; Yan, Z.; Qin, Y.; Bai, C. Desulphurization ability of blast furnace slag containing high Al_2O_3 and 5 Mass% TiO_2 at 1773 K. *Ironmak. Steelmak.* **2016**, *43*, 378–384. [CrossRef]

14. Ma, X.; Chen, M.; Xu, H.; Zhu, J.; Wang, G.; Zhao, B. Sulphide capacity of CaO–SiO_2–Al_2O_3–MgO system relevant to low MgO blast furnace slags. *ISIJ Int.* **2016**, *12*, 2126–2131. [CrossRef]

15. Ma, X.; Chen, M.; Xu, H.; Zhu, J.; Wang, G.; Zhao, B. Properties of low-MgO ironmaking blast furnace slags. *ISIJ Int.* **2018**, *8*, 1402–1405. [CrossRef]

16. Yang, Y.; McLean, A.; Sommerville, I.; Poveromo, J. The Correlation of alkali capacity with optical basicity of blast furnace slags. *Iron Steelmak.* **2000**, *27*, 103–111.

17. Shi, C.; Yang, X.; Jiao, J.; Li, C.; Guo, H. A sulphide capacity prediction model of CaO–SiO_2–MgO–Al_2O_3 ironmaking slags based on the ion and molecule coexistence theory. *ISIJ Int.* **2010**, *50*, 1362–1372. [CrossRef]

18. Hack, K. Computational thermodynamics: A mature scientific tool for industry and academia. *Pure Appl. Chem.* **2011**, *83*, 1031–1044. [CrossRef]

19. Lukas, H.; Fries, S.G.; Sundman, B. *Computation Thermodynamics the Calphad Method*; Cambridge University Press: Cambridge, UK, 2007; p. 324.

20. Bale, C.W.; Bélisle, E.P.; Chartrand, S.A.; Decterov, G.; Eriksson, A.E.; Gheribi, K.; Hack, I.H.; Kang, Y.B.; Melançon, J.; Pelton, A.D.; et al. Reprint of: FactSage thermochemical software and databases, 2010–2016. *Calphad* **2016**, *55*, 1–19. [CrossRef]

 *metals*

Article

Predictive Modeling of Blast Furnace Gas Utilization Rate Using Different Data Pre-Processing Methods

Dewen Jiang [1], Zhenyang Wang [1,*], Kejiang Li [1], Jianliang Zhang [1,2], Le Ju [3] and Liangyuan Hao [4]

[1] School of Metallurgical and Ecological Engineering, University of Science and Technology Beijing, Beijing 100083, China; csujdw@163.com (D.J.); likejiang@ustb.edu.cn (K.L.); zhang.jianliang@hotmail.com (J.Z.)

[2] School of Chemical Engineering, The University of Queensland, St. Lucia, QLD 4072, Australia

[3] School of Mathematics, Rose-Hulman Institute of Technology, Terre Haute, IN 47803, USA; 6363899jdw@gmail.com

[4] He Steel Group Co., Ltd., Shijiazhuang 050024, China; d202110163@xs.ustb.edu.cn

* Correspondence: wangzhenyang@ustb.edu.cn

Abstract: The gas utilization rate (GUR) is an important indicator parameter for reflecting the energy consumption and smooth operation of a blast furnace (BF). In this study, the original data of a BF are pre-processed by two methods, i.e., box plot and 3σ criterion, and two data sets are obtained. Then, support vector regression (SVR) is used to construct a prediction model based on the two data sets, respectively. The state parameters of a BF are selected as input parameters of the model. Gas utilization after one hour (GUR-1h), two hours (GUR-2h), and three hours (GUR-3h) are selected as output parameters, respectively. The simulation result demonstrates that using the 3σ criterion to pre-process the raw data leads to better prediction of the model compared to using the box plot. Moreover, the model has the best predictive effect when the output parameter is selected as GUR-1h.

Keywords: blast furnace; data pre-processing; extreme outlier; gas utilization rate; support vector regression

Citation: Jiang, D.; Wang, Z.; Li, K.; Zhang, J.; Ju, L.; Hao, L. Predictive Modeling of Blast Furnace Gas Utilization Rate Using Different Data Pre-Processing Methods. *Metals* **2022**, *12*, 535. https://doi.org/10.3390/met12040535

Academic Editor: Pasquale Cavaliere

Received: 17 February 2022
Accepted: 19 March 2022
Published: 22 March 2022

Publisher's Note: MDPI stays neutral with regard to jurisdictional claims in published maps and institutional affiliations.

1. Introduction

At present, the process of blast furnace (BF) ironmaking is mainly that in which iron ore and coke are fed into a BF from the top in a certain proportion, and then pig iron is smelted. After the iron ore and coke flow through a distribution device, reactions take place inside the BF. Meanwhile, hot air and pulverized coal of the tuyeres is blown into the BF to promote the chemical reaction inside the BF to form an upward airflow [1]; iron ore reacts with carbon monoxide under high-temperature and high-pressure conditions inside the BF to yield products such as molten iron, slag, and gas [2]. The molten iron flows out from the bottom of the BF and is pretreated and sent to the steel plant. BF gas is collected at the top of the furnace and can be recycled. As one of the main products of BF ironmaking, BF gas carries a huge amount of thermal and chemical energy, which can provide heat for the chemical reactions inside the BF and facilitate the reactions inside the BF.

The main components of BF gas are carbon monoxide, carbon dioxide, nitrogen, and hydrogen. In the field of metallurgy, the gas utilization rate (GUR) of BF is defined as the ratio of the carbon dioxide content to the total content of carbon monoxide and carbon dioxide. With the rapid development of industry, steel companies are making efforts to improve the gas utilization rate. On the one hand, more and more countries are paying attention to emission reduction and energy conservation [3]. On the other hand, GUR reflects the reduction and utilization of the main raw materials for BF production. It represents the level of BF energy consumption, the rationality of the gas flow distribution, and the smelting state of a BF [4,5]. Most importantly, it is an important index for reducing consumption, evaluating the quality of pig iron, and increasing the production of a BF [6].

However, a BF acts as a huge black box for ironmaking reactions with a time lag, dynamics, and complexity for the production of pig iron in the modern metallurgical industry [7]. These characteristics of the BF have caused the BF operators to be unable to grasp the information of the GUR, gas distribution, and other state parameters in time. Faced with the above realistic conditions, scholars have predicted and studied BF parameters such as GUR and gas distribution based on the following three methods. ① A geometric model based on mechanism analysis (conventional solution theory and metallurgical theory). For example, Meng investigated the relationship between the temperature in the hot preparation zone and the utilization rate of blast furnace gas, and performed thermodynamic and kinetic analyses of the reduction reaction in the thermal reserve zone using standard Gibbs free energy calculations and unreacted shrinkage core models, respectively [8]. ② A computational simulation method of simulation software. For instance, a computational fluid dynamics numerical simulation model of the cooling stave of BF based on a one-dimensional heat transfer mechanism was established. The model could reduce the influence of frequent forming and shedding of the slag crusts on the BF by analyzing the effects of the cooling stave material, the volume distribution of the cooling water pipes, and the nano-polymer in the cooling stave [9]; through mathematical modeling and energy exchange analysis, Guo conducted a numerical simulation to analyze the effect of natural gas injected through the tuyere on GUR [10]. Shen developed a three-dimensional CFX-based mathematical model, which predicts the in-furnace distributions of key performance indicators such as the gas utilization rate [11]. ③ A regression model on the basis of data-driven methods and machine learning. In the past few years, some new technologies such as machine learning, deep learning, etc., have been used in the field of BF ironmaking. Under the support of these technologies, some regression models were established and used to predict and improve GUR, BF state parameters, and production indicators. For example, the influence of BF top pressure on improving the gas flow distribution is obtained based on the fuzzy theory [12]. A strategy for burden distribution was constructed for improving the GUR [13]. An online sequential extreme learning model was proposed to predict GUR [14]. Prof. Wu used BF operating parameters to predict BF GUR [15]. An proposed a multi-time-scale fusion method to predict the gas utilization rate of a blast furnace [4]. Shi presented a method for recognizing the distribution features of the blast furnace gas flow center based on infrared image processing [16]. Jiang proposed a model based on the multi-layer perceptron to predict the gas utilization rate after 1, 2, and 3 h, respectively [17]. Zhang presented a model based on a TS fuzzy neural network and the particle swarm algorithm for predicting the gas utilization rate [18].

From the above analysis, the first and second methods have contributed to GUR prediction and optimization. However, the calculation and formulation of these two methods are based on some assumptions. Their boundary conditions are difficult to be determined and these operating parameters have a certain degree of lag and measurement error. Furthermore, the calculation is complicated. In recent years, with the advancement of sensors and detectors, a large amount of production data has been collected by steel plants. GUR and other parameters can be more accurately predicted and provide more reliable guidance for BF using data-driven and machine learning technologies.

When a model is built using data-driven methods, the reliability of data must be guaranteed. However, BF is a huge reactor in ironmaking, and the limitations of measurements imposed by the adverse operating conditions (high temperature and pressure) result in missing values and outliers in the collected data. Data pre-processing is essential as an important step in the process of building a model to ensure consistency and accuracy. It is worth noting that the hysteresis in the BF ironmaking process must be considered and few scholars have performed a comparative study on the prediction of blast furnace-related parameters by different data pre-processing methods. Meanwhile, few scholars have previously focused on the duration of the effect of the current BF condition on the GUR. Therefore, in this study, two data pre-processing methods were used to build two prediction models based on support vector regression (SVR) for forecasting GUR after one hour (GUR-1h), GUR after two hours (GUR-2h), and GUR after three hours (GUR-3h). Furthermore, the impact of the two data pre-processing methods on the prediction was analyzed, which is a fundamental and important step for improving the operating level and energy utilization of the BF.

The rest of this article is organized as follows. The original data are analyzed and pre-processed by two methods, the box plot and 3σ criterion, in Section 2. Section 3 calculates the correlation of each feature, obtains the best input parameters, and describes the algorithm used in this article. Section 4 shows the prediction results of the models based on two different data sets. Section 5 compares and evaluates the prediction effects of the models. Eventually, the conclusions are summarized in Section 6.

2. Pre-Processing of Raw Data

The data involved in this paper are from a BF in China with a working space of 4150 m^3. The 35,198 sets of data were collected from the BF and the sampling interval was 1 h. The collected parameters of each data sample are shown in Table 1. Because these parameters are general expressions in BF ironmaking and have been described in many papers, the specific meaning of each parameter will not be repeated here [17–21].

Table 1. The related parameters involved in the research.

Actual Physical Meaning	Parameter Name	Unit
Gas utilization rate	GUR	%
Blast volume	BV	Nm^3/min
Wind pressure	WP	kpa
Top pressure	TP	kpa
Pressure differential	PD	kpa
Air permeability resistance coefficient	ARC	-
Wind temperature	WT	°C
Ventilating index	VI	$\text{m}^3/\text{min·kpa}$
Oxygen content	OC	Nm^3/h
Coal injection quantity	CIQ	t
Content of carbon dioxide	COCD	%
Content of carbon monoxide	COCM	%
Center strength	Z	-
Edge strength	W	-
Intensity ratio (z/w)	IR	-
Temperature of cross-temperature measuring	TOCTM	°C
Marginal mean	MM	-
Bosh gas index	BGI	-
Theoretical combustion temperature	TCT	°C
(Blast) Kinetic energy	KE	J/s
Wind velocity	WV	m/s
Inlet water temperature	IWT	°C
Outlet water temperature	OWT	°C

In the above table, the calculation method of the GUR (η_{CO}) is shown as

$$\eta_{CO} = \frac{V_{CO2}}{V_{CO} + V_{CO2}} \tag{1}$$

where V_{CO2} and V_{CO} represent the amount of carbon dioxide and carbon monoxide, respectively.

Ironmaking is a complex reaction process involving coking, sintering, pelleting, and ironmaking. Many related reactions are carried out under high temperature and pressure. Therefore, the collected data have a certain number of outliers and missing values. Generally, two methods are used to judge the outliers and extreme outliers: the box plot method and the 3σ criterion method [22,23].

The feature data are arranged from small to large, and Q1 and Q3 are the first quartile and the third quartile of each feature parameter in a box plot, respectively. IQR is the difference between Q3 and Q1. In this context, data within the range of Q1 + 1.5IQR or minimum to Q3-1.5IQR or maximum for each data feature in the box plot are retained, and data outside this range are considered to be outliers. Data outside of (Q1 − 3IQR, Q3 + 3IQR) are considered to be extreme outliers. Data outside the range of (μ − 3σ, μ + 3σ) are judged as extreme outliers, where μ is the mathematical expectation and σ is the standard deviation in the 3σ criterion. The values of each feature are almost all concentrated in the interval (μ − 3σ, μ + 3σ). The possibility of exceeding this range is less than 0.3%. Therefore, the data outside this range can be considered as extreme outliers. The abnormal conditions of a BF must be considered when performing predictive modeling of BF. Therefore, when the collected data are pre-processed, only extreme outliers are removed and substituted with interpolated estimates.

In order to ensure the continuity of time, extreme outliers and vacancy data are usually filled instead of completely deleted. Extreme outliers are replaced with missing values in this article. The linear interpolation method is selected, which is to construct a straight line to approximate the missing value.

The distribution of each variable can be characterized by a violin plot. It is roughly judged whether each feature has an outlier from the overall distribution of the data. It is very similar to a box plot, but it can gain insight into the distribution density of each variable. At the same time, the violin plot is particularly suitable for situations where the amount of data is huge and individual observations cannot be displayed, which is consistent with the data used in this article. Figures 1–4 are comparison diagrams of the violin plot after replacing the extreme values in the original data with the box plot and the 3σ criterion. In Figures 1–4, (x) − 1, (x) − 2, and (x) − 3(x = a, b, c, . . . , f) represent the data before a feature is processed, after it is processed by the box plot, and after it is processed by the 3σ criterion, respectively. Each feature is expressed by the same color.

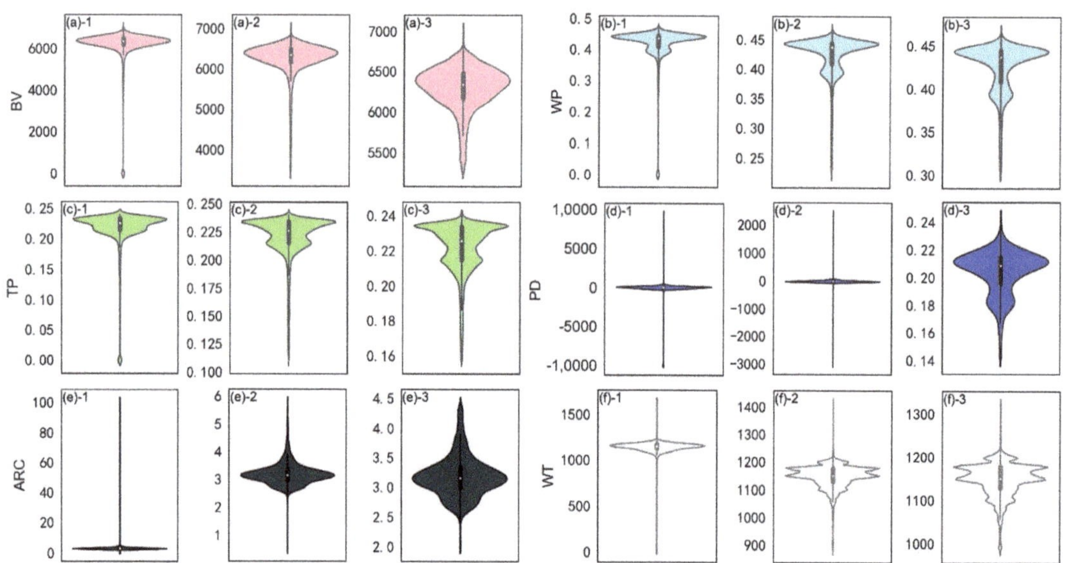

Figure 1. (a–f) Comparison of original data and processed data (1), i (i = 1, 2, 3) represents the data before a feature is processed, after it is processed by the box plot, and after it is processed by the 3σ criterion, respectively.

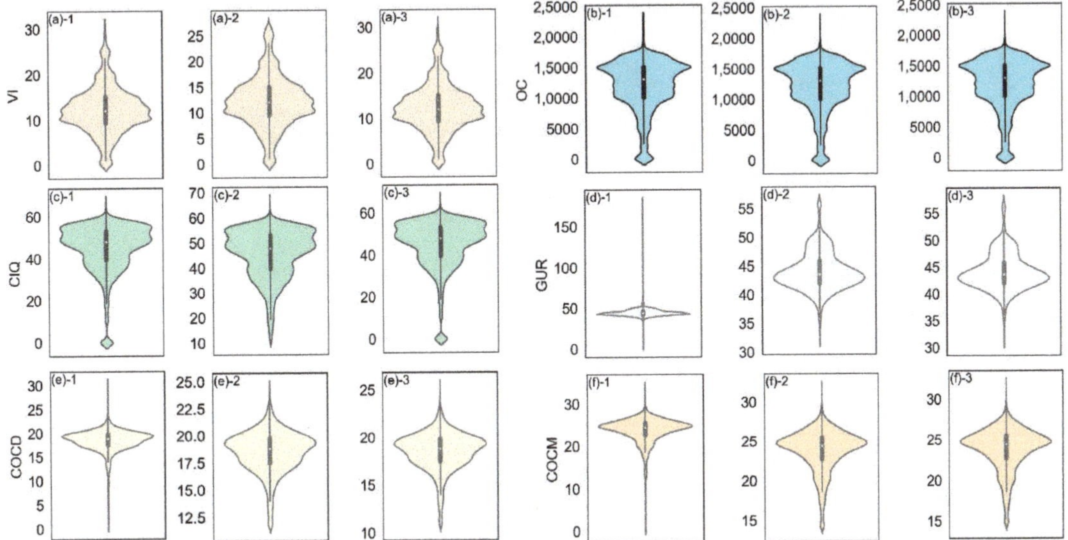

Figure 2. (**a–f**) Comparison of original data and processed data (2), i (i = 1, 2, 3) represents the data before a feature is processed, after it is processed by the box plot, and after it is processed by the 3σ criterion, respectively.

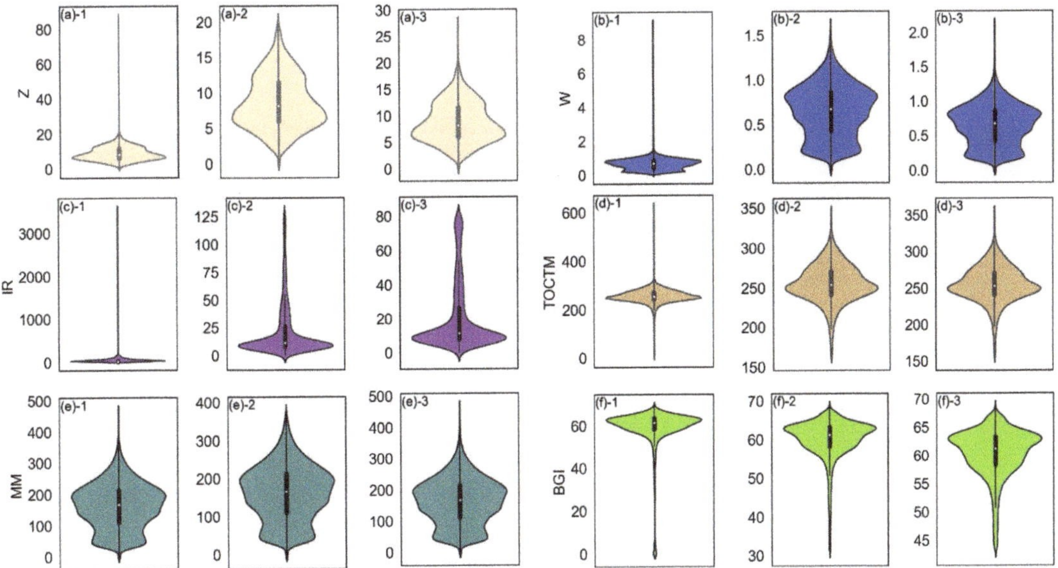

Figure 3. (**a–f**) Comparison of original data and processed data (3), i (i = 1, 2, 3) represents the data before a feature is processed, after it is processed by the box plot, and after it is processed by the 3σ criterion, respectively.

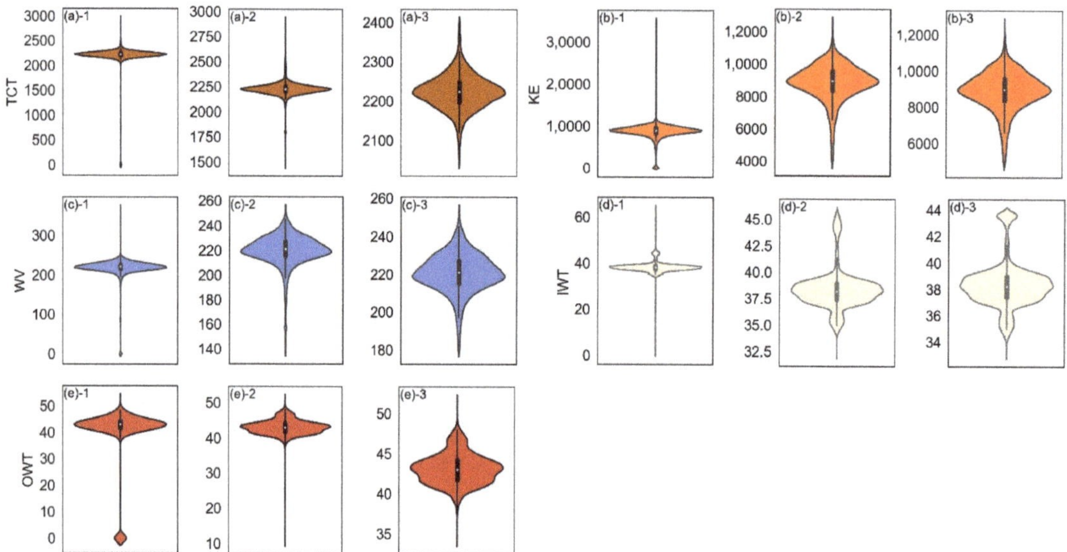

Figure 4. (**a–e**) Comparison of original data and processed data (4), i (i = 1, 2, 3) represents the data before a feature is processed, after it is processed by the box plot, and after it is processed by the 3σ criterion, respectively.

The comparisons of each parameter before and after pre-processing are reflected in Figures 1–4, where the box in each figure is the box plot. The thin black line is the whisker, and the external shape is the kernel density estimate. There are missing values, outliers, and extreme outliers in the original data, as shown in Figures 1–4. The distribution of each characteristic parameter is uneven, and the degree of discretization of the data is relatively large. Meanwhile, Figures 1–4 indicate that, compared to the data pre-processed by the box plot, the distribution of the data after being pre-processed by the 3σ criterion is more uniform. After the extreme outliers were pre-processed using the box plot and 3σ criterion, two different data sets were formed, which are called the box data set (BDS) and normal data set (NDS).

3. Model Construction

3.1. Feature Selection

The selection of input parameters in the modeling process can be determined according to the practical experience of on-site operators and the correlation between the collected parameters and GUR. For the measurement of correlation, the maximum information coefficient (MIC) is used for characterization in this paper.

If there is an association between two variables, and the scatter plots composed of these two variables are meshed, a partitioning method can always be found to describe their relevance. The correlation between two consecutive variables can be described by the MIC [24]. It mines nonlinear correlations by performing unequal interval discretization optimization on continuous variables and further makes $MIC(X, Y) \in [0, 1]$ through standardized correction with the help of this normalization function.

$$y = \frac{(y_{max} - y_{min})(x - x_{min})}{x_{max} - x_{min}} + y_{min} \tag{2}$$

where $y_{min} = -1$ and $y_{max} = 1$. After input and output variables are standardized, the mutual information between the variables is calculated as follows:

$$MIC(X;Y) = \max_{|X||Y|<B(X;Y)} \frac{I(X;Y)}{\log_2(\min\{|X|,|Y|\})} \tag{3}$$

$I(X; Y)$ represents the mutual information of X and Y in Formula (3). Moreover, $|X|$ and $|Y|$, respectively, represent the number of segments in which the variables X and Y are divided into the mesh division process. The value of B is generally set to 0.6 or 0.55. In this paper, the value of B is 0.6. The mutual information is calculated as follows:

$$I(X;Y) = \sum_{x=X}\sum_{y=Y} p(X,Y) \log_2 \frac{p(X,Y)}{p(X)p(Y)} \tag{4}$$

X and Y are two connected random variables, and $p(X, Y)$ is the joint probability density distribution function in Formula (2). The MICs of the initially selected input parameters for the BDS and NDS are shown in Figure 5.

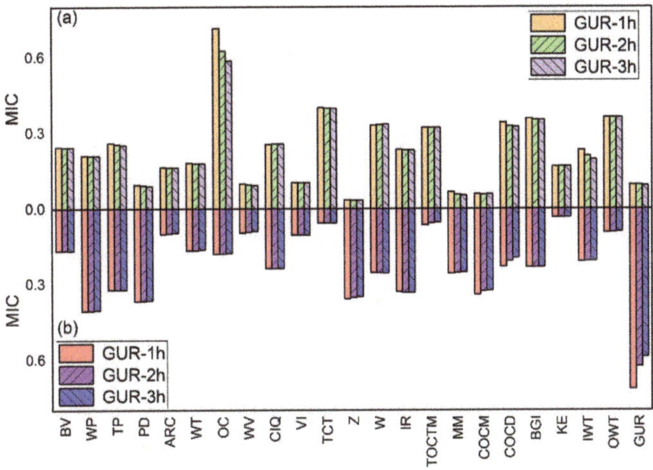

Figure 5. The values of MIC between the selected input parameters and the output parameters for BDS (**a**) and NDS (**b**).

The MICs between the feature parameters and GUR-1h, GUR-2h, and GUR-3h in the BDS and NDS data sets are shown in Figure 5a,b, respectively. The characteristic parameters with the values of MIC greater than 0.15 are finally selected as input variables based on expert experience and the above calculation results. Therefore, 16 input parameters are finally selected, respectively, for the BDS and NDS, and the selected parameters are shown in Table 2.

Table 2. Input parameters of the two models.

Common Parameters		Specific Parameters of BDS	Specific Parameters of NDS
BV	IWT	TOCTM	Z
TP	CIQ	ARC	GUR
WP	W	TCT	MM
WT	COCD	KE	COCM
OC	IR	OWT	PD
BGI			

3.2. Method of Model Construction

Traditional regression prediction generally attempts to obtain a function such as a form $f(x) = w^T x + b$ for a data set: $D = \{(x_m, y_m)$. The optimization process is to reduce directly the difference between the predicted value ($f(x)$) and the true value (y). The loss function is shown in Equation (5).

$$J(\theta) = \frac{1}{2}\sum_{i=1}^{m}\left(h\theta(x_i) - y_i\right)^2 \tag{5}$$

where y_i represents the actual value, and $h_\theta(x_i)$ means the predicted value.

The loss is calculated when $|f(x)\text{-}y| > \varepsilon$ for the SVR algorithm. If the feature vector mapping x from the low-dimensional space to the high-dimensional space is expressed as $\Phi(x)$, then the hyperplane model divided by the high-dimensional space is as shown in Equation (6):

$$f(x) = w^T \Phi(x) + b \tag{6}$$

where w is the normal vector and b is the displacement term., An interval band with a width of 2ε is constructed by taking $f(x)$ as the center. If the training sample falls into this interval band, the prediction is considered to be correct. Then, the kernel function [18] is introduced into SVR as shown in Equation (7).

$$f(x) = \sum_{i=1}^{m}\left(\hat{a}i - ai\right)\cdot k(x_i, x) + b \tag{7}$$

$$b = yi + \varepsilon - \sum_{i=1}^{m}\left(\hat{a}i - ai\right)\cdot k(xi, xj) \tag{8}$$

where $k(x_i, x_j)$ is a kernel function.

The choice of the kernel function, such as the *rbf* function, linear kernel function, polynomial kernel function, or sigmoid kernel function, is very important for the prediction result of SVR. The SVR prediction model can be finally determined by the grid search [25].

The actual data of the BF have different magnitudes, which has a significant impact on the predictive performance of a model. Therefore, in addition to handling extreme outliers, the data also need to be standardized before modeling. The process of data normalization ensures that each feature has an average of 0 and a variance of 1. It makes all features in the same magnitude, which also reduces the impact of anomaly data on the built model. At the same time, the output of the model can be restored to the original output parameters through the de-standardization method.

4. Comparison of the Prediction Results Based on Two Data Sets

In order to achieve an accurate prediction, the SVR model was established in this paper. After the model is optimized through the grid search, the best hyperparameters of the model are obtained, as shown in Table 3. C is the penalty parameter and γ is a parameter in the RBF kernel function [25,26].

Table 3. The best hyperparameters of the SVR model based on the two data sets.

Model			SVR Model		
Model parameters			Kernel function	C	γ
	GUR-1h	BDS	RBF	10	0.1
		NDS	RBF	10	0.01
Output parameters	GUR-2h	BDS	RBF	10	0.1
		NDS	RBF	100	0.01
	GUR-3h	BDS	RBF	10	0.1
		NDS	RBF	10	0.1

Due to the massive volume of test data, the 100 sets of measured values in the test set of BDS and NDS are marked in black and the correspondingly predicted values are marked in red and blue when the predicted parameter is the GUR-1h. The results are expressed in Figure 6a,b, respectively.

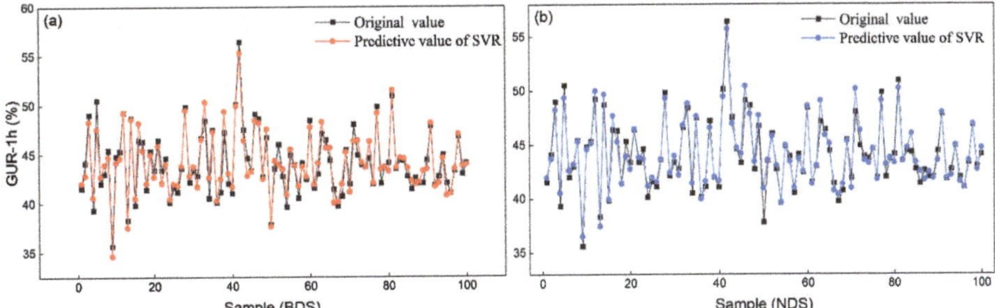

Figure 6. The prediction result of the SVR model based on BDS (**a**) and NDS (**b**) (the forecasting parameter is the GUR-1h).

Figure 6a,b show the comparison among the original values and the predicted values of the SVR model based on BDS and NDS when the predicted parameter is GUR-1h, respectively. Figure 6 shows that the black points basically coincide with the red points and blue points, which indicates that the predicted values of the SVR model based on BDS and NDS are not of much difference from the true values. In the process of testing, the 8800 sets of data, which are 25% of the total sample, are selected as the test data. After sorting according to the actual values, the forecasting bias of the SVR model is as shown in Figure 7.

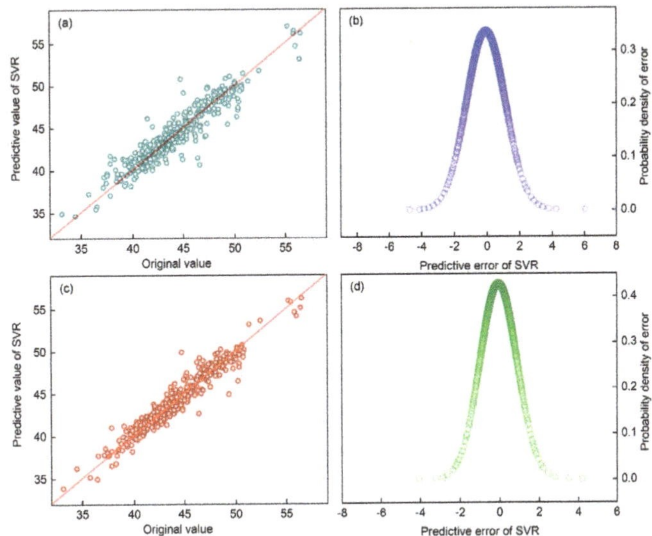

Figure 7. The comparison of the prediction error of the SVR model (the forecasting parameter is the GUR-1h). Prediction bias (**a**) and probability density of prediction errors (**b**) for the SVR model based on the BDS. Prediction bias (**c**) and probability density of prediction errors (**d**) for the SVR model based on the NDS.

Figure 7a,c show the degree of the prediction deviation of the SVR models constructed based on BDS and NDS when the predicted parameter is the GUR-1h, respectively.

Figure 7b,d are the images of the predicted deviation probability density function of the SVR models constructed based on BDS and NDS when the predicted parameter is GUR-1h, respectively. The horizontal axis is the actual value, and the vertical axis is the predictive value by the SVR models constructed based on BDS and NDS in Figure 7a,c. If the original value is very close to the predicted value, the image is in complete and exact accordance with the diagonal. It is observed that the prediction results of the SVR models constructed based on BDS and NDS fluctuate around the diagonal line, and the fluctuation range is narrow. Figure 7b,d indicate that the values of the predicted error of the two models are basically within the range of ± 2.

When the output parameter is the GUR-2h, the predicted results of the SVR model constructed using the BDS and the NDS, respectively, are as shown in Figure 8.

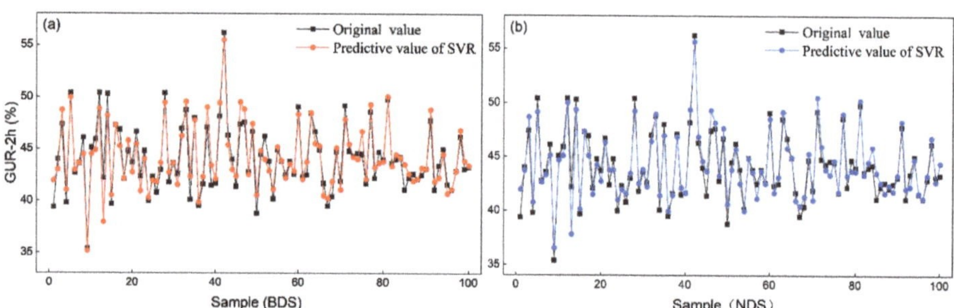

Figure 8. The prediction results of the SVR model using the BDS (**a**) and NDS (**b**) (the forecasting parameter is the GUR-2h).

Figure 8a,b show the comparisons among the original values and the predicted values of the SVR model based on BDS and NDS when the predicted parameter is the GUR-2h, respectively. The predictive errors of the SVR model constructed using the BDS and the NDS are shown in Figure 9.

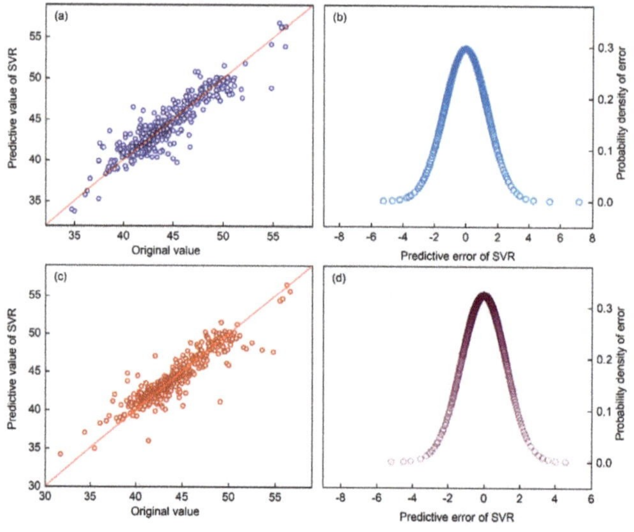

Figure 9. The comparison of the predicted errors of the SVR model (the forecasting parameter is the GUR-2h). Prediction bias (**a**) and probability density of prediction errors (**b**) for the SVR model based on the BDS. Prediction bias (**c**) and probability density of prediction errors (**d**) for the SVR model based on the NDS.

Compared to Figure 9c, the data in Figure 9a fluctuate slightly. In Figure 9b,d, the range of prediction errors gradually expands. When the output parameter is the GUR-3h, the prediction results of the SVR model constructed using the BDS and the NDS are as shown in Figure 10.

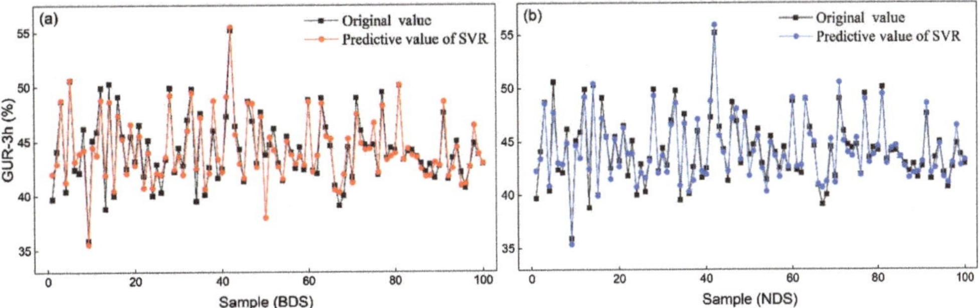

Figure 10. The prediction results of the SVR model separately based on BDS (**a**) and NDS (**b**) (the forecasting parameter is the GUR-3h).

In Figure 10a,b, the fit between the predicted values and the true value gradually deteriorates. Figure 11 represents the range of errors between the predicted values and the actual values. Compared to Figures 7 and 9, the prediction errors in Figure 11 are significantly larger.

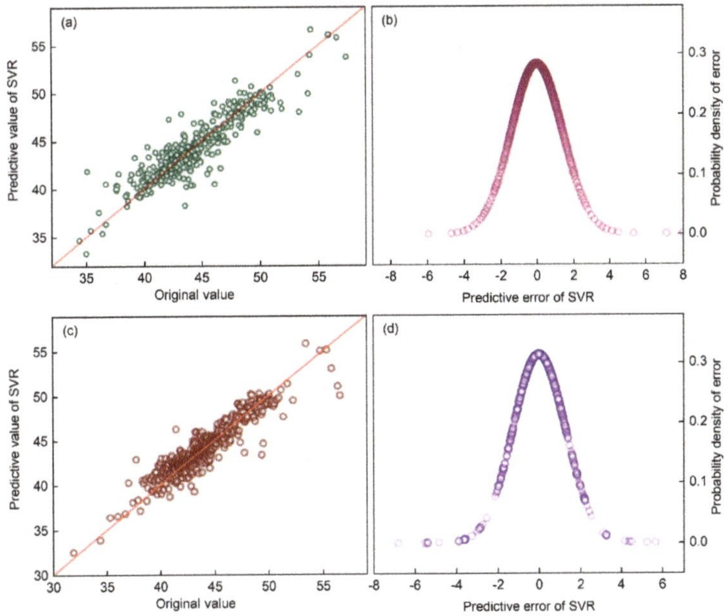

Figure 11. The comparison of the predicted errors of the SVR model (the forecasting parameter is the GUR-3h). Prediction bias (**a**) and probability density of prediction errors (**b**) for the SVR model based on the BDS. Prediction bias (**c**) and probability density of prediction errors (**d**) for the SVR model based on the NDS.

5. Evaluation Indicators and Analysis

In total, 35,198 sets of data collected by the online detection system of BF in China are used to predict the model in this paper. Moreover, 75% of the data are used for training the model and the remaining data are used for testing the model [19–21]. The evaluation of the prediction results of a model should be characterized by multiple aspects and multiple scales [27]. Generally, the characterization index is mainly the coefficient of determination (R^2), mean absolute error (MAE), root mean square error (RMSE), or hit rate (HR) [19–21,27]. The reliability of the model can be represented by these parameters within the acceptable range of the processing process, and the calculation methods are shown in Equations (9)–(12), respectively:

$$R^2 = 1 - \sum_{i=1}^{n} \left(h(x_i) - y_i \right)^2 / \sum_{i=1}^{n} \left(\overline{y} - y_i \right)^2 \tag{9}$$

$$MAE = \frac{1}{n} \cdot \sum_{i=1}^{n} |h(x_i) - y_i| \tag{10}$$

$$RMSE = \sqrt{\frac{1}{n} \cdot \sum_{i=1}^{n} (h(x_i) - y_i)^2} \tag{11}$$

$$\begin{cases} HR = \frac{1}{n} \cdot \sum_{i=1}^{n} HR_i \times 100\% \\ HR_i = \begin{cases} 1, & |h(x_i) - y_i| \leq c \\ 0, & |h(x_i) - y_i| > c \end{cases} \end{cases} \tag{12}$$

The range of R^2 is [0, 1]. In general, the larger the result, the better the fitting effect of the model. n is the total number of samples in the test set, and $h(x_i)$ and y_i are the predicted and original values of the output parameters, respectively. c is the boundary value of the hit rate. In this paper, the value of c is selected as 2%. At this time, the R^2 and HR of the two models are as shown in Figure 12a, and the MAE and RMSE of the two models are shown in Figure 12b.

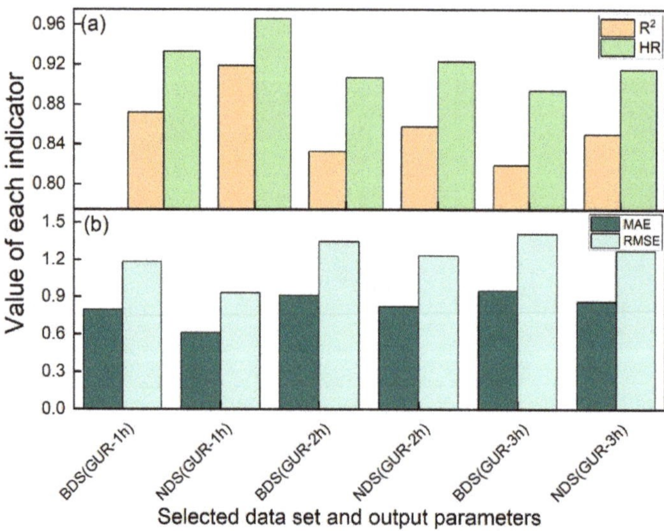

Figure 12. R^2 and HR values of the two models (**a**), and MAE and RMSE of the two models (**b**).

High values of R^2 and HR, and low values of MAE and RMSE, represent higher prediction accuracy of the model, as in Figure 12. Compared with the other two cases, when the output parameter is selected as GUR-1h, for NDS and BDS, the predicted accuracy of SVR is always the highest. When the selected data set is NDS and the output parameter is GUR-1h, the predictive accuracy and the hit rate of the SVR model are 91.9% and 96.6%, respectively. In this case, the SVR model obtains the best prediction effect. Moreover, regardless of whether the data are advanced using the 3σ criterion or the box plot, the predictive effect of the SVR model is strong.

6. Conclusions

GUR is an important indicator reflecting the energy consumption and smooth operation of the BF. This paper analyzes the impact of two data processing methods, the box plot and 3σ criterion, in predicting the blast furnace gas utilization rate. The box plot and 3σ criterion are selected to judge extreme outliers in this article, and linear interpolation is used to process extreme outliers and missing values. The simulations show that the prediction model using the SVR algorithm is more accurate based on the processed blast furnace data with the 3σ criterion. Hysteresis in blast furnace smelting must be taken into account, and the GUR-1h, GUR-2h, and GUR-3h are selected as output parameters, respectively. The experimental results show that the prediction of the gas utilization rate after one hour is most accurate using the parameters in the current state in the blast furnace smelting process. Moreover, as the time interval between predictions becomes longer, the prediction accuracy decreases.

This study is a first step; there are several avenues for further exploration. One natural extension is missing value handling. Other methods could be considered for replacing missing values. Another avenue for future work is extension to supply side applications, such as the development of a blast furnace gas utilization rate forecasting system that can be applied to actual production, to reduce energy consumption for blast furnace production, and to provide ancillary services for subsequent processes.

Author Contributions: Formal analysis, K.L.; Investigation, J.Z., L.J. and L.H.; Validation, Z.W.; Writing—original draft, D.J. All authors have read and agreed to the published version of the manuscript.

Funding: This work was supported by the National Natural Science Foundation of China (Project No.: 51904026) and China Postdoctoral Science Foundation (Project No.: BX20200045 and 2021M690370).

Conflicts of Interest: The authors declare no conflict of interest.

References

1. Ling, J.; Chuanhou, G.; Zhonghang, X. Constructing Multiple Kernel Learning Framework for Blast Furnace Automation. *IEEE Trans. Autom. Sci. Eng.* **2012**, *9*, 763–777. [CrossRef]
2. Jianqi, A.; Jialiang, Z.; Min, W.; Jinhua, S.; Takao, T. Soft-sensing method for slag-crust state of blast furnace based on two-dimensional decision fusion. *Neurocomputing* **2018**, *315*, 405–411.
3. Hao, P.; Haibin, Y.; Mingzhe, Y. Modeling and analysis of energy using efficiency of the blast furnace. *Manuf. Autom.* **2011**, *33*, 142–144.
4. Jianqi, A.; Xiaoling, S.; Min, W.; Jinhua, S. A multi-time-scale fusion prediction model for the gas utilization rate in a blast furnace. *Control Eng. Pract.* **2019**, *92*, 104120.
5. Limin, Z.; Changchun, H.; Junpeng, L.; Xinping, G. Operation status prediction based on top gas system analysis for blast furnace. *IEEE Trans. Control Syst. Technol.* **2017**, *25*, 262–269. [CrossRef]
6. Ling, J.; Chuanhou, G. Binary coding SVMs for the multiclass problem of blast furnace system. *IEEE Trans. Ind. Electron.* **2013**, *60*, 3846–3856. [CrossRef]
7. Ray, H.; Pal, S. Simple method for theoretical estimation of viscosity of oxide melts using optical basicity. *Ironmak. Steelmak.* **2004**, *31*, 125–130. [CrossRef]
8. Meng, F.; ZOU, Z. Influence of thermal reserve zone temperature in blast furnace on gas utilization rate. *J. Northeast. Univ.* **2018**, *39*, 985.
9. Weiching, C.; Wentung, C. Numerical simulation on forced convective heat transfer of titanium dioxide/water nanofluid in the cooling stave of blast furnace. *Int. Commun. Heat Mass Transf.* **2016**, *71*, 208–215.

10. Tonglai, G.; Mansheng, C.; Zhenggen, L.; Jue, T.; JunIchiro, Y. Mathematical modeling and exergy analysis of blast furnace operation with natural gas injection. *Steel Res. Int.* **2013**, *84*, 333–343. [CrossRef]

11. Shen, Y.; Guo, B.; Chew, S.; Austin, P.; Yu, A. Three-dimensional modeling of flow and thermochemical behavior in a blast furnace. *Metall. Mater. Trans. B* **2015**, *46*, 432–448. [CrossRef]

12. Jianqi, A.; Junyu, Y.; Min, W.; Jinhua, S.; Takao, T. Decoupling control method with fuzzy theory for top pressure of blast furnace. *IEEE Trans. Control Syst. Technol.* **2019**, *27*, 2735–2742. [CrossRef]

13. Min, W.; Kexin, Z.; Jianqi, A.; Jinhua, S.; Kangzhi, L. An energy efficient decision-making strategy of burden distribution for blast furnace. *Control Eng. Pract.* **2018**, *78*, 186–195.

14. Yanjiao, L.; Sen, Z.; Yixin, Y.; Wendong, X.; Jie, Z. A novel online sequential extreme learning machine for gas utilization ratio prediction in blast furnaces. *Sensors* **2017**, *17*, 1847.

15. Guimei, C.; Anwei, C.; Xiang, M. An identification method of center gas-flow distribution pattern based on sensed infrared image processing. *Inf. Control* **2014**, *43*, 110–115. [CrossRef]

16. Lin, S.; You-bin, W.; Guangsheng, Z.; Tao, Y. Recognition of blast furnace gas flow center distribution based on infrared image processing. *J. Iron Steel Res. Int.* **2016**, *23*, 203–209.

17. Jiang, D.; Wang, Z.; Zhang, J.; Jiang, D.; Li, K.; Liu, F. Machine Learning Modeling of Gas Utilization Rate in Blast Furnace. *JOM* **2022**, 1–8. [CrossRef]

18. Zhang, S.; Jiang, H.; Yin, Y.; Xiao, W.; Zhao, B. The prediction of the gas utilization ratio based on TS fuzzy neural network and particle swarm optimization. *Sensors* **2018**, *18*, 625. [CrossRef]

19. David, S.F.; David, F.F.; Machado, M. Artificial neural network model for predict of silicon content in hot metal blast furnace. In *Materials Science Forum*; Trans Tech Publications Ltd.: Bäch, Switzerland, 2016; pp. 572–577.

20. Zhang, X.; Kano, M.; Matsuzaki, S. A comparative study of deep and shallow predictive techniques for hot metal temperature prediction in blast furnace ironmaking. *Comput. Chem. Eng.* **2019**, *130*, 106575. [CrossRef]

21. Kuang, S.; Li, Z.; Yu, A. Review on modeling and simulation of blast furnace. *Steel Res. Int.* **2018**, *89*, 1700071. [CrossRef]

22. Li, L.; Wen, Z.; Wang, Z. Outlier detection and correction during the process of groundwater lever monitoring base on pauta criterion with self-learning and smooth processing. In *Theory, Methodology, Tools and Applications for Modeling and Simulation of Complex Systems*; Springer: Berlin/Heidelberg, Germany, 2016; pp. 497–503.

23. Schwertman, N.C.; Owens, M.A.; Adnan, R. A simple more general boxplot method for identifying outliers. *Comput. Stat. Data Anal.* **2004**, *47*, 165–174. [CrossRef]

24. Reshef, D.N.; Reshef, Y.A.; Finucane, H.K.; Grossman, S.R.; McVean, G.; Turnbaugh, P.J.; Lander, E.S.; Mitzenmacher, M.; Sabeti, P.C. Detecting Novel Associations in Large Data Sets. *Science* **2011**, *334*, 1518. [CrossRef] [PubMed]

25. Smola, A.J.; Schölkopf, B. A tutorial on support vector regression. *Stat. Comput.* **2004**, *14*, 199–222. [CrossRef]

26. Zhang, H.; Zhang, S.; Yin, Y.; Chen, X. Prediction of the hot metal silicon content in blast furnace based on extreme learning machine. *Int. J. Mach. Learn. Cybern.* **2018**, *9*, 1697–1706. [CrossRef]

27. Tunckaya, Y. Performance assessment of permeability index prediction in an ironmaking process via soft computing techniques. *Proc. Inst. Mech. Eng. Part E J. Process Mech. Eng.* **2017**, *231*, 1101–1113. [CrossRef]

Article

A Metallurgical Dynamics-Based Method for Production State Characterization and End-Point Time Prediction of Basic Oxygen Furnace Steelmaking

Qingting Qian [1], Qianqian Dong [1], Jinwu Xu [1], Wei Zhao [2] and Min Li [1,*]

[1] Collaborative Innovation Center of Steel Technology, University of Science and Technology Beijing, Beijing 100083, China
[2] Angang Steel Company Limited, Anshan 114021, China
* Correspondence: limin@ustb.edu.cn

Abstract: Basic Oxygen Furnace (BOF) steelmaking is an important way for steel production. Correctly recognizing different blowing periods and abnormal refining states is significant to ensure normal production process, while accurately predicting the end-point time helps to increase the first-time qualification rate of molten steel. Since the decarburization products CO and CO_2 are the main compositions of off-gas, information of off-gas is explored for BOF steelmaking control. However, the problem is that most of the existing research directly gave the proportions of CO and CO_2 as model input but barely considered the variation information of off-gas to describe the production state. At the same time, the off-gas information can be expected to recognize the last blowing period and predict the end-point time earlier than the existing methods that are based on sub-lance or furnace flame image, but little literature makes an attempt. Therefore, this work proposes a new method based on functional data analysis (FDA) and phase plane (PP), defined as FDA-PP, to describe and predict the BOF steelmaking process from the metallurgical dynamics viewpoint. This method extracts the total proportion of CO and CO_2 and its first-order derivative as dynamics features of steelmaking process via FDA, which indicate the reaction velocity and acceleration of decarburization reaction, and describes the evolution of dynamics features via PP. Then, the FDA-PP method extracts the features of phase trajectories for production state recognition and end-point time prediction. Experiments on a real production dataset demonstrate that the FDA-PP method has higher production state recognition accuracy than the classical phase space, SVM, and BP methods, which is 87.78% for blowing periods of normal batches, 90.94% for splashing anomaly, and 81.29% for drying anomaly, respectively. At the same time, the FDA-PP method decreases the mean relative prediction error (MRE) of the end-point time prediction for abnormal batches by about 10% compared with the SVM and BP methods.

Keywords: basic Oxygen Furnace steelmaking; intelligent manufacturing; functional data analysis; phase plane; blowing period recognition; anomaly monitoring; end-point time prediction

Citation: Qian, Q.; Dong, Q.; Xu, J.; Zhao, W.; Li, M. A Metallurgical Dynamics-Based Method for Production State Characterization and End-Point Time Prediction of Basic Oxygen Furnace Steelmaking. *Metals* **2023**, *13*, 2. https://doi.org/10.3390/met13010002

Academic Editor: Alexander McLean

Received: 12 November 2022
Revised: 18 December 2022
Accepted: 19 December 2022
Published: 20 December 2022

1. Introduction

Basic Oxygen Furnace (BOF) steelmaking is important in the iron and steel industry, through which over 70% of crude steel is refined all over the world [1]. A successful BOF steelmaking process should have a normal production process, accurate end-point time, and stable product quality. In reality, the process control often relies on the experience and skill of operators, so that the abnormal production state, insufficient refining time, and unqualified molten steel sometimes appear, which would threaten production safety and increase energy and resource consumption. Therefore, lots of models [2–4] are studied to recognize different blowing periods and abnormal refining states in order to optimize process parameters and ensure normal production process, as well as to predict the end-point time or oxygen blowing volume (dividing the oxygen blowing volume by the oxygen flow can obtain the end-point time) in order to increase the first-time qualification rate of molten steel and avoid re-blowing operation.

With the trend of intelligent manufacturing in the iron and steel industry [5,6], data-based methods [7–9] are developing for the above targets relying on the abundant production data, including: the sub-lance measurements, the furnace flame image, and the off-gas. Han Min et al. [10] and He Fei et al. [11] employed the adaptive-network-based fuzzy inference system (ANFIS) and the back-propagation neural network (BP) respectively to predict the oxygen blowing volume, where the model inputs were the metal's carbon content and temperature measured by TSC sub-lance. But the sub-lance measurements are unavailable for the production state recognition. Meanwhile, the sub-lance detection is always carried out within the last 1 min, and needs to suspend the process operations, which breaks the production rhythm. The furnace flame image supports both the production state characterization and the end-point prediction and can be collected without production interruption. Prof. Li Ailian's team studied the ResNet network [12] and the improved DenseNet network [13] to recognize different blowing periods relying on the furnace flame image. Wen Hongyuan [14] extracted the features of the furnace flame image and used the multiple linear regression (MLR) model and the BP model to predict the end-point time. But due to that, the end-point time prediction relied on the feature mutation of the furnace flame image which occurred very late, the time margin given by the existing model was only 20~40 s [14], too small to optimize operation parameters and adjust the steel quality.

The off-gas also supports both the production state characterization and the end-point prediction. The industry has been exploring the use of off-gas information for process control. Mats Bramming [15] built a multiway partial least squares (MPLS) model using the off-gas information and other process data to explain and predict the splashing anomaly. Our previous works [16] proposed a Mahalanobis distance-based functional derivative support vector data description (MD-FDSVDD) model to identify the splashing anomaly from the amplitude variations of off-gas. Compared with the furnace flame image, the off-gas data are sensitive to the change of decarburization rate since the decarburization products CO and CO_2 are the main compositions of off-gas. Thus, the off-gas data can be expected to predict the ending time with a larger time margin as the decarburization rate changes when the last blowing period begins.

However, for the existing research, most of the data-based models directly gave the amplitude of off-gas composition as input to a model and then obtained the output of the production state. This end-to-end calculation ignored part of the metallurgical dynamics features of BOF steelmaking, i.e., the variation information of decarburization rate, leading to a hard determination of production state with similar amplitude of off-gas composition (i.e., similar decarburization rate). As the BOF steelmaking process contains complex multi-phase physicochemical reactions, lots of operation parameters will affect the decarburization rate [17–20], and different parameters have diverse effects. On this basis, employing the variation information of off-gas composition is significant to explain how the decarburization rate is going to change with the influence of operation parameters, so that to provide an accurate description of the production state.

Therefore, this work proposes a new data-based method from the viewpoint of metallurgical dynamics to characterize the steelmaking production state and predict the end-point time. This method is based on the theories of functional data analysis (FDA) and phase plane (PP), defined as FDA-PP. The FDA-PP method firstly smooths the discrete sequences of the total proportion of CO and CO_2 in the off-gas via FDA to obtain their continuous functions and first-order derivative functions. Then, the phase plane is constructed by the fitting functions and derivative functions. The phase trajectory characterizes the decarburization rate as well as the influence of the operation parameters on the decarburization rate, describing the evolution of the production state. Next, a boundary that indicates a stable production state is estimated on the phase plane via support vector data description (SVDD) to recognize different blowing periods and abnormal production state. When the phase trajectory indicates that the last blowing period begins, the future phase trajectory is predicted via case-based reasoning (CBR), further estimating the end-point time.

Based on the above, the proposed FDA-PP method has four aspects of contributions and advantages. First, the FDA-PP method takes comprehensive consideration of the decarburization rate as well as its variation trend and establishes a metallurgical dynamics-based model for production state characterization. Second, this work is an attempt that using the off-gas data to characterize different blowing periods and predict the end-point time. Although the evolution of off-gas compositions was used to successfully recognize the splashing and drying anomalies, little literature tried to use it to identify different blowing periods. At the same time, to the best of our knowledge, the off-gas information hasn't been used for the end-point time/oxygen blowing volume prediction. Third, the FDA-PP method can be expected to predict the end-point time with 2~4 min in advance since the prediction is implemented at the beginning of the last blowing period. Fourth, based on the FDA, problems of irregular data such as uneven refining durations, different sampling frequencies, noise, and missing values, can be settled, and the evolution characteristics of the time-series data can be retained in the smoothed curves.

The remainder of this article is organized as follows: Section 2 gives an overview of BOF steelmaking and illustrates the connection between the metallurgical dynamics features with the phase plane; Section 3 describes the procedures to realize the FDA-PP method; Section 4 verifies the FDA-PP method and compares it with some commonly used methods; Section 5 makes a conclusion.

2. Problem Statement

This section first introduces the BOF steelmaking process and then analyzes the connection between the metallurgical dynamics of BOF steelmaking with the phase plane.

2.1. Overview of BOF Steelmaking

BOF steelmaking is an important way for steel production and its primary purposes are to remove the carbon content of molten steel from 4~5% to around 0.04~0.06% and increase the temperature from around 1250 °C to around 1680 °C [21]. Figure 1 is a schematic representation of the blowing process. The molten iron and the steel scrap are the main materials. The oxygen blowing from the top provides an oxidizing condition, while the nitrogen or argon blowing from the bottom acts as a stir. The auxiliary materials are added to help control the steel quality. Through controlling the oxygen lance height, the bottom blowing flow, and the weight of auxiliary materials, the molten steel is obtained [20].

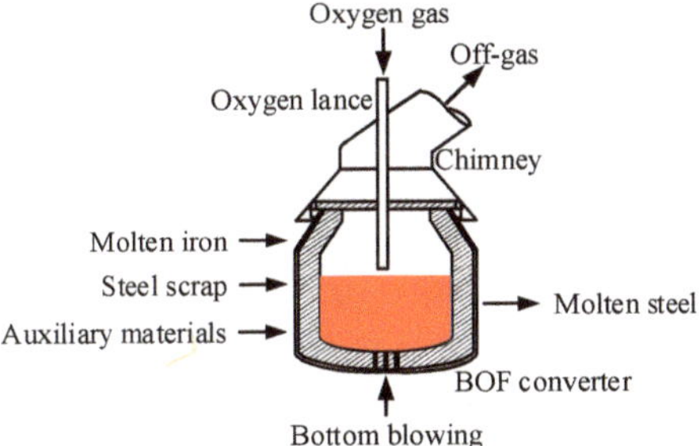

Figure 1. The schematic representation of BOF steelmaking.

The main reactions during the blowing process are the decarburization reactions, where the carbon of molten iron is oxidized by the oxygen blowing through $[C] + [O] = CO \uparrow$ and $[C] + 2[O] = CO_2 \uparrow$. Since carbon monoxide and carbon dioxide are released with the off-gas, the off-gas compositions reflect the decarburization rate of the steelmaking process [20]. According to the change of decarburization rate, the oxygen blowing can be separated to three blowing periods [22]: the first blowing period, the main blowing period, and the end blowing period, as shown in Figure 2a. Correspondingly, the change of the total proportion of CO and CO_2 in the off-gas also shows three periods, as can be seen in Figure 2b. In the first blowing period, the total proportion of CO and CO_2 in the off-gas gradually rises with increasing decarburization reaction. In the main blowing period, the decarburization rate keeps on a high value, so the total proportion of CO and CO_2 holds on a large percentage. In the last blowing period, the total proportion of CO and CO_2 decreases as the decarburization rate decreases.

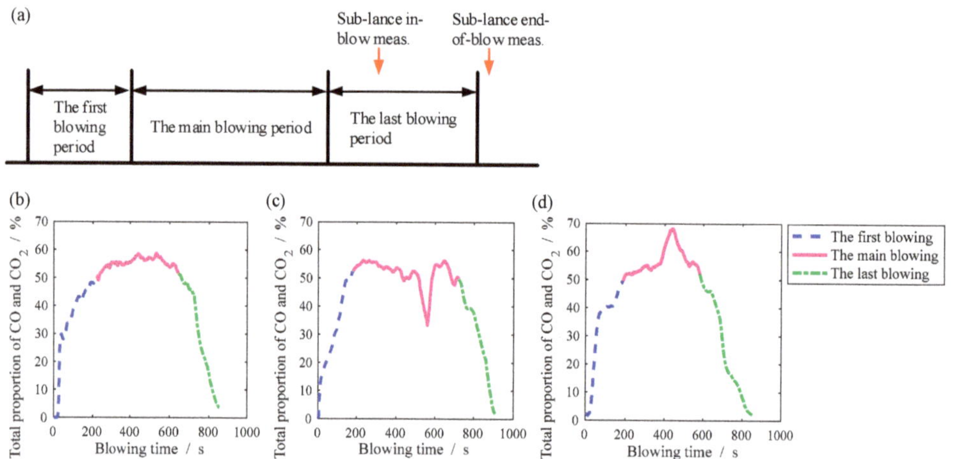

Figure 2. (**a**) The schematic representation of different blowing periods; (**b**) the off-gas compositions of a normal batch; (**c**) the off-gas compositions of a splashing batch; (**d**) the off-gas compositions of a drying batch.

The splashing and drying anomalies usually happen in the main blowing period and their influences on the total proportion of CO and CO_2 are shown in Figure 2c,d. When the splashing anomaly happens, the total proportion of CO and CO_2 drops first and then rises up. When the drying anomaly happens, the trends of CO and CO_2 increase first and then decrease, showing opposite evolutions to the splashing batch.

2.2. Metallurgical Dynamics and Phase Plane of the BOF Steelmaking

The phase plane is a method that can describe the evolution of a system's dynamics state [23,24], usually plotted with the first-order derivative and the second-order derivative. In this work, we take the decarburization reaction as the objective and use the phase plane to describe the evolution of the production state during the steelmaking process.

Since the total proportion of CO and CO_2 in the off-gas reflects the reaction velocity of the decarburization reaction in the metal pool [20], its first-order derivative indicates the reaction acceleration of the decarburization reaction, which is extracted to construct a phase plane in order to describe the dynamics evolution of decarburization reaction, i.e., the production state of BOF steelmaking. Figure 3a is the phase plane of the normal steelmaking process. We can see the phase trajectory begins at the origin point, develops clockwise from the first quadrant, holds on in a narrow region at the transverse axis, and goes back to the origin point from the fourth quadrant. The evolution of the phase trajectory is accordant to the dynamic characteristics of BOF steelmaking. In detail:

- Within the first blowing period, the phase trajectory shows increasing decarburization velocity and decreasing decarburization acceleration. The reason is that in the beginning, the silicon and manganese in the metal pool are preferentially oxidized and the carbon is following. With the silicon and manganese gradually consumed, the decarburization rate is enhanced. At the same time, as the carbon-oxygen reactions gradually reach their balances, the decarburization acceleration decreases.

- Within the main blowing period, the phase trajectory shows a stable state with the max decarburization velocity and nearly zero decarburization acceleration. The reason is that when the carbon concentration is high enough so that the transportation rate of oxygen limits the decarburization rate; thus, under a certain oxygen blowing, the decarburization reactions reach their balance condition and remain at their highest reaction velocities.
- Within the last blowing period, the phase trajectory shows decreasing decarburization velocity and a first decreasing then slightly increasing decarburization acceleration. The reason is that when the carbon concentration of the metal pool is too low so that the transportation rate of carbon limits the decarburization rate, with the unchanged oxygen blowing, the decarburization acceleration would be negative and decrease rapidly, further leading to decreasing decarburization velocity. With the carbon concentration tending to be zero, the decarburization acceleration and velocity are close to zero.

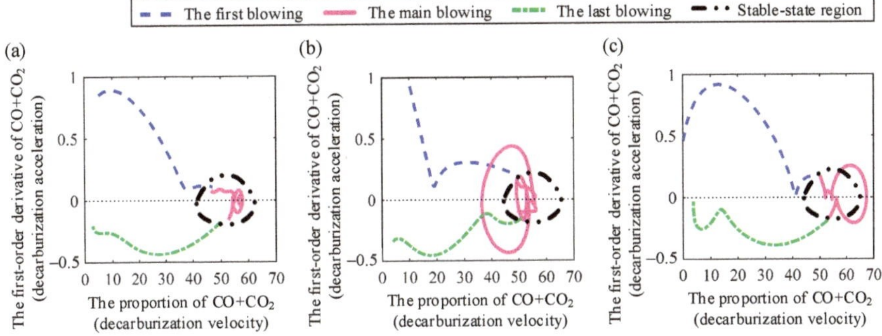

Figure 3. The phase plane of (**a**) normal batch; (**b**) splashing batch; (**c**) drying batch.

Figure 3b,c are the phase planes with splashing and drying anomalies respectively. Because of the abnormal fluidity of the slag emulsion [22], the splashing anomaly happens when the off-gas hardly traverses the slag emulsion and accumulates in the converter, whereas the drying anomaly happens when the off-gas rapidly traverses the slag and is released from the converter. Corresponding to this mechanism, when the splashing anomaly happens, the accumulated oxidation products CO and CO_2 prevent the decarburization reaction, so the decarburization velocity and acceleration rapidly decrease; after the off-gas erupts, the decarburization reaction recovers, so the decarburization velocity and acceleration rapidly increase. On this basis, the phase trajectory of the splashing anomaly exceeds the region of the stable state, develops clockwise from the fourth quadrant to the first quadrant, and finally goes back to the stable-state region. As for the drying anomaly, the decarburization reaction is promoted first because of the rapidly released off-gas and then returns to the normal level, so the phase trajectory exceeds the stable-state region and develops clockwise from the first quadrant to the fourth quadrant before returning to the stable-state region.

Based on these evolution features of the phase trajectory, different blowing periods and abnormal production states can be recognized.

3. Method Description

3.1. Extraction and Characterization of Dynamics Features

In order to describe and predict the production state of BOF steelmaking, the smoothed function and the first-order derivative function of off-gas compositions, i.e., the total proportion of CO and CO_2, are extracted by the FDA and then used to construct the phase plane to characterize the dynamics features of the decarburization reaction.

Functional data analysis [25] is a theory that regards a set of discrete observations as the sampling of a continuous function. Suppose $x(t)$ is a continuous function and $t = \begin{bmatrix} t_1 & t_2 & \ldots & t_K \end{bmatrix}^{\mathrm{T}}$ is the sampling instant, then the discrete observations $\hat{x} = \begin{bmatrix} \hat{x}_1 & \hat{x}_2 & \ldots & \hat{x}_K \end{bmatrix}^{\mathrm{T}}$ can be acquired as:

$$\hat{x} = x(t) + e = \begin{bmatrix} x(t_1) & x(t_2) & \ldots & x(t_K) \end{bmatrix}^{\mathrm{T}} + e \tag{1}$$

where K is the number of sampling points and e is the noise matrix. In order to express the continuous function $x(t)$, the FDA theory employs a linear combination of basis functions $\boldsymbol{\phi}(t) = \begin{bmatrix} \phi_1(t) & \phi_2(t) & \ldots & \phi_N(t) \end{bmatrix}^{\mathrm{T}}$ to approximate the discrete observations $\hat{x}$, which is as follows:

$$x(t) = \boldsymbol{\phi}(t)^{\mathrm{T}} c = \hat{x} - e \tag{2}$$

where $c = \begin{bmatrix} c_1 & c_2 & \ldots & c_N \end{bmatrix}^{\mathrm{T}}$ is a coefficient vector associated with the basis function system $\boldsymbol{\phi}(t)$. Usually, the coefficients c are calculated by Least Squares with a roughness penalty item, i.e.,:

$$\min_{e} e^{\mathrm{T}} e = \min_{c} \left\{ \left(\hat{x} - \boldsymbol{\phi}(t)^{\mathrm{T}} c \right)^{\mathrm{T}} \left(\hat{x} - \boldsymbol{\phi}(t)^{\mathrm{T}} c \right) + \lambda PEN_m[x(t)] \right\} \tag{3}$$

where $\boldsymbol{\phi}(t) = \begin{bmatrix} \phi_1(t_1) & \phi_1(t_2) & \ldots & \phi_1(t_K) \\ \phi_2(t_1) & \phi_2(t_2) & \ldots & \phi_2(t_K) \\ \ldots & \ldots & \ldots & \ldots \\ \phi_N(t_1) & \phi_N(t_2) & \ldots & \phi_N(t_K) \end{bmatrix}$ is the sampling matrix of the basis function vector $\boldsymbol{\phi}(t)$, $PEN_m[x(t)] = \int [D^m x(t)]^2 dt$ is the roughness penalty that is an integrated squared linear differential operator to ensure the continuous control of the function's smoothness, m is the derivative order of the roughness penalty, λ is a parameter defining the smoothness of the fitted curve. According to Equation (3), the basis function coefficients are inferred to be:

$$c = \left[\boldsymbol{\phi}(t) \boldsymbol{\phi}(t)^{\mathrm{T}} + \lambda R \right]^{-1} \boldsymbol{\phi}(t) \hat{x} \tag{4}$$

where $R = \int \frac{d^m \boldsymbol{\phi}(t)}{dt^m} \frac{d^m \boldsymbol{\phi}(t)}{dt^m}^{\mathrm{T}} dt$. Substituting Equation (4) into Equation (2) will obtain the continuous function $x(t)$. The details to determine the tuning parameters, including the order of basis functions, the number of basis functions N, the derivative order of roughness penalty m, and the coefficient of roughness penalty λ, can be found in our previous work [16]. Then, with the continuous function obtained, we can extract their derivatives to reveal their underlying dynamical features. For example, the first-order derivative function can be calculated by the following equation to characterize the function's instantaneous change rate over time.

$$x'(t) = \frac{d[x(t)]}{dt} = \frac{d\left[\boldsymbol{\phi}(t)^{\mathrm{T}} \right]}{dt} c \tag{5}$$

After the continuous function and the first-order derivative function of the total proportion of CO and CO_2 are obtained based on the theory of FDA, the phase plane can be constructed to characterize the dynamics features of BOF steelmaking. Here, the transverse axis is the total proportion of CO and CO_2, evaluating the reaction velocity of the decarburization reaction, and the longitudinal axis is the first-order derivative of the total proportion of CO and CO_2, evaluating the reaction acceleration of decarburization reaction. For other applications, other derivatives even some integration can be used to explain the system's dynamics state.

After the phase plane is constructed, the evolution features of the phase trajectory are used to recognize different blowing periods and anomalies, as well as to predict end-point time.

3.2. Production State Recognition with Dynamics Features

According to the evolution features of phase trajectory described in Section 2.2, the boundary of the stable-state region is estimated via the SVDD [16] to determine the production state, where the phase trajectories of $CO + CO_2$ of normal batches in the main blowing period are used as the training dataset. The calculation of SVDD can be found in Appendix A. Since the phase trajectory of the splashing anomaly exceeds the stable-state region from the fourth quadrant as the phase trajectory of the last blowing period does, the phase plane of CO_2 is constructed to distinguish the splashing anomaly from the last blowing period. The boundary of the stable-state region of the CO_2 phase plane is also estimated via the SVDD.

Then, based on the phase trajectories and their boundaries, the production state is determined at each moment by successively checking whether the phase trajectory of $CO + CO_2$ exceeds the boundary of the stable-state region, whether the phase trajectory of $CO + CO_2$ exceeds the boundary from the left side, and whether the phase trajectory of CO_2 exceeds the boundary of the stable-state region. The check procedure is given in Figure 4. For example, if the previous production state is the main blowing period and the current phase trajectory of $CO + CO_2$ exceeds the boundary, the location of the phase trajectory of $CO + CO_2$ needs to be checked. If the phase trajectory of $CO + CO_2$ is on the left of the boundary, a further check of the phase trajectory of CO_2 is necessary to distinguish the splashing anomaly from the last blowing period; otherwise, we can infer that the drying anomaly happens.

```
Input:    Phase trajectory of total proportion of CO and CO₂, phase trajectory of CO₂,
          boundaries of the two phase planes
Initialize:  t = 1; production_state = 'the first blowing period';
While  1
        If    The phase trajectory of CO+CO₂ enters the boundary
              break
        End
        t = t + 1;
End
production_state = 'the main blowing period'; t = t + 1;
While  1
        If    The phase trajectory of CO+CO₂ exceeds the boundary
              If    The current phase point of CO+CO₂ is on the left of the boundary
                    If    The phase trajectory of CO₂ exceeds the boundary
                          production_state = 'the splashing anomaly';
                          t = t + 1;
                          continue
                    Else
                          production_state = 'the last blowing period';
                          t = t + 1;
                          break
                    End
              Else
                    production_state = 'the drying anomaly';
                    t = t + 1;
                    continue
              End
        End
        t = t + 1;
End
Output:  t; production_state.
```

Figure 4. The check procedure for production state characterization.

3.3. End-Point Time Prediction with Dynamics Features

After the phase trajectory indicates that the last blowing period begins, the end-point time is forecast by predicting the phase trajectory in the last blowing period via case-based reasoning (CBR) [26]. The statement of CBR can be found in Appendix B. Here, the slope curves of the phase trajectories instead of the phase trajectories are employed as attributes to describe the cases/samples, since different batches have different ranges of phase trajectory in the last blowing period. And a lagging window is used to extract the former part of slope curves as the condition attributes and the latter part as the solution attributes. The similar cases are found out according to their distances to the testing case in the problem space are less than a threshold, and their solution attributes are averaged as the solution to the testing case.

The whole procedure of the proposed FDA-PP method is shown in Figure 5. This method firstly smooths the sampling sequences of off-gas data to continuous functions and extracts their first-order derivative functions to construct the phase plane to characterize the metallurgical dynamics features of BOF steelmaking. Then, based on the evolution features of phase trajectories, a boundary that indicates the region of stable production state is estimated to recognize different blowing periods as well as splashing and drying anomalies. Finally, when the last blowing period begins, the end-point time is forecast through referring to the completed batches with similar slope curves of phase trajectories.

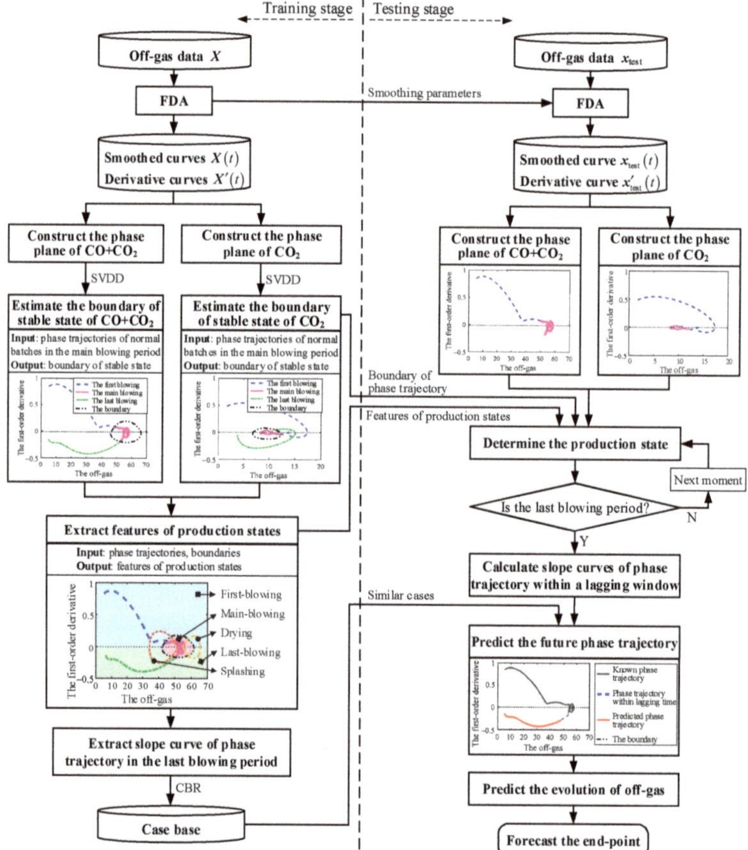

Figure 5. The schematic diagram of the proposed FDA-PP method.

4. Experiment on BOF Steelmaking

In this section, we validate the performance of the proposed FDA-PP method on the BOF steelmaking process.

4.1. Data Acquisition and Modeling Calculation

The data of BOF steelmaking are collected from a 260 tons converter in a Chinese steel plant. The proportions of CO and CO_2 in the off-gas are analyzed by the mass spectrometer with sampling frequency 0.5 Hz. Totally 476 batches/heats without re-blowing are selected, including 375 normal batches, 66 splashing batches, and 35 drying batches. The data partition can be found in Table 1. 150 normal batches are randomly selected for training and the rest 326 batches are used for testing.

Table 1. Data partition of the steelmaking data.

	Normal	Abnormal		Total
		Splashing	Drying	
Training	150	/	/	150
Testing	225	66	35	326
Total	375	66	35	476

In addition, the true end-point time is collected in Figure 6, which is calculated through subtracting the lagging time (30 s in this work) from the duration of the last blowing period. The true end-point time reflects how long in advance the FDA-PP method could predict the end-point. As seen in Figure 6, the true end-point time of over 90% batches is 100~240 s. Thus, the FDA-PP can be expected to predict the end-point and provide a time margin of about 2~4 min for operation optimization.

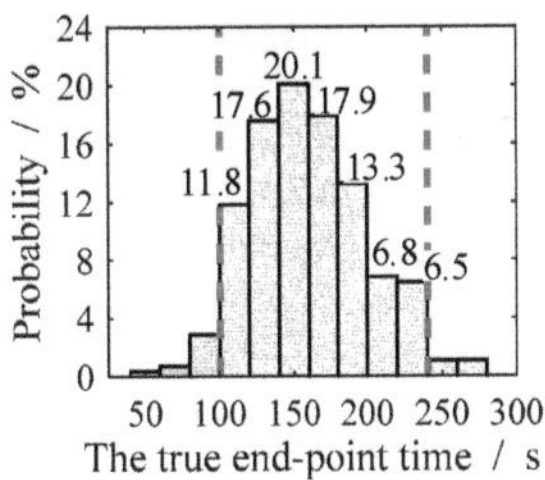

Figure 6. Distribution of the true end-point time.

Following the calculation procedure in Figure 5, firstly, 30 B-splines basis functions with order 5 are used to smooth the off-gas compositions data and extract their first-order derivatives, where the roughness penalty is the integrated squared third derivative operator with $\lambda = 10^5$. It should be noted that the time range of the fitting curve is equal to the actual blowing time of each batch. Based on the fitting curves and the first-order derivative curves, the phase plane of the total proportion of CO and CO_2 is built to characterize the production state, and the phase plane of CO_2 is built to help distinguish the splashing anomaly from the last blowing period. Then, the phase trajectories in the main blowing period are used to estimate the boundaries of the stable-state region for the production state recognition, where the bandwidths of RBF kernel are $\sigma = 3$ for the phase plane of the total proportion of CO and CO_2 and $\sigma = 1.5$ for the phase plane of CO_2. After that, the slope curves of phase trajectory in the last blowing period are recorded, where the slope curves within the lagging time window of 30 s (i.e., 15 sampling points) compose the condition attributes base and the slope curves after the lagging time window compose the solution attributes base. The criterion for similar case selection is that the Euclidean distance with the target batch ranks in the former 15%. Based on the predicted phase trajectory, the end-point time is finally estimated.

In order to further illustrate the performance of the proposed FDA-PP method, the classical phase space (PS) is calculated as comparisons, as well as two commonly used methods in BOF steelmaking, the SVM and the back-propagation neural network (BP). The details of these methods are summarized as follows:

- PS: This method sets the embedding dimension as 2 and the delay time as $\tau = 5$ to construct the phase space, and uses the phase trajectories to complete the rest calculation procedures as the FDA-PP method does.
- SVM: This method is calculated with 13 process variables, including the information of raw materials (i.e., the temperature of molten iron, the carbon content, silicon content,

manganese content, phosphorus content of molten iron, the weight of molten iron, the weight of pig iron, the weight of scrap steel), the cumulative weight of auxiliary materials, and the operation parameters (i.e., the total proportion of CO and CO_2, the height of oxygen lance, the cumulative volume of oxygen blowing, the flow of bottom blowing). Among them, the information of auxiliary materials and operation parameters is recorded as sequences. Since the SVM is a supervised classification method, 10 normal batches, 10 splashing batches, and 10 drying batches are trained for production state recognition. For the end-point time prediction, still 150 normal batches are used for training.

- BP: This method is carried out with the same variables and data partition as the SVM method does, where the size of the hidden layer is set as 60 for the production state recognition and 5 for the end-point prediction.

To avoid randomness of individual results, the FDA-PP method as well as all the compared methods are tested 50 times based on the random data partition. The results are discussed in the following sections.

4.2. Results of the Production State Recognition

Table 2 shows the accuracy of the production state recognition. For the normal batches, we collect the recognition accuracies of the first blowing period, the main blowing period, and the last blowing period, where correctly identifying the blowing periods not only provides guidance for process control but also gives a start signal for the end-point time prediction. For the abnormal batches, we collect the recognition accuracies of the splashing and drying anomalies, where correctly alarming the anomalies can help workers take action to restrain the anomalies.

Table 2. Accuracy of production state recognition.

Method	Blowing Periods Recognition of Normal Batches			Anomaly Identification of Abnormal Batches	
	First-Blowing	Main-Blowing	Last-Blowing	Splashing Anomaly	Drying Anomaly
FDA-PP	99.16%	87.78%	88.49%	90.94%	81.29%
PS	98.22%	82.33%	84.13%	67.00%	74.29%
SVM	87.28%	66.67%	71.93%	62.12%	54.55%
BP	65.59%	43.21%	50.26%	30.30%	41.67%

Table 2 illustrates that the proposed FDA-PP method has the best accuracy on the production state recognition, where the blowing period recognition accuracy of normal batches is 87.78%, the anomaly identification accuracies of splashing and drying batches are 90.94% and 81.29% respectively. For the compared methods, the PS method has lower accuracy than the FDA-PP method because the difference calculation in the PS method would bring errors in expressing the off-gas composition's variation and cannot isolate the noise interference. The SVM and BP methods are inferior to the FDA-PP method and the PS method although they utilize 13 variables to take the information of raw materials, auxiliary materials, and operation parameters into consideration. This result demonstrates that the dynamics features expressed by the phase plane and phase space are significant in determining a system's operating condition. For BOF steelmaking, different blowing periods and anomalies would generate similar production data. The SVM and BP methods analyze the measurements at an individual moment and miss the evolution features of the sequence within an interval, so easily confuses different production states. The FDA-PP and PS methods not only indicate the decarburization rate but also describe how the decarburization rate is going to change with the influence of operation parameters, so have better performance on the production state recognition.

We take a splashing batch and a drying batch as examples to show the results of production state recognition, as can be seen in Figure 7. The FDA-PP method successfully recognizes the three blowing periods as well as the splashing and drying anomalies. The PS method successfully identifies the drying batch but confuses the splashing anomaly with the last blowing period. The SVM method identifies different blowing periods but misses catching the drying anomaly, and it gives a false alarm before the splashing anomaly. The BP method nearly cannot judge the production state of the splashing batch and falsely regards the drying as the splashing.

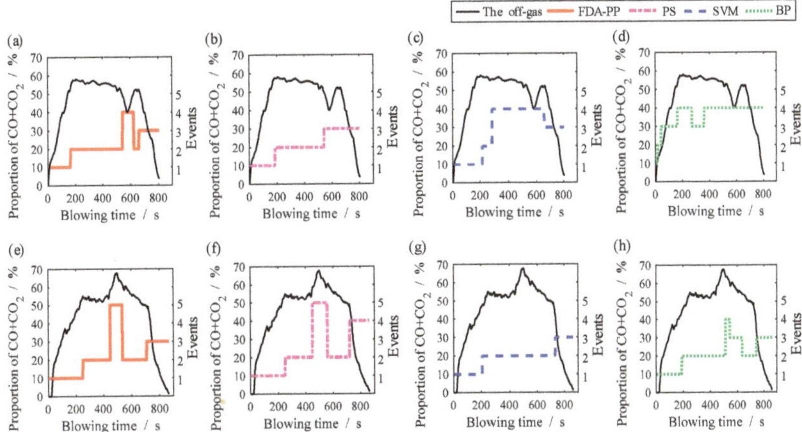

Figure 7. (**a–d**) Production state recognition of a splashing batch by different methods; (**e–h**) production state recognition of a drying batch by different methods. Here, the events indicate that: 1-the first blowing period, 2-the main blowing period, 3-the main blowing period, 4-the splashing anomaly, 5-the drying anomaly.

Furthermore, as the production state is recognized at each sampling point, the FDA-PP method is able to be implemented on the online characterization, where the production state at sampling point t is recognized using the sequence from the start point to the current point.

4.3. Results of the End-Point Time Prediction

Based on the production state recognition, the end-point time is going to be predicted as the last blowing period begins. Thus, the ending time prediction is a forecast ahead of time and can be available for the online application, where the end-point can be predicted using the sequence within the lagging time during the last period. Table 3 collects the mean relative prediction error (MRPE) of the end-point time, which is calculated as follows:

$$\delta = \frac{1}{N} \cdot \frac{\left|\Delta \hat{t}_n - \Delta t_n\right|}{\Delta t_n} \tag{6}$$

where Δt_n is the true value of the end-point time of batch n which is calculated through subtracting the lagging time (30 s in this work) from the duration of the last blowing period, $\Delta \hat{t}_n$ is the predicted value of the end-point time of batch n, and N is the number of batches. For the SVM and BP methods, since Section 4.2 indicates that they can hardly identify different blowing periods, the beginning moment of the last blowing period in the SVM and BP methods is provided by the FDA-PP method.

Table 3. MRPE of end-point time prediction.

Method	Normal Batches	Splashing Batches	Drying Batches	Mean Value
FDA-PP	18.07%	19.31%	17.01%	18.13%
PS	41.36%	49.79%	46.91%	46.02%
SVM	19.53%	31.28%	26.78%	25.86%
BP	19.44%	32.58%	29.94%	27.32%

From Table 3 we can see that the proposed FDA-PP method is much more accurate than the PS method and more accurate on the abnormal batches than the SVM and BP methods. The FDA-PP method can predict the ending time with MRPE 18.07%. Especially when the splashing or drying anomaly happens, the MRPE of the end-point time can be controlled within 20%. As for the SVM and BP methods, although they achieve similar MRPE on the normal batches as the FDA-PP method, their prediction errors on the abnormal batches are about 10% larger than the FDA-PP method. The superiority of FDA-PP on the abnormal batches is due to that the FDA-PP method predicts the end-point through estimating the evolution of dynamics features (the phase trajectory) but the SVM and BP methods only utilize the measurements of production data. When the anomalies happen, the dynamics features would change, and the decarburization reaction rate would be influenced, bringing difficulties in end-point prediction. Therefore, the SVM and BP methods have larger prediction errors on the abnormal batches.

Figure 8 shows the distribution of the absolute deviation of the predicted end-point time, i.e., the numerator in Equation (6). As can be seen in Figure 8, using the FDA-PP method, the end-point time prediction of more batches can be controlled in smaller absolute deviations than the PS, SVM, and BP methods, especially of the abnormal batches. The FDA-PP method can ensure the absolute deviation of the predicted end-point time on 20% splashing batches and 27% drying batches no more than 10 s, and on 60% splashing batches and 61% drying batches no more than 30 s. As comparison, the PS, SVM, and BP methods can hardly control the absolute deviation on abnormal batches to be in 10 s, and their ratios of abnormal batches whose absolute deviations are no more than 30 s are less than the FDA-PP method.

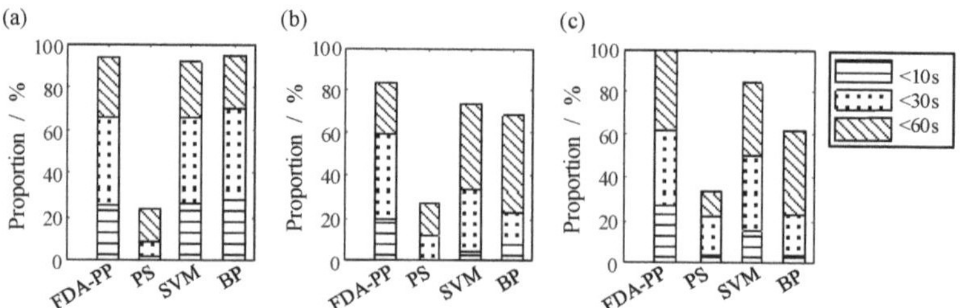

Figure 8. Absolute deviation of the predicted end-point time on (**a**) the normal batches, (**b**) the splashing batches, (**c**) the drying batches.

More physical insights into the production state prediction of BOF steelmaking are that, first, with the predicted phase trajectory, the evolution of the total proportion of CO and CO_2 in the last blowing period can be reconstructed, then further studies such as the steel compositions prediction and the temperature prediction can be realized associated with information of raw materials and operation parameters. Second, although the FDA-PP method only utilizes the off-gas data, it realizes similar even better performance than the SVM and BP methods that process variables; thus, the off-gas can be expected to replace the sub-lance to control the BOF steelmaking so that avoiding the production process suspension.

5. Conclusions

This work aims at the production state recognition and the end-point time prediction, and proposes a new data-based method associated with the off-gas data, defined as FDA-PP. This method describes the evolution of the production state from the metallurgical dynamics viewpoint and takes a comprehensive consideration of the decarburization rate as well as its variation trend. As the change of off-gas compositions was only used to recognize the splashing and drying anomalies in the existing research, this work is an attempt that using the off-gas data to distinguish different blowing periods and predict the end-point time.

In order to evaluate the performance of the proposed method, the real production data from a steel plant are studied. The results show that the FDA-PP method achieves the production state recognition accuracy at 87.78% for normal batches, 90.94% for splashing batches, and 81.29% for drying batches, respectively, more accurate than the PS, SVM, and BP methods. At the same time, the MRPE of end-point time prediction by the FDA-PP method is lower than the compared methods. Especially on the abnormal batches, although the FDA-PP method only employs the proportions of CO and CO_2 in the off-gas, its prediction error of ending time is about 10% lower than the SVM and BP methods that utilize 13 variables to take the information of raw materials, auxiliary materials, and operation parameters into consideration. Moreover, the FDA-PP method can predict the end-point time once the model judges that the last blowing period begins, so providing 2~4 min time margin for the operation parameters optimization.

Following this work, the predicted sequences of the total proportion of CO and CO_2 by the FDA-PP method give a possibility to predict the steel compositions and temperature. In addition, more efforts can be devoted to exploring the use of off-gas information for production control so that replacing the sub-lance and ensuring continuous production rhythm.

Author Contributions: Conceptualization, M.L. and Q.Q.; Investigation, M.L., Q.Q., Q.D. and J.X.; Resources, M.L. and W.Z.; Methodology, Q.Q. and M.L.; Software, Q.Q. and Q.D.; Validation, Q.Q. and Q.D.; Formal Analysis, Q.Q.; Writing—Original Draft Preparation, Q.Q.; Writing—review & Editing, Q.Q. and M.L.; Supervision, M.L. and J.X.; Project Administration, M.L.; Funding Acquisition, M.L. All authors have read and agreed to the published version of the manuscript.

Funding: This work was supported by the Guangdong Province Project for Research and Development in Key Areas (Grant No.2020B0101130007).

Data Availability Statement: Restrictions apply to the availability of these data. Data was obtained from Angang Steel Company Limited and are available from the corresponding author with the permission of Angang Steel Company Limited.

Conflicts of Interest: The authors declare no conflict of interest.

Appendix A. Calculation of Support Vector Data Description

The support vector data description (SVDD) is a single-class classification model. It works by estimating a minimum spherical boundary around the normal samples in a high-dimensional feature space so that the abnormal samples are outside the boundary. Taking the phase trajectories of normal batches in the main blowing period as the training dataset, the optimization problem of SVDD is as follows:

$$\min_{a,R,\xi} R^2 + C\sum_i \xi_i$$
$$s.t. \ \|g(z_i) - a\|^2 \leq R^2 + \xi_i, \ \xi_i \geq 0 \tag{A1}$$

where $z_i = [x_n(t_k) \ x'_n(t_k)]$ is the phase trajectory of batch n at moment t_k, $g(\cdot)$ is a non-linear function to settle the nonlinearity, a and R are the center and radius of the sphere respectively, C is an adjustable parameter balancing the sphere volume and model error, $\xi_i \geq 0$ is a slack variable allowing outliers in the training dataset. Based on the Lagrange multipliers and the Karush-Kuhn-Tucker (KKT) condition, the optimization problem in Equation (A1) can be converted to be

$$\max_{\gamma} \sum_i \gamma_i g(z_i)^T g(z_i) - \sum_i \sum_p \gamma_i \gamma_p g(z_i)^T g(z_p) = \max_{\gamma} \sum_i \gamma_i \kappa(z_i, z_i) - \sum_i \sum_p \gamma_i \gamma_p \kappa(z_i, z_p)$$
$$s.t. \ 0 \leq \gamma_i \leq C, \sum_i \gamma_i = 1 \tag{A2}$$

where γ_i and γ_p are Lagrange multipliers of z_i and z_p, $\kappa(z_i, z_p) = \langle g(z_i), g(z_p) \rangle = g(z_i)^T g(z_p)$ is the kernel function. Settling the above optimization problem can obtain the optimal values of Lagrange multipliers, and the training samples satisfying $\gamma_i > 0$

constitute the support vectors, i.e., $SV = \{z_i | \gamma_i > 0\}$. These support vectors determine the boundary of the stable-state region, which can be computed as:

$$R^2 = \kappa(z_l, z_l) - 2 \sum_{z_i \in SV} \gamma_i \kappa(z_i, z_l) + \sum_{z_i \in SV} \sum_{z_p \in SV} \gamma_i \gamma_p \kappa(z_i, z_p) \tag{A3}$$

where z_l is one of the support vectors. In practice, the kernel function mostly employs the radial basis function (RBF) kernel, that is:

$$\kappa(z_i, z_p) = \exp\left(-\frac{\|z_i - z_p\|^2}{\sigma^2}\right) \tag{A4}$$

where σ is the bandwidth that ensures a tight sphere. Thus, the first item in Equation (A3) can be simplified to be $\kappa(z_l, z_l) = 1$. In testing stage, whether the phase trajectory exceeds the stable-state region or not can be determined by the following equation:

$$r_{\text{test}}^2 = 1 - 2 \sum_{z_i \in SV} \gamma_i \kappa(z_i, z_{\text{test}}) + \sum_{z_i \in SV} \sum_{z_p \in SV} \gamma_i \gamma_p \kappa(z_i, z_p) > R^2 \tag{A5}$$

Appendix B. Statement of Case-Based Reasoning

The case-based reasoning (CBR) is a branch of artificial intelligence, where a case/sample is described by the condition attributes (the inputs) in problem space and the solution attributes (the outputs) in solution space. The basic idea of CBR is that similar cases in problem space are also close to each other in solution space. It consists of the following steps: case description, case retrieval, case reuse, case revise and retain.

In the case description step, appropriate condition attributes and solution attributes of cases needs to be determined. Since the phase trajectory in the last blowing period has different range for different batch, the slope curves of the phase trajectories instead of the phase trajectories are employed as the attribute to describe the case. In detail, a lagging window is used to extract the former part of slope curves as the condition attributes and the later part as the solution attributes. Furthermore, due to the blowing time needs to be predicted, the phase trajectory is resampled to transform the time-dependent curves $(x(t), x'(t))$ to the amplitude-dependent curves $[\alpha, f(\alpha)]$, where $\alpha = x(t)$ and is normalized to $[0, 1]$, $f(\alpha) = x'(t)$.

In the case retrieval step, a criterion needs to be determined to define the similar cases in problem space. In this work, the similar cases are found out according to their Euclidean distances to the testing case in the problem space with a certain threshold.

In the case reuse step, an approach needs to be settled to combine the solution attributes of the retrieval cases. In this work, the solution attributes of the similar cases, i.e., their slope curves after the lagging window, are averaged as the solution to the testing case. Thus, the phase trajectory in the last blowing period of the testing batch can be constructed, further the end-point time can be forecast.

In the case revise and retain step, if the similarities between the testing case with all the cases in the case base are less than a certain threshold, the solution attributes of the testing case can be modified according to the suggested solution attributes, and the testing case can be added to the case base to expand the solution attributes. In this work, the case revise and retain step is untapped in model calculation.

References

1. World Steel Association. Fact Sheet: Steel and Raw Materials. Available online: https://worldsteel.org/steel-topics/statistics/world-steel-in-figures-2022/#crude-steel-production-by-process-2021 (accessed on 3 November 2022).
2. De Vos, L.; Bellemans, I.; Vercruyssen, C.; Verbeken, K. Basic Oxygen Furnace: Assessment of Recent Physicochemical Models. *Metall. Mater. Trans. B* **2019**, *50*, 2647–2666. [CrossRef]
3. Heenatimulla, J.; Brooks, G.A.; Dunn, M.; Sly, D.; Snashall, R.; Leung, W. Acoustic Analysis of Slag Foaming in the BOF. *Metals* **2022**, *12*, 1142. [CrossRef]

4. Singha, P.; Shukla, A.K. Contribution of Hot-Spot Zone in Decarburization of BOF Steel-Making: Fundamental Analysis Based upon the FactSage-Macro Program. *Metals* **2022**, *12*, 638. [CrossRef]
5. Shanmugam, S.P.; Nurni, V.N.; Manjini, S.; Chandra, S.; Holappa, L.E.K. Challenges and Outlines of Steelmaking toward the Year 2030 and Beyond—Indian Perspective. *Metals* **2021**, *11*, 1654. [CrossRef]
6. Zhou, D.; Xu, K.; Lv, Z.; Yang, J.; Li, M.; He, F.; Xu, G. Intelligent Manufacturing Technology in the Steel Industry of China: A Review. *Sensors* **2022**, *22*, 8194. [CrossRef] [PubMed]
7. Nenchev, B.; Panwisawas, C.; Yang, X.; Fu, J.; Dong, Z.; Tao, Q.; Gebelin, J.-C.; Dunsmore, A.; Dong, H.; Li, M.; et al. Metallurgical Data Science for Steel Industry: A Case Study on Basic Oxygen Furnace. *Steel Res. Int.* **2022**, *93*, 2100813. [CrossRef]
8. Wang, R.; Mohanty, I.; Srivastava, A.; Roy, T.K.; Gupta, P.; Chattopadhyay, K. Hybrid Method for Endpoint Prediction in a Basic Oxygen Furnace. *Metals* **2022**, *12*, 801. [CrossRef]
9. Zou, Y.; Yang, L.; Li, B.; Yan, Z.; Li, Z.; Wang, S.; Guo, Y. Prediction Model of End-Point Phosphorus Content in EAF Steelmaking Based on BP Neural Network with Periodical Data Optimization. *Metals* **2022**, *12*, 1519. [CrossRef]
10. Han, M.; Zhao, Y. Dynamic control model of BOF steelmaking process based on ANFIS and robust relevance vector machine. *Expert Syst. Appl.* **2011**, *38*, 14786–14798. [CrossRef]
11. Fei, H.; Xianyi, C.; Zhenghai, Z. Prediction of oxygen-blowing volume in BOF steelmaking process based on BP neural network and incremental learning. *High Temp. Mater. Process.* **2022**, *41*, 403–416. [CrossRef]
12. Zhao, D. Research on Prediction of Converter Endpoint Based on Image Processing. Master's Thesis, Inner Mongolia University of Science and Technology, Baotou, China, 2020.
13. Li, A.; Li, C.; Wang, Y.; Cui, G.; Xie, S. Recognition of converter blowing periods based on improved DenseNet Network. *J. Univ. Jinan* **2022**, *36*, 273–277. [CrossRef]
14. Wen, H. Research on Modeling and Prediction of BOF End-Point Based on the Furnace Mouth Radiation Information. Ph.D. Thesis, Nanjing University of Science and Technology, Nanjing, China, 2009.
15. Brämming, M.; Bjorkman, B.; Samuelsson, C. BOF process control and slopping prediction based on multivariate data analysis. *Steel Res. Int.* **2016**, *87*, 301–310. [CrossRef]
16. Qian, Q.; Fang, X.; Xu, J.; Li, M. Multichannel profile-based anomaly detection and its application in the monitoring of basic oxygen furnace steelmaking process. *J. Manuf. Syst.* **2021**, *61*, 375–390. [CrossRef]
17. Li, G.-H.; Wang, B.; Liu, Q.; Tian, X.-Z.; Zhu, R.; Hu, L.-N.; Cheng, G.-G. A process model for BOF process based on bath mixing degree. *Int. J. Miner. Metall. Mater.* **2010**, *17*, 715–722. [CrossRef]
18. Wang, Z. Study on the Control of Steelmaking Process and Blowing End-Point for Medium-High Carbon Steel Melting by Converter. Ph.D. Thesis, University of Science and Technology Beijing, Beijing, China, 2016.
19. Li, N.; Lin, W.; Cao, L.; Liu, Q.; Sun, L.; Liao, S. Carbon prediction model for basic oxygen furnace off-gas analysis based on bath mixing degree. *Chin. J. Eng.* **2018**, *40*, 1244–1250. [CrossRef]
20. Rahnama, A.; Li, Z.; Sridhar, S. Machine Learning-Based Prediction of a BOS Reactor Performance from Operating Parameters. *Processes* **2020**, *8*, 371. [CrossRef]
21. Dering, D.; Swartz, C.L.; Dogan, N. A dynamic optimization framework for basic oxygen furnace operation. *Chem. Eng. Sci.* **2021**, *241*, 116653. [CrossRef]
22. Deo, B.; Overbosch, A.; Snoeijer, B.; Das, D.; Srinivas, K. Control of Slag Formation, Foaming, Slopping, and Chaos in BOF. *Trans. Indian Inst. Met.* **2013**, *66*, 543–554. [CrossRef]
23. Bae, S.; Mo, D.Y. Dynamical analysis of an anemoscope in the phase plane. *Appl. Math. Model.* **2010**, *34*, 1884–1891. [CrossRef]
24. Han, Z.; Xu, N.; Chen, H.; Huang, Y.; Zhao, B. Energy-efficient control of electric vehicles based on linear quadratic regulator and phase plane analysis. *Appl. Energy* **2018**, *213*, 639–657. [CrossRef]
25. Ramsay, J.O.; Silverman, B.W. *Functional Data Analysis*, 2nd ed.; Springer: New York, NY, USA, 2005.
26. Han, M.; Li, Y.; Cao, Z. Hybrid intelligent control of BOF oxygen volume and coolant addition. *Neurocomputing* **2014**, *123*, 415–423. [CrossRef]

 *metals*

Article

The Formation Mechanisms and Evolution of Multi-Phase Inclusions in Ti-Ca Deoxidized Offshore Structural Steel

Zhe Rong [1],* , Hongbo Liu [2], Peng Zhang [2], Feng Wang [2], Geoff Wang [1], Baojun Zhao [1,3], Fengqiu Tang [1] and Xiaodong Ma [1],*

[1] Sustainable Minerals Institute, The University of Queensland, Brisbane 4072, Australia; gxwang@uq.edu.au (G.W.); baojun@uq.edu.au (B.Z.); f.tang@uq.edu.au (F.T.)
[2] HBIS Group Co., Ltd., Shijiazhuang 050000, China; ustbliuhongbo@163.com (H.L.); zhangpengysu@126.com (P.Z.); wangfeng02@hbisco.com (F.W.)
[3] Faculty of Materials Metallurgy and Chemistry, Jiangxi University of Science and Technology, Ganzhou 341000, China
* Correspondence: z.rong@uq.edu.au (Z.R.); x.ma@uq.edu.au (X.M.)

Abstract: To understand and clarify the formation mechanisms and evolution of complex inclusions in Ti-Ca deoxidized offshore structural steel, inclusions in industrial steel were systematically investigated. The number density of total inclusions generally decreased from Ladle Furnace (LF), Vacuum Degassing (VD), Tundish to the final product except for Ti and Ca addition. The major inclusions during the refining process were $CaO-Al_2O_3-SiO_2-(MgO)-TiO_x$ and $CaO-Al_2O_3-SiO_2$. $CaO-Al_2O_3-SiO_2-(MgO)-TiO_x$ inclusion initially originated from the combination of $CaO-SiO_2-(MgO)$ in refining slag or refractory and deoxidization product Al_2O_3 and TiO_2. With the refining process proceeding and Ca addition, the Al_2O_3 concentration in the $CaO-Al_2O_3-SiO_2-(MgO)-TiO_x$ inclusions gradually dropped while the CaO and TiO_2 concentrations gradually increased. The $CaO-Al_2O_3-SiO_2$ inclusions originally came from refining slag, existing as $2CaO \cdot Al_2O_3 \cdot SiO_2$, and maintained a liquid state during the early stage of LF. After Ca treatment, it was gradually transferred to $2CaO \cdot SiO_2$ due to Al_2O_3 continuously being reduced by Ca. The liquidus of $2CaO \cdot SiO_2$ inclusion was higher than that of molten steel, so they presented as a solid-state during the refining process. After welding thermal simulation, $CaO-Al_2O_3-SiO_2-(MgO)-TiO_x$ inclusions were proven effective for inducing intragranular acicular ferrite (IAF) while $CaO-Al_2O_3-SiO_2$ was inert for IAF promotion. Additionally, Al_2O_3-MgO spinel in multiphase $CaO-Al_2O_3-SiO_2-(MgO)-TiO_x$ inclusion has different formation mechanisms: (1) initial formation as individual Al_2O_3-MgO spinel as a solid-state in molten steel; (2) and it presented as a part of liquid inclusion $CaO-Al_2O_3-SiO_2-(MgO)-TiO_x$ and firstly precipitated due to its low solubility.

Keywords: oxide metallurgy; high heat input welding; Ti-Ca deoxidation; inclusion evolution; formation mechanism

Citation: Rong, Z.; Liu, H.; Zhang, P.; Wang, F.; Wang, G.; Zhao, B.; Tang, F.; Ma, X. The Formation Mechanisms and Evolution of Multi-Phase Inclusions in Ti-Ca Deoxidized Offshore Structural Steel. *Metals* **2022**, *12*, 511. https://doi.org/10.3390/met12030511

Academic Editor: Mark E. Schlesinger

Received: 17 February 2022
Accepted: 14 March 2022
Published: 17 March 2022

Publisher's Note: MDPI stays neutral with regard to jurisdictional claims in published maps and institutional affiliations.

1. Introduction

With the development of large-scale engineering structures, high heat input welding has been widely used in shipbuilding, marine engineering, and oil container fields due to its advantages, such as high efficiency and high stability. However, the increment in the heat input will result in the coarsening of grains in the heat affect zone (HAZ), the origin of a tiny cleavage crack, and a heterogeneous microstructure, thereby significantly deteriorating the toughness [1].

To address these problems, researchers have started to focus on inclusions that are inevitable in metal products. Some non-metallic inclusions can have positive impacts on the steel microstructure and mechanical properties [2–4]. Initially, in the 1970s, nanosized TiN particles were found to be effective at inhibiting the growth of austenite grains and beneficial for the mechanical property of steel [5]. They were formed during the cooling process after solidification due to their relatively low solubility. In the 1990s, titanium oxide-containing inclusions, formed in the molten steel and solidification process, were reported to act effectively as intragranular nucleation sites for acicular ferrite, contributing to a smaller grain size and consequently improving the weldability of steel. These practical inclusions were formed during the second refining process and had a relatively high liquidus, so they had excellent thermostability. The technology of titanium oxides and their positive effect on the steel microstructure and phase transformation behaviour was proposed and termed "oxide metallurgy" [6–9]. Since then, the understanding of inclusions in the steelmaking industry has been rejuvenated. Harmful inclusions, such as large-sized and Al_2O_3 inclusions that may cause clogging, should be removed as much as possible while those with a small size and particular chemical composition can be maintained and utilized to achieve outstanding mechanical properties [10,11]. Oxide metallurgy has been one of the most effective methods used to address the toughness problem in HAZ.

In recent decades, the use of strong deoxidizers, such as Ca, Mg, Zr, and rare Earth metals (REMs) [10,12–15], has attracted considerable research interest due to their stronger affinity with O and S and higher thermal stability of inclusions. Traditionally, Ca treatment is mostly used for modifying Al_2O_3 inclusions with a high melting point into CaO-Al_2O_3 with a lower melting point, thereby solving the nozzle clogging problem [16–18]. Kobe Steel also reported that modified inclusion of CaO-Al_2O_3 could help prohibit the coarse TiN precipitating on oxides and facilitate the formation and dispersive distribution of fine TiN particles [19]. So far, research has mainly focused on Ca treatment in Ti-bearing Al-killed steel. Wang et al. [20] compared Al-Ca deoxidized steel and Ti-Ca deoxidized steel in the lab, pointing out that CaO-Al_2O_3-TiO_2-MnS-related inclusions were dominant and active in facilitating fine IAF formation [21,22]. CaO-Al_2O_3-TiO_2 will act as a nucleation core while MnS will precipitate around the core during the cooling process and form a manganese depletion zone (MDZ) to induce IAF formation [23–25].

Although several investigations regarding Ti-Ca deoxidized steel have been reported, they mainly focused on the effect of different deoxidizers on inclusions in final products or the effect of inclusions on the microstructure in HAZ. However, few studies harnessing the integrated understanding of inclusion evolution in each step of actual industry practice are available. In this paper, the inclusion characteristics regarding categories, number density, and evolution were systematically investigated to understand and clarify the formation mechanisms and evolution of complex inclusions in Ti-Ca-treated offshore structural steel, which will benefit the translation Ti-Ca oxide metallurgy of offshore steels.

2. Materials and Methods

Steel samples were obtained from an industrial process for offshore structural steel, which followed the sequential steps through a basic oxygen furnace (BOF)→ladle furnace (LF)→vacuum degassing (VD)→continuous casting (CC)→thermomechanical control process (TMCP) in Wusteel, HBIS. Ferrosilicon (Si $\geq$ 75.0%, Al $\leq$ 1.5%, Fe $\geq$ 21.0%) and ferromanganese (65.0–72.0% Mn, Si $\leq$ 2.5%, C $\leq$ 7.0%) alloys together with lime were added to the molten steel during BOF tapping. Then, the molten steel was transferred into LF for refining and bottom argon blowing was adopted for homogenizing the steel composition and accelerating the removal of large-sized inclusions. LF refractories were composed of MgO-C bricks. After refining in LF at 1650 °C, the molten steel was transferred to VD at 1600 °C and was eventually transported into tundish for continuous casting. In particular, samples were collected at the beginning of LF 1 min after Ti addition (Ti-Fe: 25.0–35.0%Ti, Al $\leq$ 8.0%, Si $\leq$ 4.5%), 1 min after Ca addition, VD soft blowing, tundish, and hot-rolled plate (final products). The chemical compositions of the steel samples from different stages are shown in Table 1. The product sample underwent heat treatment as welding thermal simulation at a peak temperature of 1400 °C. The measured temperature curve of the welding thermal simulation is shown in Figure 1. The welding samples were held at 1400 °C for 180 s, and the cooling rates of the 1400–800 °C and 800–500 °C range were approximately 3 and 0.75 °C/s (t8/5 = 400 s), respectively.

Table 1. Chemical composition of samples taken from different stages (wt%).

No.	C	Si	Mn	P	S	Nb	V	Al	O	Ti	Ca
LF begin	0.025	0.089	1.29	0.013	0.025	-	-	0.0062	0.0110	-	-
Ti added	0.032	0.201	1.31	0.013	0.0070	0.02	0.039	0.0065	0.0068	0.0042	-
Ca added	0.048	0.200	1.43	0.013	0.0034	0.02	0.039	0.0058	0.0031	0.015	0.0015
VD	0.066	0.214	1.47	0.013	0.0022	0.022	0.043	0.0052	0.0035	0.010	0.0026
Tundish	0.065	0.216	1.47	0.013	0.0021	0.023	0.043	0.0047	0.0032	0.010	0.0016
TMCP	0.069	0.222	1.52	0.013	0.0021	0.023	0.044	0.0040	0.0031	0.014	0.0014

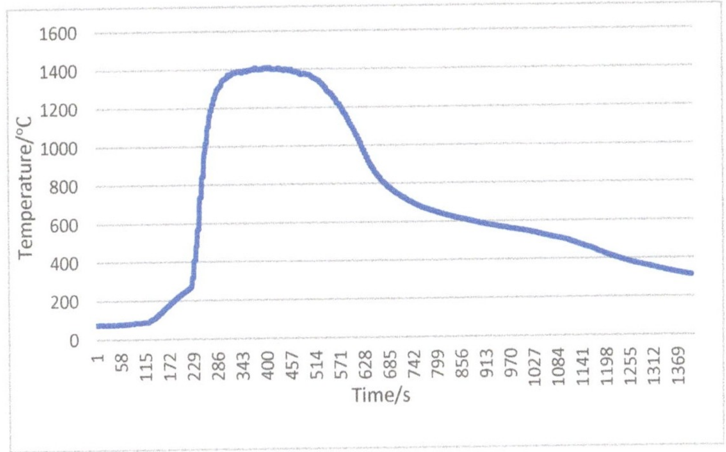

Figure 1. The measured temperature curve of welding thermal simulation.

Steel samples were mounted in resin and ground, polished, and etched for inclusion composition analysis and microstructure observation. Composition analysis of inclusions was conducted by using a JXA 8200 electron probe microanalyzer (EPMA, JEOL, Akishima, Japan) with wavelength-dispersive spectrometers (WDSs).

The EPMA operating details are as follows: the standard samples are $CaSiO_3$ for Ca and Si, pure MgO for Mg, pure Al_2O_3 for Al, pure TiO_2 for Ti, Spessartine for Mn, Fe_2O_3 for Fe, and FeS for S. An acceleration voltage of 15 kV, probe current of 15 nA, probe diameter of "zero" (the smallest operating probe diameter achieved by the focused electron beams), peak measuring time of 30 s, and background measuring time of 10 s were used. Backscattered electron imaging and EDS analysis were utilized to recognize the non-metallic inclusions, especially those complex inclusions with multiple phases. Then, WDS was conducted on the inclusions to analyze the inclusion composition accurately. EPMA can accurately determine the composition of inclusions exceeding 1.5 μm. Field emission scanning electron microscopy (FE-SEM) was used to analyze inclusions sized 1.5 μm or less by element spot analysis, line scanning, and map scanning to demonstrate the elementary distribution of multiphase inclusions.

To describe the number densities of inclusions, 20 photos were taken randomly under 500 magnifications for each sample to ensure reliable statistics. The area of each photo was 0.246 mm × 0.196 mm = 0.0482 mm^2, so the total observation area of each sample was 0.964 mm^2.

3. Results and Discussions

3.1. Characteristics of the Inclusions at Different Stages

Due to complex alloy systems including Si, Mn, Al, Ti, and Ca related to the deoxidization process, the inclusions can be classified into nine groups: Al_2O_3-MnO, Al_2O_3-

SiO_2-MnO, CaO-Al_2O_3-SiO_2-MnO, CaO-SiO_2, CaO-Al_2O_3-SiO_2, CaO-Al_2O_3-SiO_2-TiO_x, Al_2O_3-MnO-TiO_x, SiO_2-MnO, and CaO-TiO_x, as shown in Table 2. Generally, several valences of Ti, such as TiO, TiO_2, Ti_2O_3, and Ti_3O_5, exist in steel, and it is assumed that Ti-oxide was mainly TiO_2 when analyzing the inclusion composition, so a TiO_2 standard sample was used during the EPMA analysis. Among these inclusions, Al_2O_3-MnO, Al_2O_3-SiO_2-MnO, and CaO-Al_2O_3-SiO_2-MnO are either primary deoxidization products or those that contacted with lime, which was added during the BOF tapping process. CaO-SiO_2 and CaO-Al_2O_3-SiO_2 are grouped into slag entrapment products (large-sized inclusions) and reduction products (small-sized inclusions). CaO-Al_2O_3-SiO_2-TiO_x and CaO-TiO_x belong to the refining products. Al_2O_3-MnO-TiO_x and SiO_2-MnO inclusions are considered to be secondary oxidization products during the vacuum break of VD and Tundish processes.

Table 2. Classification of inclusions based on the chemical compositions (wt%).

Inclusions	Chemical Composition					
	CaO	Al_2O_3	SiO_2	MgO	TiO_2	MnO
Al_2O_3-MnO	<10%	>60%	<10%	<10%	<10%	>10%
Al_2O_3-SiO_2-MnO	<10%	>10%	>20%	<10%	<10%	>10%
CaO-Al_2O_3-SiO_2-MnO	>10%	>10%	>10%	<10%	<10%	>10%
Al_2O_3-MnO-TiO_x	<10%	>20%	<10%	<10%	>30%	>10%
SiO_2-MnO	<10%	>30%	<10%	<10%	<10%	>30%
CaO-SiO_2	>40%	<10%	>10%	<10%	<10%	<10%
CaO-Al_2O_3-SiO_2	>20%	>20%	>20%	<10%	<10%	<10%
CaO-Al_2O_3-SiO_2-TiO_x	>20%	>10%	>10%	<10%	>10%	<10%
CaO-TiO_x	>30%	<10%	<10%	<10%	>30%	<10%

The initial inclusions in the LF-entry sample include Al_2O_3-MnO, Al_2O_3-SiO_2-MnO, CaO-Al_2O_3-SiO_2-MnO, and CaO-SiO_2. Among these inclusions, Al_2O_3-MnO and Al_2O_3-SiO_2-MnO are primary deoxidization products with small sizes generally less than 3 μm. Although Fe-Si and Fe-Mn ferroalloys are mainly used as pre-deoxidizer in industrial practice, inevitably a small amount of Al impurities exist in alloys. As Al has a stronger ability to combine with oxygen than Si and Mn, almost all the primary deoxidization products contained Al_2O_3. CaO-Al_2O_3-SiO_2-MnO can be considered as the coalescence between primary deoxidization products and CaO from lime and CaO-SiO_2 from refining slag that was added to molten steel during BOF tapping. CaO-SiO_2 inclusions in the LF entry stage were generally larger than 10 μm, but the proportion of CaO-SiO_2 inclusions in this stage was very low. They originated from slag entrapment. After Ti was added to LF, the main inclusion types were CaO-Al_2O_3-SiO_2 and CaO-Al_2O_3-SiO_2-TiO_x 1–10 μm in size. The dissolved aluminium coming from continuous reaction of ferroalloys with CaO-SiO_2 from refining slag then formed CaO-Al_2O_3-SiO_2 inclusions. The additive Ti entered molten steel, reacting with CaO-Al_2O_3-SiO_2 inclusions and forming CaO-Al_2O_3-SiO_2-TiO_2 inclusions. After Ca was added to the molten steel, a considerable proportion of CaO-SiO_2 inclusions was observed apart from CaO-Al_2O_3-SiO_2 and CaO-Al_2O_3-SiO_2-TiO_x inclusions due to the severe slag entrapment caused by the splashing of the Ca addition, and the reduction of Al_2O_3 by additive calcium. In the following VD and Tundish, until the final products, CaO-SiO_2, CaO-Al_2O_3-SiO_2, and CaO-Al_2O_3-SiO_2-TiO_x inclusions remained the main types. The most significant difference was that Al_2O_3-MnO-TiO_x and SiO_2-MnO inclusions were observed after the vacuum break of VD and Tundish. In general, these were typical re-oxidization products due to the excessive surface turbulence during molten steel teeming into the Tundish ladle. However, these kinds of re-oxidization products were seldom found in the final products, which indicated that most of them floated upwards into the mold flux and were removed effectively. In the final products, CaO-TiO_x inclusions were found apart from CaO-SiO$_2$, CaO-Al$_2$O$_3$-SiO$_2$, and CaO-Al$_2$O$_3$-SiO$_2$-TiO$_x$ inclusions. They may be generated from either the enrichment of the CaO component and the reduction of Al$_2$O$_3$ and SiO$_2$, or the precipitation of CaO-TiO$_x$ from CaO-Al$_2$O$_3$-SiO$_2$-TiO$_x$ inclusions during the cooling process.

3.2. Number Density and Size Distribution of Inclusions

The number density of all inclusions in industrial steel is shown in Figure 2. The primary inclusions in the LF entry sample were generally primary deoxidization products, such as Al$_2$O$_3$-MnO and Al$_2$O$_3$-SiO$_2$-MnO, featuring a high proportion of fine inclusions that were less than 1 μm and had a number density of 1560/mm^2. With ladle furnace refining processing, when Fe-Ti alloy was added to the molten steel, the total amount of inclusions decreased dramatically to about 300/mm^2, especially those with a small size (<1 μm). This is due to the coalescence of small inclusions, floating upward into the refining slag. When Si-Ca wire was added to the molten steel and the end of LF refining, the total amount of inclusions increased slightly to higher than 400/mm^2 owing to the severe splashing during Si-Ca wire feeding. It also featured a higher proportion of 1–2 μm and 2–5 μm inclusions, and this can be explained by inclusion coalescence and growth to be larger ones. In the VD section, the total amount of inclusions dropped gradually to around 200 mm^2, small-sizd inclusions of less than 1 μm. This is because small-sized inclusions kept coalescing and gathering during the vacuum break and soft blowing process of VD. From VD to Tundish and products, the total amount of inclusions decreased steadily, and the main factor was the decrease in the inclusions with small sizes less than 1 μm.

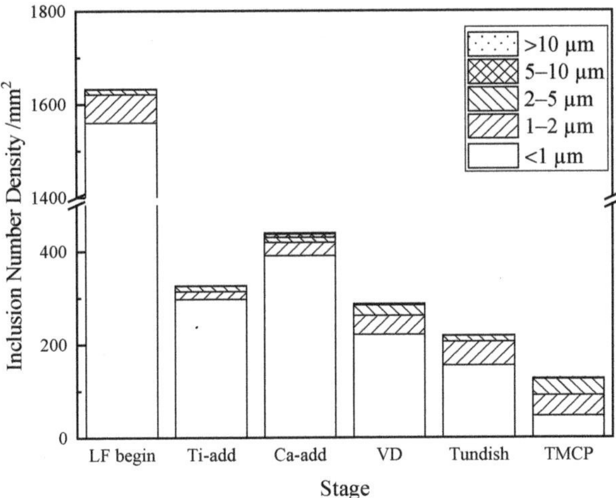

Figure 2. The number density of inclusions at different stages.

The oxide components in inclusions at different stages are shown in Figure 3. As can be seen, Al$_2$O$_3$ and MnO, as primary deoxidization products, accounted for the majority initially when LF began. With the LF refining processing, when Fe-Ti and Si-Ca alloys were added to the molten steel, CaO and SiO$_2$ were substituted for Al$_2$O$_3$ and MnO, accounting for more than 50% and 30%, respectively. Therefore, the content of Al$_2$O$_3$ and MnO dramatically decreased to less than 10%. From the Ca addition to vacuum degassing, the CaO content dropped a little to 40% due to the evaporation effect of Ca, and accordingly, the Al$_2$O$_3$ content increased a little to about 20%. However, the CaO content increased gradually to about 50% after the vacuum break of VD, which was due to extra Si-Ca alloy being added to the molten steel. The TiO$_2$ concentration fluctuated from 10% to 15% after Ti-Fe addition.

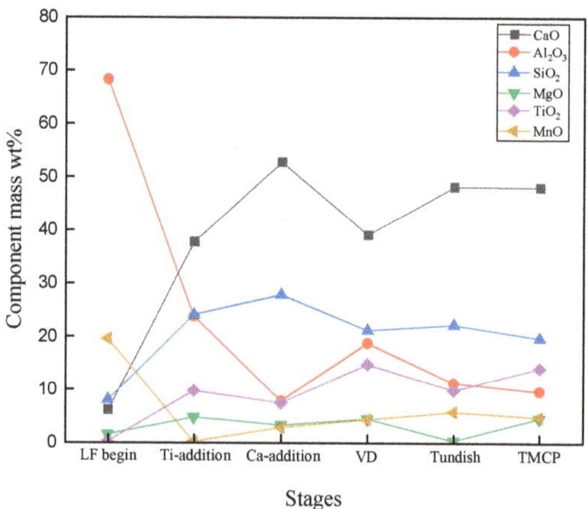

Figure 3. Oxide component in the inclusions at different stages.

CaO-SiO$_2$, CaO-Al$_2$O$_3$-SiO$_2$, and CaO-Al$_2$O$_3$-SiO$_2$-(MgO)-TiO$_x$ inclusions remained the main types among all the processes after Ti addition. Particularly, CaO-SiO$_2$ and CaO-Al$_2$O$_3$-SiO$_2$ can be described in a CaO-Al$_2$O$_3$-SiO$_2$ system, thus CaO-Al$_2$O$_3$-SiO$_2$-(MgO)-TiO$_x$ and CaO-Al$_2$O$_3$-SiO$_2$ inclusions were focused on and their formation mechanisms and evolutions were investigated and are discussed here.

3.3. Formation Mechanism and Evolution of CaO-Al$_2$O$_3$-SiO$_2$-(MgO)-TiO$_x$ Inclusion

The initial CaO-Al$_2$O$_3$-SiO$_2$-(MgO)-TiO$_x$ (MgO was marked here for discussing the formation mechanism) inclusion was found after Ti addition. Three typical inclusions (size ranged from 1 to above 10 μm) with different sizes are shown in Figure 4. The size of the three inclusions were 10.8, 5.1, and 2.5 μm, respectively. All of them belonged to CaO-Al$_2$O$_3$-SiO$_2$-(MgO)-TiO$_x$ inclusions, among which the MgO content was relatively low at about 5%. The detailed chemical composition is shown in Figure 5 to demonstrate the varying trend of each oxide component with different sizes.

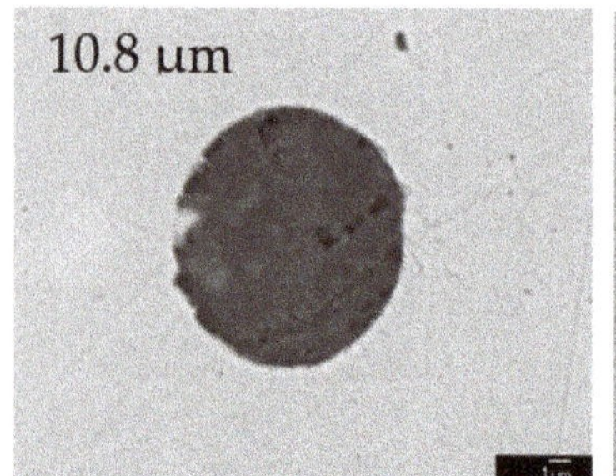

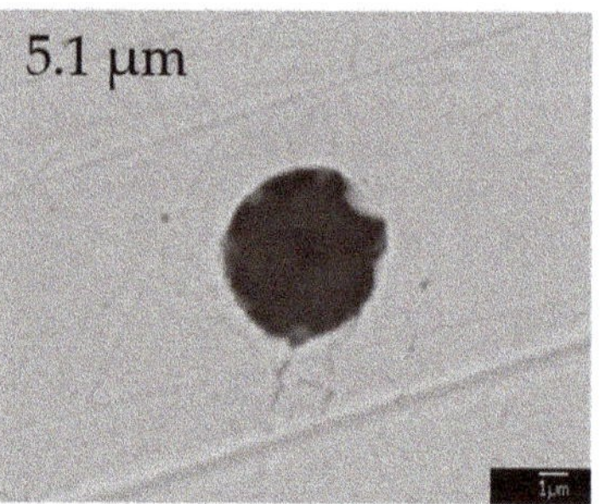

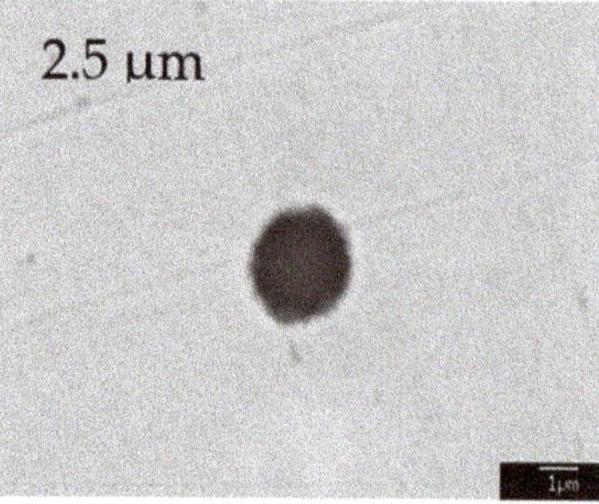

Figure 4. Morphologies of typical inclusions with different sizes after Ti addition.

It is obvious from Figure 5 that the mass proportion of CaO and SiO$_2$ steadily dropped with the decrease in the inclusion size. The mass proportion of MgO remained stable at a relatively low level. Adversely, the mass proportion of Al$_2$O$_3$ and TiO$_2$ showed a gradually increasing trend. CaO and SiO$_2$ were generally the main contents of the refining slag, and MgO caneither come from refining slag or the corrosion of MgO-C refractory and Al$_2$O$_3$, and TiO$_x$ was a typical deoxidization product of Ti-Ca killed steel. Generally, deoxidization products are small in size while the size of particles from slag and refractory can range widely. When small-sized deoxidization products are combined with large-sized

CaO-SiO$_2$(-MgO) particles, the CaO-SiO$_2$(-MgO) content in the average composition will be high, and Al$_2$O$_3$-TiO$_2$ will be low. On the other hand, when a small-sized deoxidization product is combined with small- or medium-sized CaO-SiO$_2$(-MgO) particles, the CaO-SiO$_2$(-MgO) content in the average composition will be relatively lower, and Al$_2$O$_3$-TiO$_2$ will be moderately higher. As a result, it can be concluded from Figure 5 that the CaO-Al$_2$O$_3$-SiO$_2$-(MgO)-TiO$_x$ inclusion in LF originated from the reaction between CaO-SiO$_2$ from slag, MgO from refractory, and the deoxidization products Al$_2$O$_3$ and TiO$_x$.

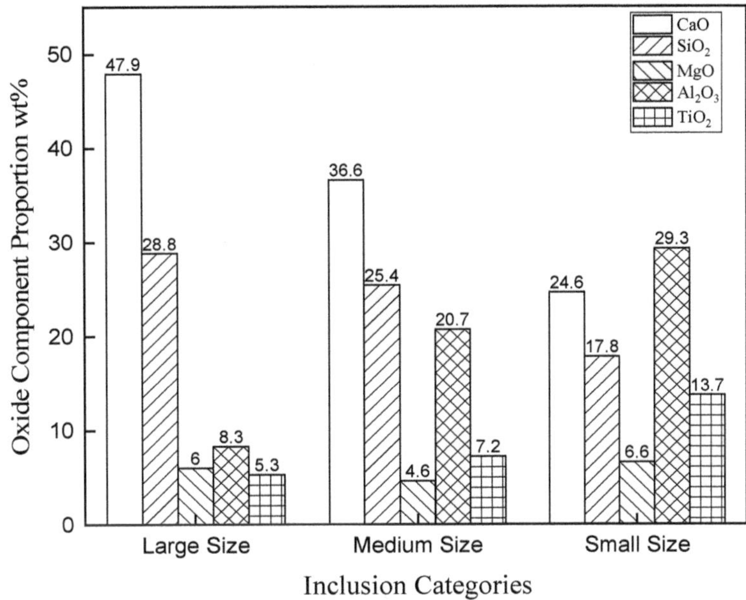

Figure 5. Oxide component proportion in different sized CaO-Al$_2$O$_3$-SiO$_2$-(MgO)-TiO$_x$ inclusions.

The CaO-Al$_2$O$_3$-SiO$_2$-MgO-TiO$_x$ inclusions' composition at different stages was projected into pseudo-ternary phase diagrams of CaO-Al$_2$O$_3$-TiO$_x$ at fixed 20%SiO$_2$ (14.8–23.2% SiO$_2$) and 5%MgO (2.7–7.8%MgO), as shown in Figure 6. In Figure 6a, all the inclusions' composition is projected, and relatively dispersive. Accordingly, the average composition of each stage was calculated and is shown in Figure 6b. After the addition of Ti in the LF refining, the main inclusions are generally located in the regions of CaO-Al$_2$O$_3$-MgO and melilite, and with Ca addition in LF and following VD, the main inclusions are basically located in melilite and Ca$_2$SiO$_4$ regions, and the inclusion composition in final product is eventually located in the CaTiO$_3$ region.

The morphologies of CaO-Al$_2$O$_3$-SiO$_2$-(MgO)-TiO$_x$ at different stages are shown in Figure 7. Generally, they were spherical. From the liquidus temperature in the pseudo-ternary phase diagram, the compositions of inclusions at different stages are located in the 1400–1500 °C liquidus region, which means they were in a liquid state in molten steel and thereby presented a sphere shape. The morphologies shown in Figure 7 indicate relatively fast cooling and the precipitation of different components; thus, the morphologies reflect the formation mechanism of inclusion. The component with a light grey colour is TiO$_2$, and the area proportion of TiO$_x$ continuously increased with the refining process proceeding, which was in accordance with the TiO$_x$ component trend shown in the phase diagram. During the early stage of LF refining after Ti addition, large amounts of Al-Ti-oxide particles started to gather around and stick to the CaO-SiO$_2$-(MgO) inclusion on its surface. Particularly, Al$_2$O$_3$ was from the oxidization of residual Al in the ferroalloys, and TiO$_x$ was from the oxidization of Ti in the. Ti-Fe alloy. These small deoxidization products did not have enough time to dissolve into the inclusion from the refining slag, so they stayed on the surface. With the refining process proceeding, Ti-oxide continuously increased and had sufficient time to gradually dissolve into the inclusion from the refining slag and formed new inclusions as a whole. As can be seen in the phase diagrams shown in Figure 6, the liquidus of the average composition of the inclusions was in the range from 1400 to 1450 °C. So, these new inclusions were mainly liquid inclusions in molten steel. During the cooling of the sample-taking process, different phases precipitated and formed multiphase inclusions, as shown in Figure 7. As a result, the morphology after cooling presented, such as the Ti-oxide component, "entered" the inclusion and formed the core of the multiphase inclusions. Meanwhile, the Al_2O_3 and SiO_2 component generally decreased due to the reduction of Ca, so the average content of CaO showed an increasing trend when compared with that of Al_2O_3 and SiO_2, as shown in Figure 6. The formation mechanism of $CaO-Al_2O_3-SiO_2-(MgO)-TiO_x$ is in accordance with that of the $CaO-Al_2O_3-SiO_2-(MgO)$ inclusions in Si-Mn killed steel with a limited aluminum content [21].

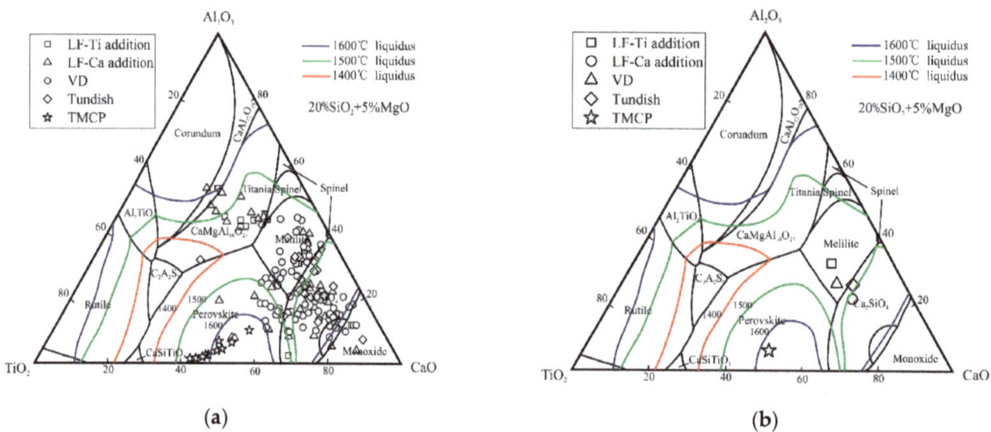

Figure 6. Pseudo-ternary phase diagrams $CaO-Al_2O_3-TiO_2$ at fixed 20%SiO_2 (14.8–23.2% SiO_2) and 5%MgO (2.7–7.8%MgO); (**a**) experimental data; (**b**) average data.

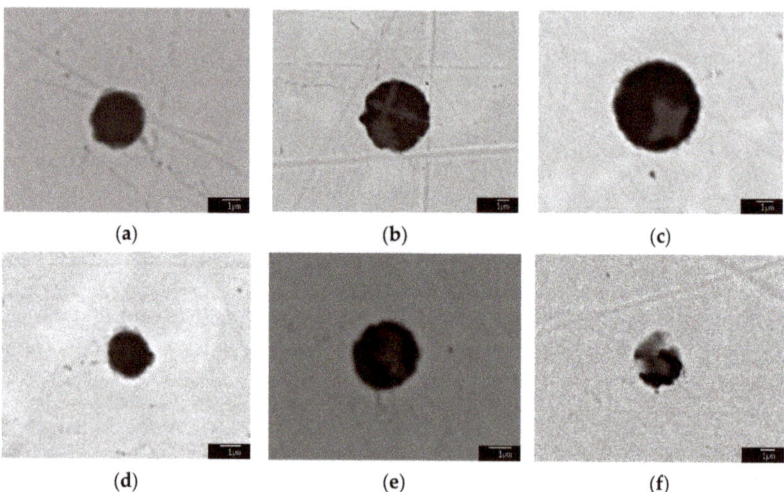

Figure 7. The morphologies of $CaO-Al_2O_3-SiO_2-(MgO)-TiO_x$ at different stages: (**a**) Ti addition; (**b**) Ca additon; (**c**) LF end; (**d**) VD; (**e**) Tundish; (**f**) product.

In summary, the formation and evolution mechanism of CaO-Al$_2$O$_3$-SiO$_2$-(MgO)-TiO$_x$ inclusion during the whole refining process can be described in a schematic diagram, as shown in Figure 8.

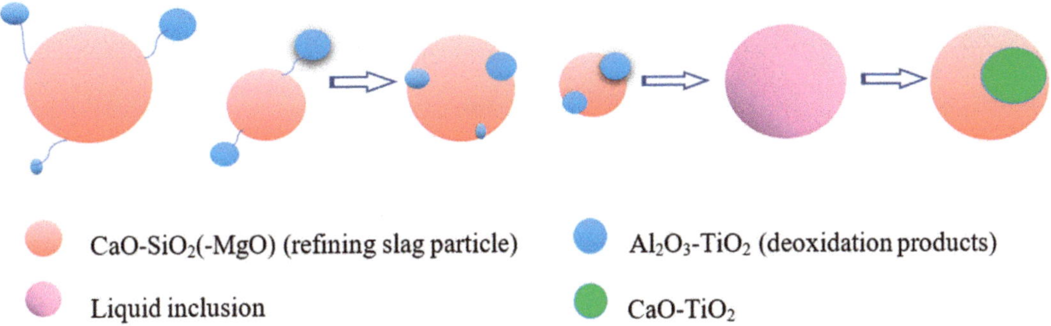

Figure 8. The formation and evolution mechanism of CaO-Al$_2$O$_3$-SiO$_2$-(MgO)-TiO$_2$ inclusion during the whole refining process.

3.4. Formation Mechanism and Evolution of CaO-Al$_2$O$_3$-SiO$_2$ Inclusion

There are generally two morphology types of CaO-SiO$_2$-Al$_2$O$_3$ inclusions as shown in Figure 9. One is a single-phase inclusion with a smooth surface, and the other has a rough surface. These two types of CaO-SiO$_2$-Al$_2$O$_3$ inclusions have different chemical compositions as shown in Table 3. The basicity CaO/SiO$_2$ of them is similar at about 1.9–2.0. The difference is mainly the component of Al$_2$O$_3$: Al$_2$O$_3$ of the inclusion in Figure 9a is 24.5% while that of the inclusion in Figure 9b is only 2.8%, so the inclusion in Figure 9b can be defined as CaO-SiO$_2$. The chemical composition distribution of CaO-SiO$_2$-Al$_2$O$_3$ inclusions in steel samples taken from different stages of industrial manufacturing was projected into CaO-Al$_2$O$_3$-SiO$_2$ ternary phase diagrams, as shown in Figure 10.

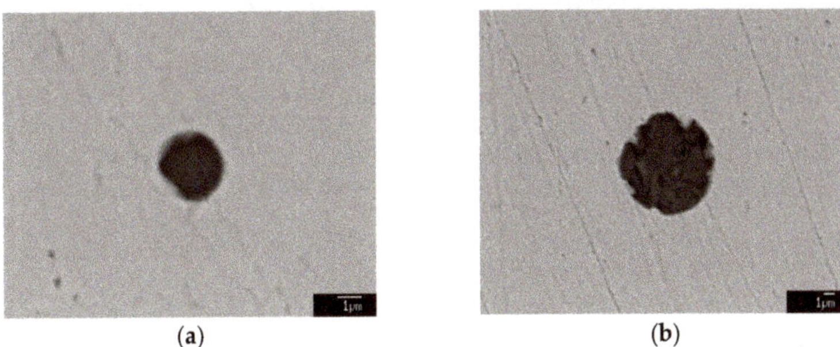

(a)　　　　　　　　　　(b)

Figure 9. Two morphology types of CaO-Al$_2$O$_3$-SiO$_2$ inclusions. (**a**) CaO-Al$_2$O$_3$-SiO$_2$, (**b**) CaO-SiO$_2$.

Table 3. Chemical compositions of CaO-Al$_2$O$_3$-SiO$_2$ inclusions.

No.	CaO	Al$_2$O$_3$	SiO$_2$	MgO	TiO$_2$	MnO	FeO	S
a	40.6	25.5	20.5	2.3	2.9	0.2	7.0	0.8
b	59.2	2.8	31.3	0.3	0.9	0.2	5.1	0.1

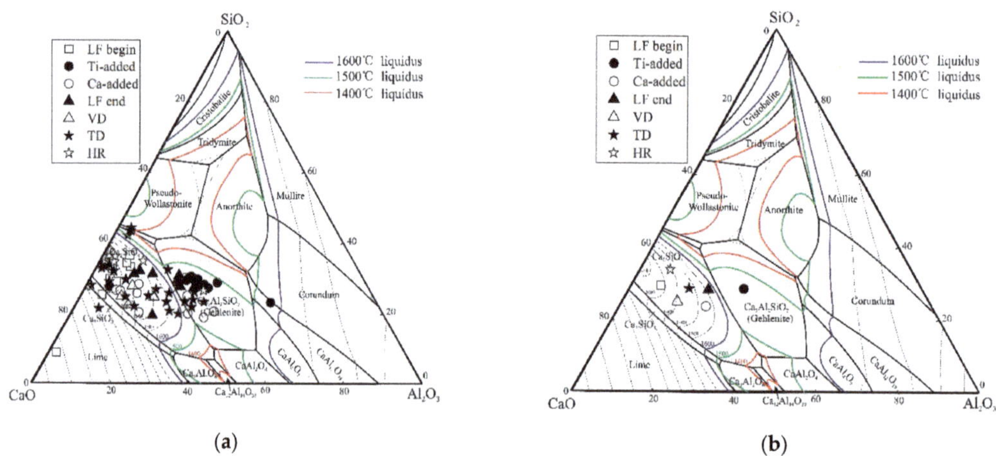

Figure 10. CaO-Al$_2$O$_3$-SiO$_2$ ternary phase diagrams.(**a**) experimental data, (**b**) average data.

The CaO-Al$_2$O$_3$-SiO$_2$ and CaO-SiO$_2$ inclusions' compositions at different stages were projected into ternary phase diagrams of CaO-Al$_2$O$_3$-SiO$_2$, as shown in Figure 10. Since the inclusion composition of the industrial sample is distributed dispersedly (Figure 10a), the average composition of inclusions during different stages is shown in Figure 10b. The initial composition of the LF entry is located in the 2CaO· SiO$_2$ region, and with the refining processing from LF→VD→TD→final products, the average composition moved from the 2CaO· SiO$_2$ region to the 2CaO· Al$_2$O$_3$· SiO$_2$ region (Gehlenite) and eventually returned to the 2CaO· SiO$_2$ region (Ca$_2$SiO$_4$). First, additive lime during the BOF tapping process brought CaO into molten steel, and it met the primary deoxidization product SiO$_2$ and formed 2CaO· SiO$_2$. When Ti-Fe alloy was added, residual Al enterted the molten steel and formed the deoxidization product Al$_2$O$_3$. The combination of Al$_2$O$_3$ and Ca$_2$SiO$_4$ formed 2CaO· Al$_2$O$_3$· SiO$_2$, and significantly decreased the liquidus of inclusion. Then, Ca treatment led to the reduction of Al$_2$O$_3$ and SiO$_2$ in 2CaO· Al$_2$O$_3$· SiO$_2$, so the Al$_2$O$_3$ and SiO$_2$ content kept dropping while the CaO content gradually increased. When LF began, the average inclusion composition located in the 2CaO· SiO$_2$ region and the CaO/SiO$_2$ ratio was about 1.7–1.8, and its liquidus was around 2000 °C. After the Ti addition, the average inclusion composition was located in the Gehlenite region, and the liquidus significantly decreased to about 1500–1600 °C due to the increment of Al$_2$O$_3$. The inclusions after the Ti addition had a spherical and smooth shape because they remained liquid in molten steel. After Ca treatment, the CaO/SiO$_2$ ratio tended to be slightly higher than the initial CaO/SiO$_2$ ratio, and the average composition entered the Ca$_2$SiO$_4$ region, the liquidus of which was about 1700–1800 °C. As a result, the CaO-SiO$_2$ inclusions from VD, tundish, and the final product presented rough surfaces, which indicated that they remained in a solid state in molten steel as shown in Figure 11. The reactions during these processes can be described as Formula (1)–(3):

$$2CaO + SiO_2 \rightarrow 2CaO{\cdot}SiO_2 \tag{1}$$

$$2CaO{\cdot}SiO_2 + [Al] + [O] \rightarrow 2CaO{\cdot}Al_2O_3{\cdot}SiO_2 \tag{2}$$

$$2CaO{\cdot}Al_2O_3{\cdot}SiO_2 + [Ca] \rightarrow CaO - Al_2O_3 - SiO_2 + [Al] + [Si] \tag{3}$$

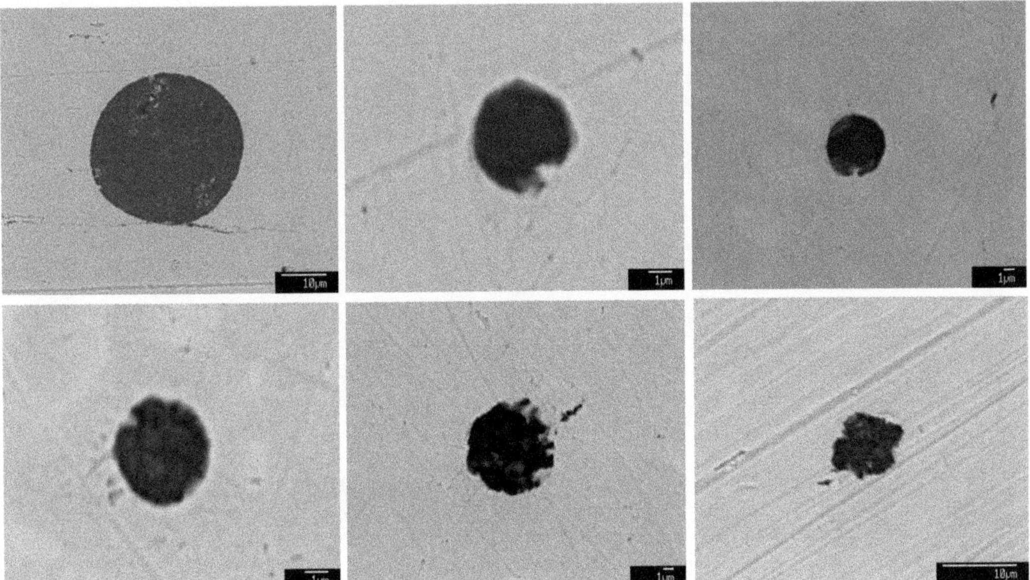

Figure 11. The morphologies of CaO-Al$_2$O$_3$-SiO$_2$ inclusions at different stages.

3.5. Effect of Inclusions on IAF Formation

As many researchers have reported, titanium oxide is the key part of "oxide metallurgy". Generally, they perform as an effective and stable nucleation site for the induction of intragranular acicular ferrite (IAF). As discussed above, the main categories of inclusion in Ti-Ca industrial steel are CaO-Al$_2$O$_3$-SiO$_2$-(MgO)-TiO$_x$ and CaO-Al$_2$O$_3$-SiO$_2$ (CaO-SiO$_2$ can be defined as one of CaO-Al$_2$O$_3$-SiO$_2$). However, CaO-Al$_2$O$_3$-SiO$_2$ may not be an effective inclusion for IAF promotion while CaO-Al$_2$O$_3$-SiO$_2$-(MgO)-TiO$_x$ can be a potential nucleation site for IAF. As a result, the chemical composition of some typical inclusions was investigated using EDS mapping scanning of FE-SEM to clarify the element distribution in multiphase inclusions. Figure 12 shows the morphologies of typical Ti-oxide-containing inclusions in final products, and their chemical composition is shown in Table 4. They were generally CaO-Al$_2$O$_3$-(SiO$_2$-MgO)-TiO$_x$ inclusions, with MnS precipitating around them.

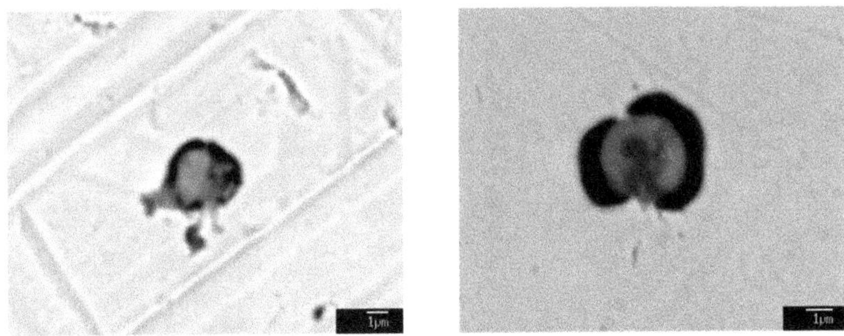

Figure 12. The morphologies of typical Ti-oxide-containing inclusions in the final products.

Table 4. The chemical composition of typical Ti-oxide-containing inclusions in the final products.

No.	CaO	Al_2O_3	SiO_2	MgO	TiO_2	MnO	FeO	S
1	22.7	15.7	0.7	6.6	15.3	16.1	12.1	11.0
2	27.7	19.2	6.9	9.2	22.8	2.8	6.6	4.7

Additionally, element mapping scanning was used to identify the chemical composition of different phases in multiphase inclusions. The element distribution of 2 typical multiphase inclusions is shown in Figures 13 and 14. The size of the inclusion in Figure 13 is about 2 μm. The shape of the inclusion shows different parts combined, and it mainly consists of two parts: MgO-Al_2O_3 spinel formed the core with two small MgO + MgS particles attached to it. The core was half surrounded by the CaO-TiO_2 layer, and the interface between the core and surrounding layer was relatively clear. The size of the inclusion in Figure 14 is about 1.5 μm. The shape looks like a whole, and it contains two phases: the left part is MgO-Al_2O_3 spinel, and the right part is CaO-TiO_x and (Ca, Mn) S. Generally, MgO-Al_2O_3 spinel is formed by the combination of [Mg], [Al], and [O] in molten steel and has a very high melting point. However, different mechanisms of the formation of MgO-Al_2O_3 spinel in Ti-Ca-treated industrial steel may exist. As for the multiphase inclusion in Figure 13, it is indicated that MgO-Al_2O_3 spinel was first formed when dissolved [Al] from ferroalloy combined with [Mg] from the refractory. Then, spinel acted as a heterogeneous nucleation site, and MgS and CaO-TiO_x started to precipitate around the core with the decreasing temperature of molten steel and the solidification process. As a result, clear interfaces between each phase existed. As for the multiphase inclusion in Figure 14, the explanation of its formation may be illustrated, as shown in Figure 8: liquid inclusion CaO-Al_2O_3-SiO_2-(MgO)-TiO_x was formed in the steel refining process, but with the decreasing temperature of molten steel and the solidification process, MgO-Al_2O_3 spinel firstly crystallized and precipitated due to its highest melting point, and then other phases, such as CaO-TiO_x and (Ca, Mn) S, precipitated in succession.

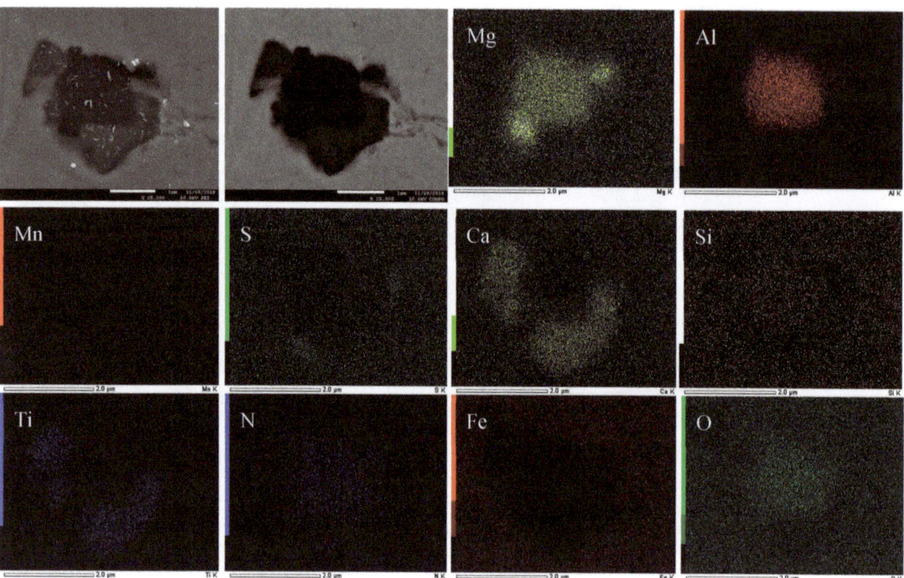

Figure 13. The element distribution of typical multiphase inclusion.

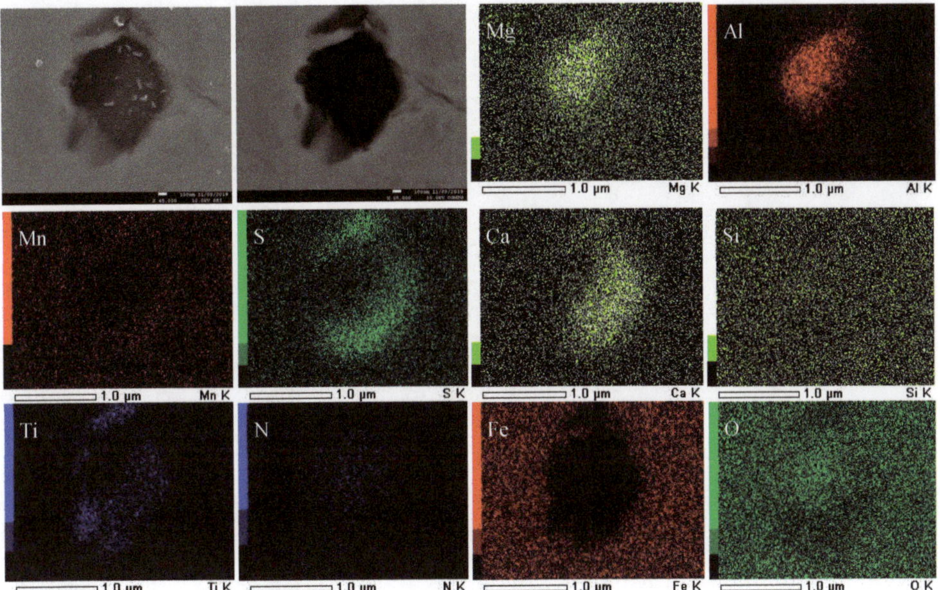

Figure 14. The element distribution of typical multiphase inclusion.

A large-sized multiphase inclusion is shown in Figure 15 and the results of EPMA analysis for each phase are listed in Table 5. The main body of this large-sized inclusion consisted of CaO-Al_2O_3-SiO_2 (red number 1) and CaO-TiO_x (red number 2). It seemed that this large-sized inclusion was crushed and separated into several parts composed of $(Ca, Mn)S$ (red number 3), CaO-Al_2O_3-MgO (red number 4) and CaO-SiO_2 (red number 5) during the rolling process of TMCP, so it spread along the rolling direction.

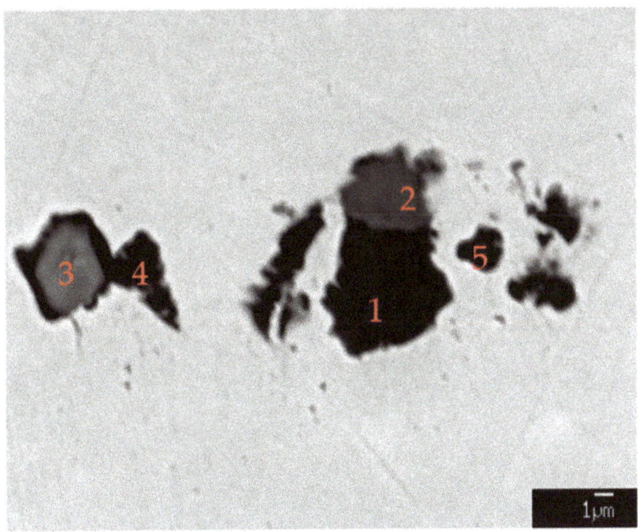

Figure 15. The morphology of a large-sized multiphase inclusion.

Table 5. The chemical composition of each part of the large-sized multiphase inclusion.

No.	CaO	Al_2O_3	SiO_2	MgO	TiO_2	MnO	FeO	S	Composition
1	40.5	29.2	24.9	3.0	0.3	0.1	2.4	0.0	$CaO\text{-}Al_2O_3\text{-}SiO_2$
2	40.3	1.4	0.5	0.3	54.7	0.0	2.9	0.0	$CaO\text{-}TiO_x$
3	53.4	0.1	0.0	0.3	0.0	6.7	9.1	30.3	(Ca, Mn) S
4	15.1	46.2	3.2	19.6	0.9	2.8	8.1	3.6	$CaO\text{-}Al_2O_3\text{-}MgO$
5	53.4	0.2	27.6	1.9	0.1	0.5	16.4	0.1	$CaO\text{-}SiO_2$

After welding simulation, the inclusions and microstructure were investigated, and some typical inclusions that can effectively induce IAFs were observed and detected using EPMA. The morphologies and chemical composition of typical inclusions are shown in Figure 16 and Table 6, respectively. The induced IAFs are marked in red (AF1–AF6, AF1–AF3) in Figure 16. It can be confirmed that only TiO_2 containing inclusions were effective for IAF promotion while $CaO\text{-}Al_2O_3\text{-}SiO_2$ inclusions were not found to be the nucleation site of IAF formation. These effective IAF inclusions were generally multiphase, and their size ranged from 2 to 5 µm and consisted of $CaO\text{-}Al_2O_3\text{-}(SiO_2\text{-}MgO)\text{-}TiO_x$. This indicates that $CaO\text{-}Al_2O_3\text{-}(SiO_2\text{-}MgO)\text{-}TiO_x$ was essential for inducing IAFs in Ti-Ca-treated offshore structural steel. Particularly, it can be found that the concentration of Ti-oxide in these effective IAF inclusions (20–30%wt) was significantly higher than the average concentration of Ti-oxide in all inclusions, as shown in Figure 13, which was about 15%wt. The higher concentration of Ti-oxide is generally considered to increase the possibility of IAF formation, thereby enhancing the mechanical property of steel. As a result, how to increase the proportion of $CaO\text{-}Al_2O_3\text{-}(SiO_2\text{-}MgO)\text{-}TiO_x$ and how to increase the concentration of Ti-oxide can be a potential and meaningful research direction, and more work is yet to be done in the future.

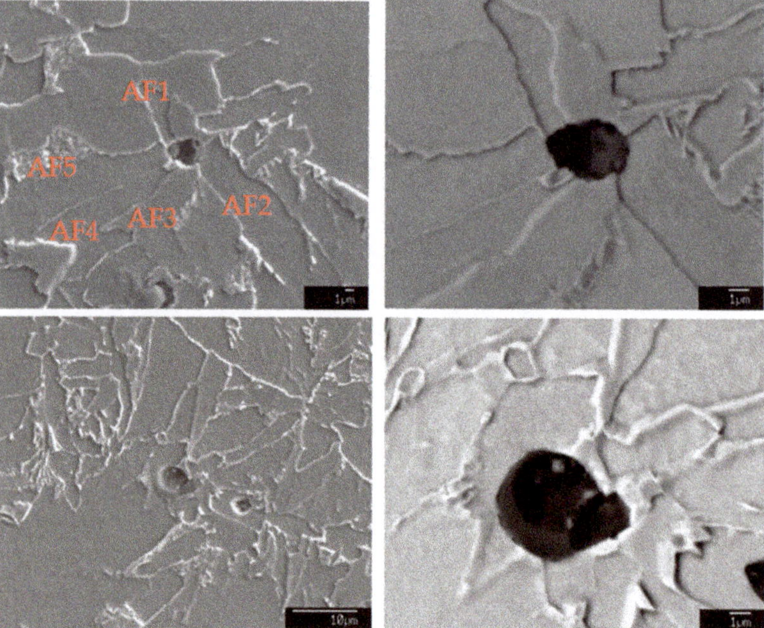

Figure 16. *Cont.*

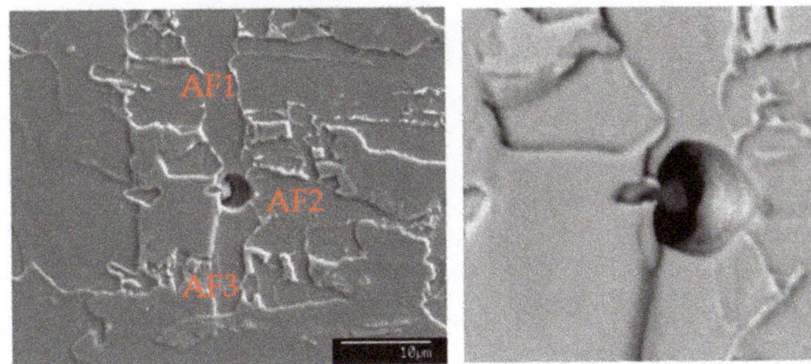

Figure 16. The morphologies of three typical inclusions after welding simulation.

Table 6. The chemical composition of typical inclusions after welding simulation.

No.	CaO	Al$_2$O$_3$	SiO$_2$	MgO	TiO$_2$	MnO	FeO	S
1	13.4	29.3	8.0	24.8	19.1	1.6	10.6	0.1
2	11.7	16.5	16.3	20.8	30.1	1.7	3.3	0.2
3	3.7	18.2	0.3	3.8	36.1	3.2	37.3	1.0

To clearly identify different phases in these multi-phased inclusions and thoroughly clarify the formation mechanisms, mapping scanning of FE-SEM was used for detecting the elementary distribution of effective inclusions that induced IAFs (inclusion A, B, and C). The results are shown in Figures 17–19.

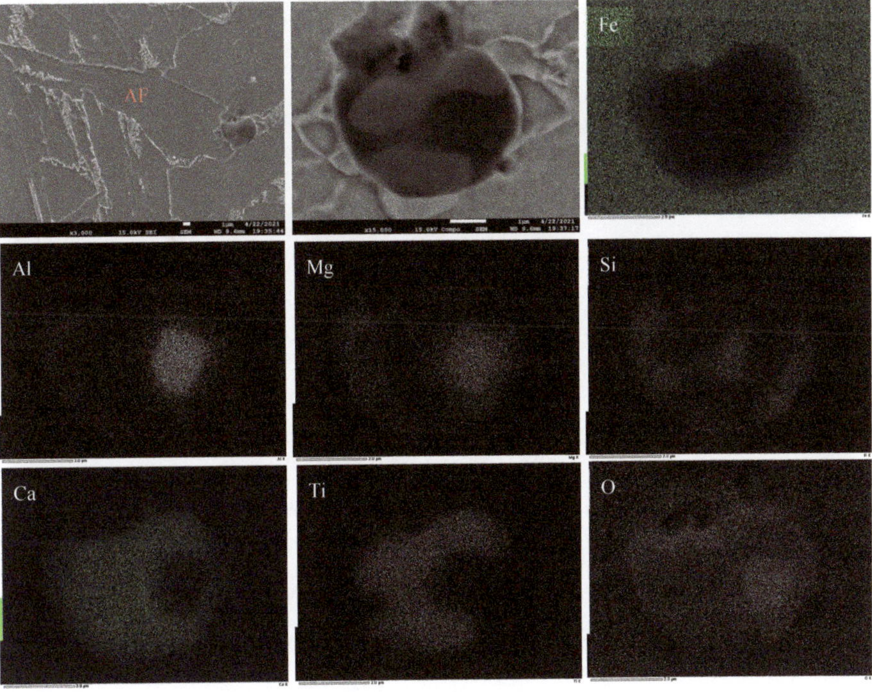

Figure 17. EDS mapping analysis of one typical inclusion A effective for IAF nucleation in Ti-Ca deoxidized steel.

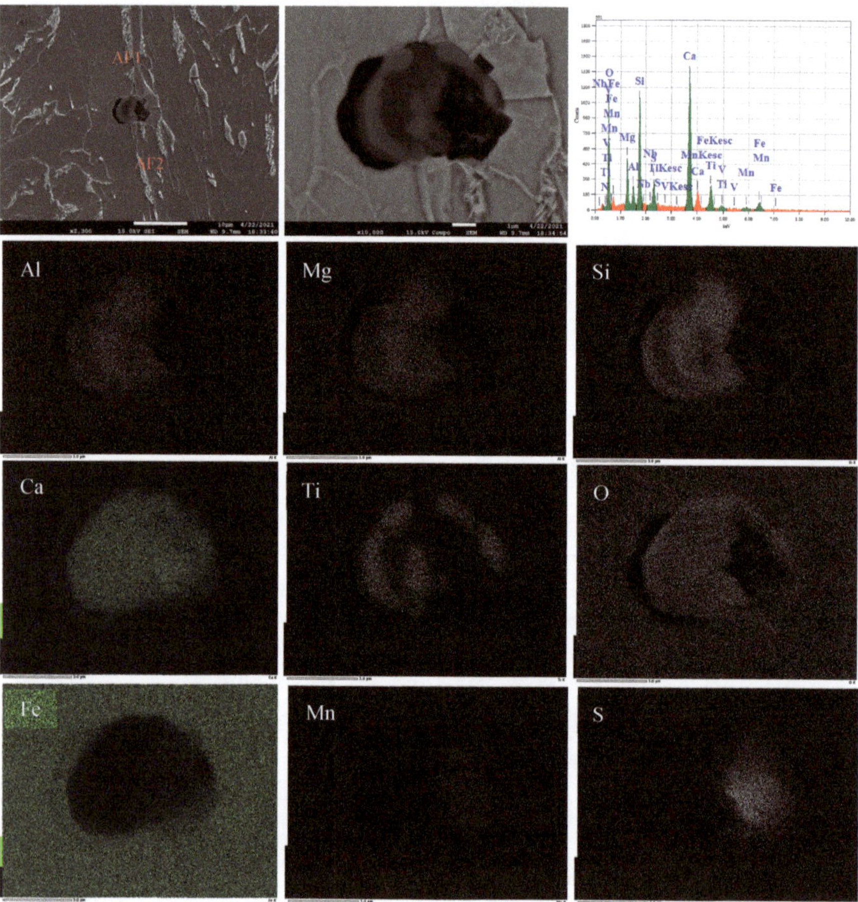

Figure 18. EDS mapping analysis of one typical inclusion B effective for IAF nucleation in Ti-Ca deoxidized steel.

As can be seen in Figure 17, the equivalent diameter of inclusion A is about 5.1 μm and one piece of acicular ferrite lath is induced by this inclusion. The morphology of the multi-phased inclusion has a spherical shape. This is a typical CaO-Al_2O_3-SiO_2-MgO-TiO_x inclusion, and its main body consists of CaO-Al_2O_3-SiO_2-MgO, Al_2O_3-MgO, and CaO-TiO_x, respectively. Particularly, CaO-Al_2O_3-SiO_2-MgO is the main content of secondary refining slag and has a low liquidus, thereby presenting a liquid state in molten steel. An Al_2O_3-MgO spinel particle 1.2 μm in size is embedded into the CaO-Al_2O_3-SiO_2-MgO liquid inclusion, and it has a very high liquidus temperature. CaO-TiO_2 also has a higher liquidus temperature than that of molten steel except for the condition when $TiO_2 = 80\%$. As a result, it can be concluded that the CaO-Al_2O_3-SiO_2-MgO-TiO_x liquid inclusion was formed in molten steel as described in Figure 8. Then, Al_2O_3-MgO spinel and CaO-TiO_x perovskite precipitated in order during the cooling process.

As shown in Figure 18, inclusion B also shows the spherical shape, and two pieces of acicular ferrite laths are induced by this inclusion. It consists of CaO-Al_2O_3-SiO_2-MgO, $(Ca, Mn) S$, and CaO-TiO_x. Particularly, the left part of the main body is CaO-Al_2O_3-SiO_2-MgO, and the right part of the main body is $(Ca, Mn) S$ while CaO-TiO_x perovskite precipitates along the edge of the above two phases.

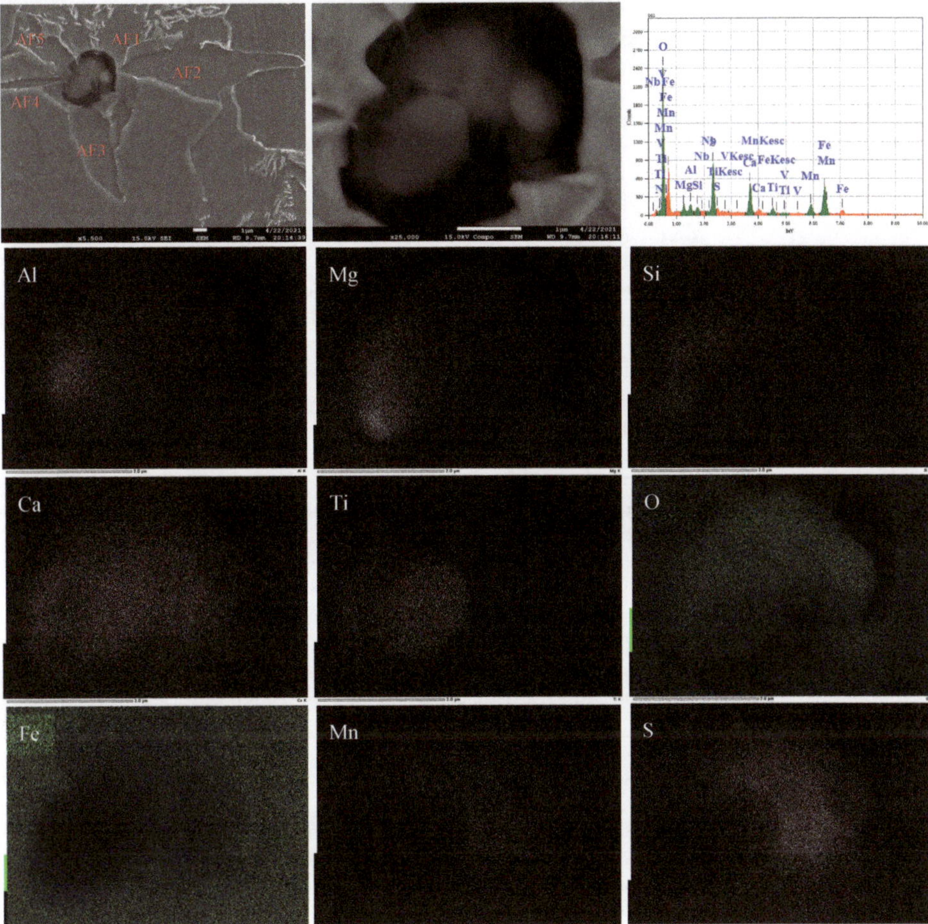

Figure 19. EDS mapping analysis of one typical inclusion C effective for IAF nucleation in Ti-Ca deoxidized steel.

In Figure 19, five pieces of acicular ferrite laths are induced by inclusion C. Different from the above two inclusions, the morphology reveals that it is formed due to the coalescence of several particles: $CaO\text{-}Al_2O_3\text{-}SiO_2\text{-}MgO$, (Ca, Mn) S, and $CaO\text{-}TiO_x$. During the secondary refining process, phases with a higher liquidus temperature, such as (Ca, Mn) S and $CaO\text{-}TiO_x$, are "captured" by liquid-phase $CaO\text{-}Al_2O_3\text{-}SiO_2\text{-}MgO$. These particles do not melt and form a new liquid inclusion but mechanically coalesce together.

Almost all the detected inclusions effective at inducing IAFs are found to be $CaO\text{-}Al_2O_3\text{-}SiO_2\text{-}MgO\text{-}TiO_x$-based inclusions. It can be confirmed that $CaO\text{-}Al_2O_3\text{-}SiO_2\text{-}MgO\text{-}TiO_x$-based inclusions are effective nucleation sites for IAF promotion while $CaO\text{-}Al_2O_3\text{-}SiO_2$-based inclusions are ineffective at inducing IAFs. The above three elementary distribution analyses also prove the formation mechanism of $CaO\text{-}Al_2O_3\text{-}SiO_2\text{-}MgO\text{-}TiO_x$-based inclusions due to the combination of particles from refining slag and deoxidization products. Then, different phases precipitate in order during the cooling process after secondary refining.

4. Conclusions

To clarify the formation mechanism and evolution of oxide inclusions in Ti-Ca-treated offshore structural steel, industrial sampling was conducted from LF, VD, and TD to final products. Continuous changes and correlations of various inclusions were explained by analysing the number density, morphology, and chemical composition using EPMA and FE-SEM. The primary findings are concluded as follows:

1. The evolution of inclusions during different stages in Ti-Ca-treated offshore structural steel is from primary deoxidization products (Al_2O_3-MnO, Al_2O_3-SiO_2-MnO) and their combination with lime (CaO-Al_2O_3-SiO_2-MnO and CaO-SiO_2)$\rightarrow$CaO-Al_2O_3-SiO_2-(MgO)-TiO_x and CaO-Al_2O_3-$SiO_2$$\rightarrow$CaO-$Al_2O_3$-$SiO_2$-(MgO)-$TiO_x$, CaO-$Al_2O_3$-$SiO_2$, and secondary deoxidization products (Al_2O_3-MnO-TiO_x and SiO_2-MnO)$\rightarrow$ CaO-Al_2O_3-SiO_2-(MgO)-TiO_x and CaO-Al_2O_3-SiO_2. The number density of the inclusions in Ti-Ca-treated industrial steel generally dropped from LF, VD, and Tundish to the final product, except for the Ti-Fe and Si-Ca addition in the LF, the number density slightly increased. The total decrease in the inclusion number density was mainly due to the significantly decreasing number density of small inclusions (<1 μm) during the refining process.

2. The formation mechanism of CaO-Al_2O_3-SiO_2-(MgO)-TiO_x inclusion was due to CaO-SiO_2-(MgO) from refining slag and refractory combining with the deoxidization product Al_2O_3 and TiO_x. With the refining process proceeding, Ti-oxide continuously increased and gradually "entered" the inclusion and formed the core of the multiphase inclusions while the Al_2O_3 component generally decreased due to the reduction of Ca, so the average content of CaO showed an adverse trend when compared with that of Al_2O_3.

3. The formation mechanism of CaO-Al_2O_3-SiO_2 inclusions is the initial $2CaO\cdot Al_2O_3\cdot SiO_2$ inclusion came from the combination of CaO-SiO_2 particles in refining slag and the deoxidization product Al_2O_3, and its liquidus was lower than that of molten steel, so it presented a liquid state in steel and had a smooth surface. After Ca addition, the initial $2CaO\cdot Al_2O_3\cdot SiO_2$ was gradually transferred to $2CaO\cdot SiO_2$ with Al_2O_3 continuously reduced by Ca. $2CaO\cdot SiO_2$ had a higher liquidus than that of molten steel, so it presented as a solid state in steel and had a rough surface.

4. In Ti-Ca-treated offshore structural steel, after welding simulation, CaO-Al_2O_3-SiO_2 inclusions were not effective at inducing IAFs while CaO-Al_2O_3-SiO_2-(MgO)-TiO_x inclusions were proven to be effective nucleation sites for promoting IAFs. The Al_2O_3-MgO spinel component in welding samples may have different formation mechanisms: one is that it formed directly in molten steel as a solid state, and other phases and inclusions, such as CaO-TiO_x and MnS, precipitated on Al_2O_3-MgO spinel, so the interface between each phase was clear. Another is that CaO-Al_2O_3-SiO_2-(MgO)-TiO_x as a whole formed in molten steel as a liquid state, and Al_2O_3-MgO spinel firstly precipitated due to its highest melting point and was followed by other phases, so the interface between each phase was not clear.

Author Contributions: Conceptualization, Z.R. and X.M.; methodology, Z.R., B.Z. and X.M.; software, Z.R.; validation, B.Z., X.M. and G.W.; formal analysis, Z.R.; investigation, Z.R. and X.M.; resources, H.L., P.Z. and F.W.; data curation, H.L., P.Z. and F.W.; writing—original draft preparation, Z.R.; writing—review and editing, Z.R, F.T.; visualization, Z.R.; supervision, X.M.; project administration, B.Z., X.M. and G.W.; funding acquisition, B.Z. and X.M. All authors have read and agreed to the published version of the manuscript.

Funding: This research was funded by Hebei Iron and Steel Group (HBIS), grant number ICSS2017-02.

Institutional Review Board Statement: Not applicable.

Informed Consent Statement: Not applicable.

Data Availability Statement: Not applicable.

Acknowledgments: Financial support for this research provided by Hebei Iron and Steel Group (HBIS) is greatly appreciated. The authors also acknowledge the facilities and the scientific and technical assistance of the Australian Microscopy& Microanalysis Research Facility at the Centre of Microscopy and Microanalysis at the University of Queensland.

Conflicts of Interest: The authors declare no conflict of interest.

References

1. Wan, X.L.; Wu, K.M.; Wang, H.H.; Li, G.Q. Applications of Oxide Metallurgy on High Heat Input Welding Steel. *China Metall.* **2015**, *25*, 6–12.
2. Villegas, R.; Redjaimia, A.; Confente, M.; Perrot-Simonetta, M.T. Fractal Nature of Acicular Ferrite, and Fine Precipitation in Medium Carbon Micro-Alloyed Forging Steels. Proc. New Developments on Metallurgy and Applications of High Strength Steels. Ed. Asoc. Argentina de Materiális. Buenos Aires. 2008. Available online: https://www.researchgate.net/publication/301840861_Proc_of_New_Developments_on_Metallurgy_and_Applications_of_High_Strength_Steels_Conference_Minerals_Wiley_John_Sons_Incorporated (accessed on 15 February 2022).
3. Sarma, D.S.; Karasev, A.V.; Jönsson, P.G. On the Role of Non-metallic Inclusions in the Nucleation of Acicular Ferrite in Steels. *ISIJ Int.* **2009**, *49*, 1063–1074. [CrossRef]
4. wan Kim, J.; koo Kim, S.; sik Kim, D.; deuk Lee, Y.; keun Yang, P. Formation Mechanism of Ca-Si-Al-Mg-Ti-O Inclusions in Type 304 Stainless Steel. *ISIJ Int.* **1996**, *36*, 140–143. [CrossRef]
5. Kanazawa, S.; Nakashima, A.; Okamoto, K.; Kanaya, K. Improved Toughness of Weld Fussion Zone by Fine TiN Particles and Development of a Steel for Large Heat Input Welding. *Tetsu Hagané* **1975**, *61*, 2589–2603. [CrossRef]
6. Takamura, J.; Mizoguchi, S. Roles of Oxides in Steels Performance. In Proceedings of the 6th International Iron and Steel Congress, ISIJ, Nagoya, Japan, 21–26 October 1990; Volume I, pp. 591–597.
7. Mizoguchi, S.; Takamura, J. Control of Oxides as Inoculants-Metallurgy of Oxides in Steels. In Proceedings of the 6th International Iron and Steel Congress, ISIJ, Nagoya, Japan, 21–26 October 1990; Volume I, pp. 598–604.
8. Sawai, T.; Wakoh, M.; Ueshima, Y.; Mizoguchi, S. Effects of Zirconium on the Precipitation of MnS in Low Carbon Steels. In Proceedings of the 6th International Iron and Steel Congress, ISIJ, Nagoya, Japan, 21–26 October 1990; Volume I, pp. 605–611.
9. Ogibayashi, S.; Yamaguchi, K.; Hirai, M.; Goto, H. The Features of Oxides in Ti-deoxidized Steel. In Proceedings of the 6th International Iron and Steel Congress, ISIJ, Nagoya, Japan, 21–26 October 1990; Volume I, pp. 612–617.
10. Kojima, A.; Yoshii, K.I.; Hada, T.; Saeki, O.; Ichikawa, K.; Yoshida, Y.; Shimura, Y.; Azuma, K. Development of High Toughness Steel Plates for Box Columns with High Heat Input Welding. *Shinnittetsu Giho.* **2004**, *90*, 39–44.
11. Kojima, A.; Kiyose, A.; Uemori, R.; Minagawa, M.; Hoshino, M.; Nakashima, T.; Ishida, K.; Yasui, H. Super High HAZ Toughness Technology with Fine Microstructure Imparted by Fine Particles. *Shinnittetsu Giho.* **2004**, *380*, 2–6.
12. Uemori, R. Analysis of the Crystallographic Direction of the Intra-granular Ferrite in Low Alloy Steel. *CAMP-ISIJ* **2001**, *14*, 1174.
13. Kojima, A.; Tomoya, K.; Noboru, H. Effect of Microstructure and Chemical Composition on Laser Weld Metal Toughness. Welded Structure Symposium 2002. *Collect. Lect. Pap.* **2002**, *11*, 327.
14. Kojima, A. High HAZ Toughness Steels with Fine Microstructure Imparted by Fine Particles. *CAMP-ISIJ* **2003**, *16*, 360.
15. Shigesato, G. Progress of High-Performance Steel Plates with Excellent HAZ Toughness. *Nippon. Steel Sumitomo Met. Tech. Rep.* **2018**, *119*, 22–25.
16. Ye, G.; Jönsson, P.; Lund, T. Thermodynamics and Kinetics of the Modification of Al2O3 Inclusions. *ISIJ Int.* **1996**, *36*, 105–108. [CrossRef]
17. Ito, Y.I.; Suda, M.; Kato, Y.; Nakato, H.; Sorimachi, K.I. Kinetics of Shape Control of Alumina Inclusions with Calcium Treatment in Line Pipe Steel for Sour Service. *ISIJ Int.* **1996**, *36*, S148–S150. [CrossRef]
18. Higuchi, Y.; Numata, M.; Fukagawa, S.; Shinme, K. Inclusion Modification by Calcium Treatment. *ISIJ Int.* **1996**, *36*, S151–S154. [CrossRef]
19. Kato, T.; Sato, S.; Ohta, H.; Shiwaku, T. Effects of Ca Addition on Formation Behavior of TiN Particles and HAZ Toughness in Large-Heat-Input-Welding. *Kobelco Technol. Rev.* **2011**, *30*, 76–79.
20. Wang, K.; Jiang, M.; Wang, X.; Wang, Y.; Zhao, H.; Cao, Z. Formation Mechanism of CaO-SiO_2-Al_2O_3-MgO Inclusions in Si-Mn-Killed Steel with Limited Aluminum Content during Low Basicity Slag Refining. *Metall. Mater. Trans. B* **2016**, *47*, 282–290. [CrossRef]
21. Wang, C.; Wang, X.; Kang, J.; Yuan, G.; Wang, G. Effect of Thermomechanical Treatment on Acicular Ferrite Formation in Ti-Ca Deoxidized Low Carbon Steel. *Metals* **2019**, *9*, 296. [CrossRef]
22. Wang, X.; Wang, C.; Kang, J.; Yuan, G.; Misra, R.D.K.; Wang, G. Improved Toughness of Double-pass Welding Heat Affected Zone by Fine Ti-Ca Oxide Inclusions for High-strength Low-alloy Steel. *Mater Sci. Eng. A* **2020**, *780*, 139198. [CrossRef]
23. Yamamoto, K.; Hasegawa, T.; Takamura, J.I. Effect of Boron on Intra-granular Ferrite Formation in Ti-Oxide Bearing Steels. *ISIJ Int.* **1996**, *36*, 80–86. [CrossRef]

24. Mabuchi, H.; Uemori, R.; Fujioka, M. The Role of Mn Depletion in Intra-Granular Ferrite Transformation in the Heat Affected Zone of Welded Joints with Large Heat Input in Strucutral Steels. *ISIJ Int.* **1996**, *36*, 1406–1412. [CrossRef]
25. Fukunaga, K.; Chijiiwa, R.; Watanabe, Y.; Kojima, A.; Nagai, Y.; Mamada, N.; Adachi, T.; Date, A.; Taniguchi, S.; Uemori, R.; et al. Advanced Titanium Oxide Steel with Excellent HAZ Toughness for Offshore Structures. In Proceedings of the ASME 2010 29th International Conference on Ocean, Offshore and Arctic Engineering OMAE2010, Shanghai, China, 6–11 June 2010; pp. 1–9.

metals

Article

Formation of Complex Inclusions in Gear Steels for Modification of Manganese Sulphide

Haseeb Ahmad [1], Baojun Zhao [2,*], Sha Lyu [3,*], Zongze Huang [4], Yingtie Xu [4], Sixin Zhao [4] and Xiaodong Ma [1,*]

[1] Sustainable Minerals Institute, The University of Queensland, Brisbane 4072, Australia; h.ahmad@uq.edu.au
[2] REEM International Institute for Innovation, Jiangxi University of Science and Technology, Ganzhou 341000, China
[3] Department of Materials Science and Engineering, Southern University of Science and Technology, Shenzhen 518055, China
[4] Baoshan Iron and Steel Co., Ltd., Shanghai 200122, China; zzhuang@baosteel.com (Z.H.); xuyingtie@baosteel.com (Y.X.); zhaosixin@baosteel.com (S.Z.)
* Correspondence: bzhao@jxust.edu.cn (B.Z.); lus@sustech.edu.cn (S.L.); x.ma@uq.edu.au (X.M.)

Abstract: Suitable MnS inclusions in gear steel can significantly improve the steel machinability and reduce the manufacturing costs. Two gear steel samples with different sulphur contents were prepared via aluminium deoxidation followed by calcium treatment. The shape, size, composition and percentage distribution of the inclusions present in the steel samples were analyzed using an electron probe micro-analysis (EPMA) technique. The average diameter of MnS precipitated on an oxide inclusion is less than 5 μm. It was found that the steel with high sulphur content contains a greater number of elongated MnS precipitates than low sulphur steel. Moreover, there are more oxide inclusions such as calcium-aluminates and spinels with a small amount of solid solution of (Ca,Mn)S in low content sulphur steel after calcium treatment, which indicates the modification of solid alumina inclusions into liquid aluminates. The typical inclusions generated in high sulphur steel are sulphide encapsulating oxide inclusions and some core oxides were observed as spinel. The formation mechanisms of complex inclusions with different sulphur and calcium contents are discussed. The results are in good agreement with thermodynamic calculations.

Keywords: gear steel; inclusions characteristics; EPMA; sulphide modification

Citation: Ahmad, H.; Zhao, B.; Lyu, S.; Huang, Z.; Xu, Y.; Zhao, S.; Ma, X. Formation of Complex Inclusions in Gear Steels for Modification of Manganese Sulphide. *Metals* **2021**, *11*, 2051. https://doi.org/10.3390/met11122051

Academic Editor: Eli Aghion

Received: 4 November 2021
Accepted: 16 December 2021
Published: 18 December 2021

Publisher's Note: MDPI stays neutral with regard to jurisdictional claims in published maps and institutional affiliations.

1. Introduction

Gears structure a fundamental part in the running of different machines and cars for transmitting power. There are a range of gears used in different areas according to the particular prerequisites, such as automotive gears, mining gears, marine gears, wind turbines, bicycle gears, instrumentation gears and conveyor systems. The automotive industry requires automotive gears for high force capacity, high fatigue strength and to convert mechanical energy efficiently and silently. However, machining of gear steels, generally, is linked with high energy consumption and decreased cutting tool life. The worsening of mechanical properties depends on the amount, size, morphology and dispersion of non-metallic inclusions. Since the machining costs (estimated ~40%) represent a large fraction of the total production expenses of most components, considerable attention has been paid to designing innovative techniques and technologies that make machinability viable with mechanical properties [1–3]. Adding a small amount of sulphur (typically from 100 to 500 ppm) is known to cause an outstanding improvement in the machinability of gear steels via manganese sulphide formation, which improves the machinability via chip embrittlement and increases tool life. Sulphides, especially MnS, with a round or shaft morphology are usually advantageous for this purpose. MnS inclusions have good ductility and can easily bend into long strip shapes after hot working, which causes anisotropy [4] in steel properties, consequently reducing the fatigue life [3] and altering the transverse properties of steel, and is furthermore destructive to the machinability of steels [5]. Most of the oxide and sulphide inclusions are endogenous, formed as the result of deoxidation of steel. However, dual phase types with or without sulphides and pure sulphide inclusions are generated during all succeeding stages of steel processing, even during solidification [1,6–10]. In contrast, alumina (Al_2O_3) is another type of inclusion generated as the result of Al deoxidation in molten steel. Large and irregular shapes of alumina inclusions mostly float up towards the slag, while smaller alumina inclusions (less than 10 μm) unable to float up get stuck in the steel and solidify, which can act as a stress raiser or become a crack source or even help the crack propagation under cyclic loading, and suppress the anti-fatigue property of steel. In addition, too much Al_2O_3 inclusions can cause nozzle clogging due to accumulation in the nozzle during the casting process, and because it is hard and brittle, it causes excessive tool wear in the machining process [7].

Normally, calcium addition is an effective way to alter the morphology and dispersion of non-metallic inclusions in the continuous casting process or in ingots. Calcium treatment is normally applied in Al-deoxidized steels to alter the solid alumina inclusions into liquid ones [11–13] which effortlessly pass through without spout impediment. In the current operations, the inclusions modification proficiency using Ca is usually not high; the surplus Ca may cause the development of high temperature stable inclusions such as CaS and $Ca \cdot Al_{12}O_{19}$, leading to spout impediment [2,10,14]. In resulphurized steel, Ca is also used to transform the MnS inclusions by reducing the relative plasticity of MnS inclusions during hot working. Blais et al. proposed the optimum Ca/S ratio to be 0.7 for effective modification of sulphide inclusions [15]. The reaction formation of CaS was reported via the formation of a CaS transient phase right after the Ca injection treatment [16] or formed after modification of Al_2O_3 to calcium aluminates [17]. Verma et al. proposed that transit CaS can react with the alumina to yield modified inclusions [18,19], while Wang et al. proposed that the reaction between Al_2O_3-CaO inclusions with the dissolved [S] and [Al] in the melt forms CaS around the inclusions [20]. Park et al. studied the mechanism of complex inclusions formation by using synthetic slag without adding Ca in the melt and predicted the critical sulphur [S] level required to precipitate CaS in the inclusions. By increasing the SiO_2 in the slag, the critical [S] level increases at a given [Al] content [21].

For uniform distribution of fine MnS precipitates, it is effective to use such oxides which have a high sulphide capacity and low melting temperature [11–13,22]. Miki [23] concluded that sulphur solubility in the inclusions increases with increasing MnO content in oxides, therefore increasing the probability of MnS precipitation on oxides. Thus, improvement of steel properties lies in the proper morphology control of oxides and sulphides. If the adverse effects of both sulphides (MnS) and pure oxides (Al_2O_3) can be minimized or even avoided via Ca treatment, the deformation of pure MnS precipitates will be easily confined, and also the oxide inclusions will not directly become the reason for steel failure. Although there is enough literature available on inclusions modification in Al-deoxidized and Ca-treated steels in lab scale models, there is very limited data available to improve the quality of industrial products because of the complex nature of reactions taking place in the ladle. To clearly define the role of Ca in modifying the sulphide and oxide inclusions in Al-killed Ca-treated gear steels, it is necessary to obtain the fundamental knowledge to understand the formation mechanism of inclusions including the composition of each phase from the viewpoint of thermodynamics and evolution during the steelmaking process on an industrial scale. Therefore, in this study we focused on the characterization and formation of complex oxide-sulphide inclusions in different sulphur-level industrial samples with an Al-deoxidation and Ca-treatment process. The formation mechanisms and thermodynamic prediction are discussed for the optimization of manganese sulphide modification.

2. Materials and Methods

To investigate the effect of sulphur and calcium on complex oxide and sulphide inclusions, two samples having different amounts of sulphur and calcium were taken from two industrial products. The chemical compositions of the samples are given in Table 1 below. Two steel grades (named GS1 and GS2) have almost the same Mn concentration: GS1 contains 290 ppm S and 14 ppm Ca, whereas GS2 contains 140 ppm S and 19 ppm Ca. The steelmaking process followed sequential steps through an electric arc furnace (EAF) → ladle (LF) refining → vacuum degassing (VD) → continuous casting (CC). Both samples underwent the same manufacturing route, i.e., the scrap metal along with hot metal from the converter was added into the electric arc furnace (Danieli & Co., Buttrio, Italy). During melting, oxygen lancing was used to reduce the carbon content to the desired level. After melting, the molten steel was tapped to the ladle furnace for further treatment, in which lumps of pure aluminum were added to reduce the oxygen content. Pure argon gas purging from the bottom was also employed to accelerate the mixing of aluminum and removal of oxide impurities in the slag. After ladle treatment, the furnace was transferred to the vacuum degassing process to ensure the complete removal of impurities. Ca in the form of CaSi was added into the vacuum degassing process to liquify the remaining impurities (in order to avoid nozzle blockage and surface defects), and finally, the furnace was transferred to the continuous casting plant to cast the steel into long bars of variable sizes. Finally, both samples were further rolled down into small diameter rods (after soaking in the furnace having a temperature ranging between 1000 and 1050 °C), i.e., GS1 was rolled to 40 mm diameter and GS2 to 25 mm.

Table 1. Chemical compositions of two gear steel samples (wt.%).

Grade	C	Si	Mn	Ni	Cr	*N	*P	*S	*Al	*Ca	*O	Fe
GS1	0.18	0.15	1.13	0.17	1.07	106	81	290	284	14	19	Bal.
GS2	0.21	0.25	1.31	0.028	1.26	155	68	140	407	19	20	Bal.

*N, P, S, Al, Ca and O are in ppm.

Analysis of Steel

Steel samples were cautiously prepared for electron probe microanalyzer (EPMA) analysis. Longitudinal sections were cut for inclusion analysis. After setting into the epoxy resin, the samples were ground up to 4000 grit size paper and slightly polished. Extensive polishing and etching can easily wash away the inclusions. A 20 × 20 mm^2 area of each sample was manually scanned for inclusions to ensure no inclusion was left behind.

A JXA 8200 electron probe X-ray microanalyzer (JEOL, Tokyo, Japan) with wavelength-dispersive detectors was used for microstructure and compositional analysis. Parameters employed for analysis include an accelerating voltage of 15 kV, 15 nA current, zero probe diameter, peak measuring time of 30 s, background position of ±5 μm and background measuring time of 5 s. EDS was used to identify and locate non-metallic inclusions in the steel; WDS was performed to measure their composition. Standards used for analysis were spinel (MgO-Al$_2$O$_3$), calcium silicate (CaSiO$_3$), iron sulphide (FeS$_2$), spessartine (Mn$_3$Al$_2$Si$_3$O$_{12}$) and hematite (Fe$_2$O$_3$) for Al, Mg, Ca, Si, S, Mn and Fe, respectively. The built-in ZAF correction procedure supplied with the probe was applied. Different crystals to cover the entire X-ray spectrum were used, which were lithium fluoride (LIF), pentaerythritol (PET) and thallium acid phthalate (TAP). Assuming the valence state of each element, compositions of the phases were re-calculated. The overall accuracy of the EPMA measurement of each component is within 1 wt.%.

3. Results and Discussion

3.1. Morphology and Composition of Inclusions

Longitudinal sections of both GS1 and GS2 samples were scanned for a better understanding of the morphology and distribution of the inclusions. In GS1 and GS2 samples, there are approximately 300 and 370 inclusions respectively in the scanned area of 2 cm^2, which was measured via EMPA. The inclusions are categorized into different groups based on their major chemical compositions and morphologies including single/dual phases and spherical/elongated sulphides. The classification of inclusions is given in Table 2, which describes the overall composition range of oxides and sulphides in single phase and also in complex inclusions. The wt.% of alumina, silica, calcium and magnesium oxides in the inclusions along with sulphur and manganese can be easily seen. Figure 1 shows the typical inclusions in GS1 and GS2 samples based on their compositions and phases, i.e., single phase and dual phase.

Table 2. Classification of inclusions in both samples based on the chemical compositions and morphologies.

Inclusion Type		Composition Range (wt.%)						
		Al_2O_3	CaO	MgO	SiO_2	FeO	CaS	MnS
Single Oxides	Al_2O_3	>80	<05	<05	<01	<05	<05	<05
	Al_2O_3-CaO	>50	>40	<05	<01	<05	<05	<05
	Al_2O_3-MgO	>60	<05	>30	<01	<05	<05	<05
Sulphides	(Ca,Mn)S	<05	<05	<05	<01	<05	>45	>45
	MnS	<05	<05	<05	<01	<05	<05	>90
Dual Oxy-Sulphides	Al_2O_3-(Ca,Mn)S	>50	<05	<05	<01	<05	>20	>20
	Al_2O_3-CaO-(Ca,Mn)S	>40	>20	<05	<01	<05	>15	>05
	Al_2O_3-MgO-(Ca,Mn)S	>50	<05	>20	<01	<05	>15	>10

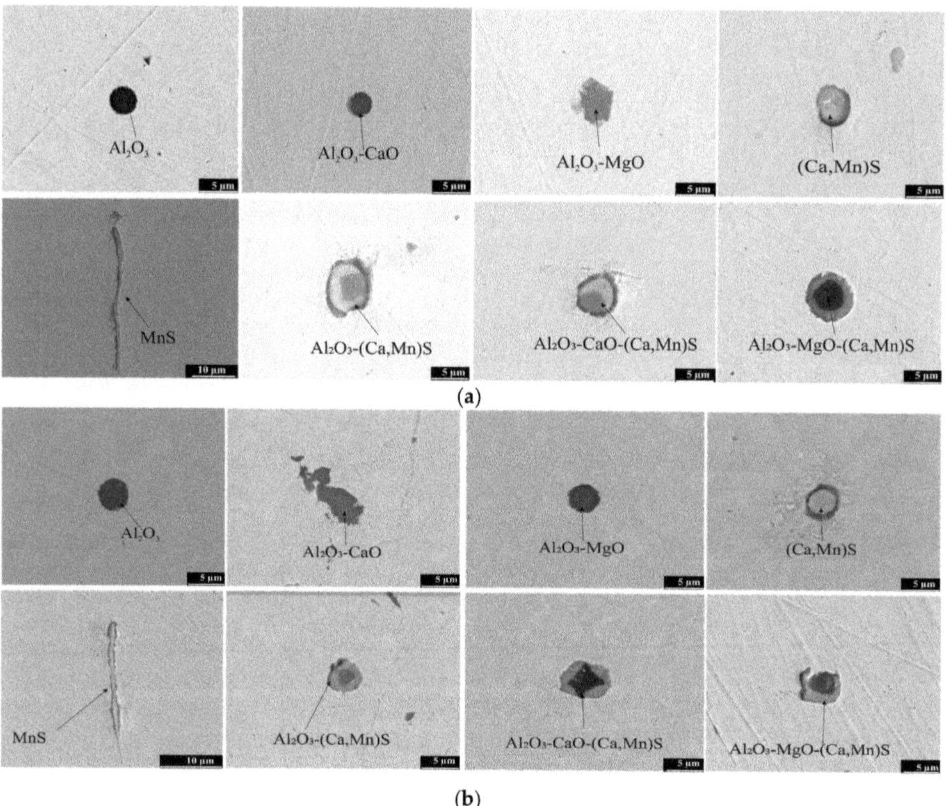

Figure 1. Typical morphology and composition of inclusions in both samples. (**a**) GS1 sample, (**b**) GS2 sample.

3.2. Number Fraction and Size Range of Inclusions

Figure 2a,b represent the detailed size and percentage distribution of individual inclusions in GS1 and GS2 samples, respectively. According to Figure 1a,b, most of the inclusions are round, while some are in irregular shapes, especially the pure MnS inclusions, which are elongated due to rolling. ImageJ (LOCI, University of Wisconsin, Madison, WI, USA) was used to measure the size of the inclusions. Most of the inclusions in the samples are regular in shape and easy to measure. For irregular inclusions morphology, the inclusion size is simply defined by measuring the longest length. Inclusions percentage is calculated by dividing the number of inclusions of each type by the total number of inclusions. In Figure 2a, there are a greater number (around 15%) of pure MnS inclusions in the range of 10–20 µm, while around 17% of the pure MnS inclusions are larger than 20 µm, a smaller number (only 5%) of pure alumina inclusions and a higher number of calcium-aluminates, and also complex oxy-sulphides, whereas in Figure 2b, there are around 16% pure alumina having a size greater than 5 µm, and the spinel (MgO-Al$_2$O$_3$) inclusions are around 14% in that size range. It is shown that the pure alumina inclusions are more in number than the complex oxides and sulphides; these inclusions are less 10 µm in size, which is in an acceptable range for the mechanical properties of gear steels. Around 2% of the calcium aluminates are less or equal to 5 µm, and about 8% of the calcium-aluminate inclusions are in the 5–10 µm range. Around 9% of the total inclusions are duplex ones in which the solid solution of Ca and Mn sulphide encapsulated the alumina and spinel inclusions. Most of the inclusions are in the acceptable size range, i.e., 5–15 µm, except the pure MnS, which was plastically deformed during rolling of the steel.

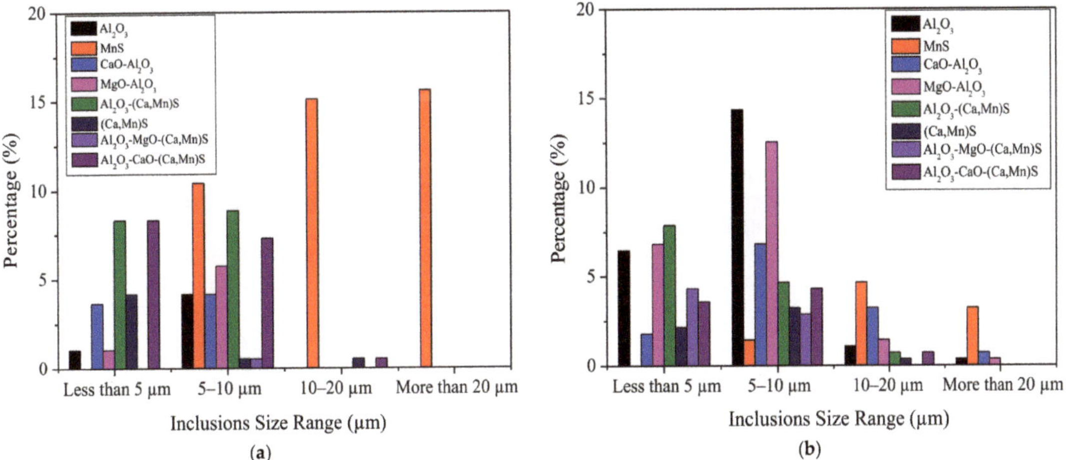

Figure 2. The size of different types of inclusions in the gear steels. (**a**) GS1 sample, (**b**) GS2 sample.

Figure 3 specifically represents the size comparison between pure MnS inclusions in both high and low sulphur samples, respectively. Steel with high sulphur content (GS1 sample) and only 14 ppm Ca shows a higher number of pure MnS inclusions which are elongated and also larger in size due to rolling of the steel because of the plastic (soft) nature of pure manganese sulphide, as has already been mentioned above. The inclusions of the GS2 sample with a relatively low sulphur content have a small number of pure MnS, also shown in Figure 3, but have a solid solution of (Ca,Mn)S. Overall, Figure 3 shows a better MnS modification result due to Ca treatment in the GS2 sample, because most of the portion of pure MnS inclusions was transformed into a solid solution of (Ca,Mn)S.

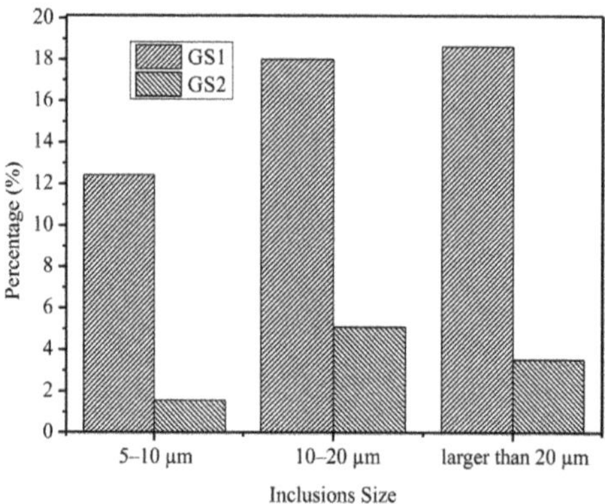

Figure 3. Pure MnS size comparison between the two samples.

4. Thermodynamics of Inclusions Formation

Chemical compositions of Al_2O_3, Al_2O_3-MgO and Al_2O_3-CaO inclusions are projected onto the Al_2O_3-MgO-CaO ternary system as shown in Figure 4a,b of both samples respectively (where M is MgO, MA is MgO-Al_2O_3, CA is CaO-Al_2O_3, CA_2 is Ca-Al_4O_7, CA_6 is Ca-$Al_{12}O_{19}$, C3A is Ca_3-Al_2O_6 and $C_{12}A_7$ is Ca_{12}-$Al_{14}O_{34}$). The liquidus temperature of inclusions can be predicted by Figure 4. The liquidus temperature of Al_2O_3, Al_2O_3-rich CaO, CaO-rich Al_2O_3 and Al_2O_3-MgO spinel inclusions is much higher than 1600 °C, which is normally the molten steel refining temperature, while some Al_2O_3-CaO inclusions are below 1600 °C. When Al was added to molten steel as a deoxidizer, Al_2O_3 inclusions were initially formed. The dissolved [Al] and [O] often reacted with traces of dissolved [Mg] or (MgO) refractory to form (MgO-Al_2O_3) spinel inclusions. These inclusions were formed through the reduction of MgO in the slag or refractory by aluminum even though Mg was not intentionally added in the steel. Mg reacted with Al_2O_3 to form MgO-Al_2O_3 inclusions. The formation of spinel inclusions through a steel-slag-refractory reaction has been discussed elsewhere in detail [21,24–26]. Ca-treatment was used to transform solid spinel and solid alumina into liquid calcium aluminates [21]. However, some Al_2O_3 inclusions were not properly modified by Ca treatment and formed solid Al_2O_3-rich CaO and CaO-rich Al_2O_3 inclusions, which have a high liquidus temperature.

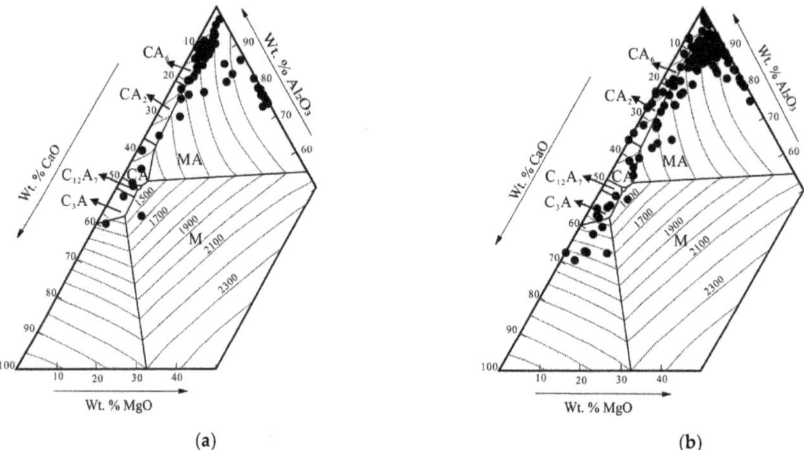

(a)   (b)

Figure 4. The composition distribution of Al_2O_3-MgO-CaO inclusions on the Al_2O_3-MgO-CaO phase diagram. (a) GS1 sample, (b) GS2 sample.

(Ca,Mn)S was observed to be associated with Al_2O_3-CaO inclusions to form Al_2O_3-CaO-(Ca, Mn)S oxysulphide inclusions. Kang et al. [27] measured the phase equilibria and calculated the Al_2O_3-CaO-CaS oxysulphide system using FactSage thermochemical software. To understand the solubility of CaS in the Al_2O_3-CaO at high temperature and to justify why pure CaS inclusions or CaS encapsulating oxide inclusions are not present in the present inclusions system, the Al_2O_3-CaO-CaS system was re-calculated using an updated version of the FactSage 7.3 (CRCT, Montreal, Canada and GTT Technologies, Aachen, Germany) phase diagram module using the FactOxide, FactPS. and FactMisc. databases with 1 atm pressure, and no other assumptions were considered [28], which is shown in Figure 5. The region of full-liquid phase at 1600 °C is delimited by the thinner lines on the diagram. Within the isothermal sections of 1600 °C, CaS solid precipitates in the molten steel during Ca-treatment can dissolve into liquid calcium aluminate inclusions in the primary phase fields of $Ca_3Al_2O_6$ and $CaAl_2O_4$. In the CaS primary phase field, the solubility of CaS in the liquid calcium aluminate is up to 8.5 wt.%, and it decreases with increasing Al_2O_3 content in the inclusions at 1600 °C. The liquid inclusions formed in GS1 and GS2 samples were calculated, which are indicated in Figure 5 as solid points. The CaS composition in liquid calcium aluminate inclusions is around 3.2 wt.%. According to Higuchi et al. the dissolved CaS will re-precipitate from the liquid Al_2O_3-CaO-CaS oxysulphide inclusions to form Al_2O_3-CaO-CaS dual phases during solidification of steel [16], but in the present case it did not happen. Precipitation of MnS on CaS at low temperature to form (Ca, Mn)S in solidified steels will be discussed in the next section.

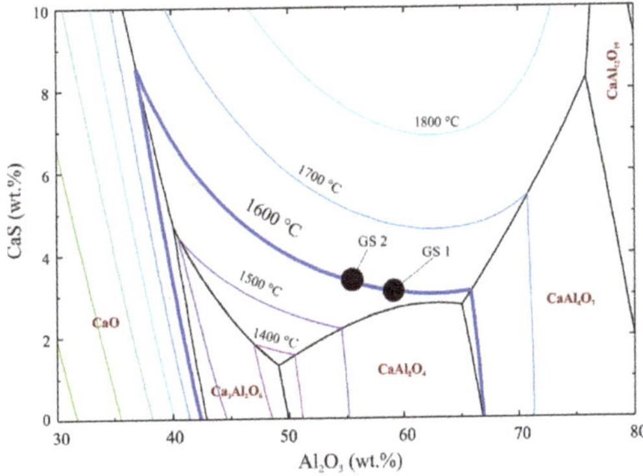

Figure 5. Calculated liquidus projection of CaO-Al_2O_3-CaS system in low CaS region.

Thermodynamics of inclusions precipitation and evolution in the equilibrium cooling of steels from 1600 °C to 1000 °C are shown in Figure 6a,b for GS1 and GS2 samples based on the chemical compositions in Table 1, respectively. Figure 6a,b show liquid calcium aluminates formed in molten GS1 and GS2 steels at 1600 °C. The ratio of CaO/Al_2O_3 in the liquid oxide inclusions mainly depends on the modification extent of Al_2O_3 by Ca and concentrations of Ca, Al and O in Al-killed steels. The liquid calcium aluminate inclusions transformed to the $CaAl_2O_4$ phase at around 1600 °C, then altered from $CaAl_4O_7$ to $CaAl_{12}O_{19}$ and finally transformed to Al_2O_3 inclusions below 1400 °C. A large number of Al_2O_3 or Al_2O_3-rich inclusions were observed in the GS1 and GS2 solidified steels, as shown in their compositions in Figure 4 above, which is consistent with the thermodynamic calculations. The Al_2O_3 or Al_2O_3-rich inclusions might be formed from the cooling of liquid inclusions or a deficient modification of Al_2O_3 due to Ca approaching equilibrium. One thing worth noting is the amount of AlN formed in both steels; GS2 has more mass % of AlN than GS1. This is due to the fact that a greater amount of Al is used for deoxidation of steel. Nitrogen is usually added in the form of a ferroalloy for grain refinement, to prevent the austenite grain coarsening during heat treatment and to enhance the mechanical properties of the steels via grain boundary pinning through aluminum-nitride precipitates

(AlN) [29]. AlN was not detected during EPMA analysis, maybe because of the detection limit of the probe, which mostly measures inclusions larger than or equal to 1 μm. AlN may have a sub-micron or nano size, and also the cooling rate in the ladle and during continuous casting is not at equilibrium, so maybe that is why AlN was unable to be detected by the probe despite the prediction during thermodynamic calculations.

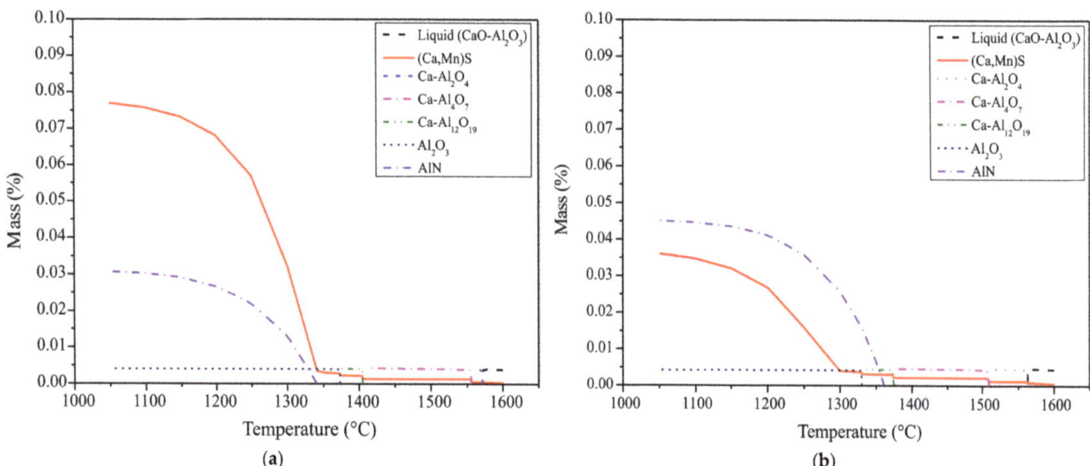

Figure 6. Equilibrium precipitation of inclusions and evolution during steel solidification. (**a**) GS1 sample, (**b**) GS2 sample.

The variation of CaS and MnS components in the (Ca, Mn)S solid solution in equilibrium solidification in both samples is specifically shown in Figure 7. According to FactSage calculations, nearly pure CaS precipitated in molten Ca-treated Al-killed GS1 and GS2 steels at 1600 °C. The formation of solid CaS is usually caused by excessive sulphur and calcium in liquid steel. At steel melting temperature, both Ca and S have high activities and the amount of CaS generated in both samples is more than 90%. In GS1, with the decrease in temperature, this amount decreases until it reaches the balance with MnS at around 1330 °C, and after that, MnS begins to grow at the expense of CaS and reaches more than 90% at around 1000 °C, whereas CaS is only less than 10% at that temperature. On the other hand, in GS2, with the decrease in temperature the amount of CaS decreases until 1270 °C. The analysis of samples showed different results in the case of CaS inclusions, as proved by Figure 5. Individual or complex oxy-CaS inclusions were not observed during EPMA analysis because CaS was dissolved in the CaO-Al$_2$O$_3$ and did not solidify as CaO-Al$_2$O$_3$-CaS. Detailed mechanism of inclusions evolution during solidification can be seen in Figure 8. Decrease in CaS in the GS2 sample at a lower temperature can provide the complete casting of several heats with no nozzle blockage and a lower amount of MnS generated.

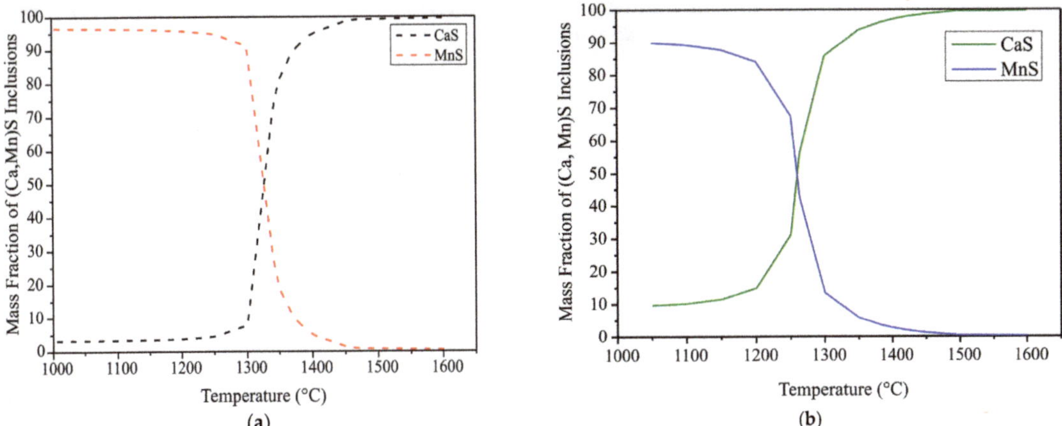

Figure 7. Mass percentage of CaS and MnS in the solid solution of (Ca,Mn)S. (**a**) GS1 sample, (**b**) GS2 Sample.

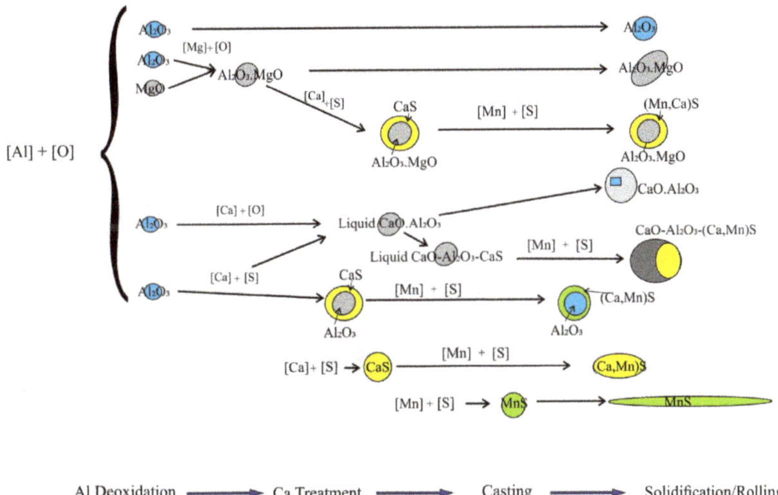

Figure 8. The schematic diagram of inclusions formation at the different stages of Al deoxidation, Ca treatment and solidification.

5. Formation Mechanism of Inclusions

The reactions during modification of alumina inclusions by calcium can be divided into the following kinetic steps:

1. Ca injection into the ladle, which immediately melts the Ca, because its melting point is much lower than the melting temperature of steel, followed by the dissolution of calcium from the gas bubbles-steel interface to the bulk of the steel and finally transfer of the dissolved calcium from the bulk steel to the steel-inclusion interface:

$$Ca(s) = Ca(l) + Ca(g) = [Ca] \tag{1}$$

2. Diffusion of calcium from the steel-inclusion interface into the alumina core followed by chemical reaction of the calcium with alumina:

$$[Ca] + [O] + (x + 1/3)(Al_2O_3) = (CaO \cdot xAl_2O_3) + 2/3[Al] \tag{2}$$

3. The sulphur content varies in both samples, consequently the extent of CaS formation changes. Ca can react with sulphur in two ways:

$$[Ca] + [S] = (CaS) \qquad (3)$$

$$(CaO) + 2/3[Al] + [S] = (CaS) + 1/3\ (Al_2O_3) \qquad (4)$$

In the above equations, square brackets denote an element dissolved in molten steel, and round brackets indicate the compounds in the inclusions. Before the calcium-alloy wire injection, mainly Al_2O_3 was produced due to the Al deoxidation of liquid steel. During the calcium treatment, the injection of Ca will lead to a severe splashing of liquid steel due to the high vapor pressure of Ca, and the added [Ca] reacts with [O] and Al_2O_3 inclusions to transform solid Al_2O_3 to liquid calcium aluminate inclusions. Some semi-solid Al_2O_3-rich core-shell inclusions might be partially modified due to kinetic diffusion of [Ca] controlling limitation [30–32]. When the [S] content is higher than the critical value for the precipitation of CaS, the CaO in the liquid inclusions reacts with sulphur [S] and aluminum [Al] in the molten steel. Therefore, the composition of the inclusions in the core is transformed to a $MgO \cdot Al_2O_3$-rich system (close to spinel), and a CaS shell is formed. On the other hand, when the [S] content is lower than the critical value for the precipitation of CaS, liquid inclusions are formed [21]. The solubility of [Mn] and [S] decreases during the solidification of steels. Segregation of [Mn] and [S] takes the primary inclusions as the heterogeneous nucleation sites for MnS precipitation, then reacts with primary CaS wrapping on the oxides to form (Ca,Mn)S solid solution [32–35]. A schematic diagram of inclusions formation is demonstrated in Figure 8 and can be visualized via SEM-EDS analysis as shown in Figure 9. Figure 9 with EDS scanning explains the formation of a crescent shaped (Ca,Mn)S layer surrounding the spinels and in the other image it shows the oval shape (Ca,Mn)S associated with the spinels.

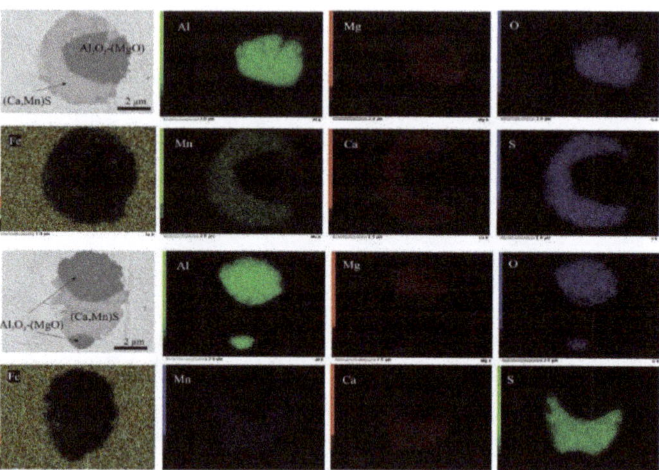

Figure 9. Duplex inclusions formed by (Ca,Mn)S precipitated on and collided with Al_2O_3-MgO inclusions.

The stable inclusions which can be formed based on the compositions of GS1 and GS2 in Table 1 by varying [Al] and [Ca] were calculated using FactSage 7.3 software with a phase diagram module with 1 atm pressure (FactOxide, FactPS and FactMisc. databases) to understand the controlling of inclusions. Figure 10 shows the stability diagram of the inclusions formed in the Fe-C-Si-Mn-Cr-S-O-Al-Ca systems at 1600 °C. The compositions of GS1 and GS2 are located close to the boundary of CaS + liquid and totally liquid inclusions region, which indicates the amount of CaS is relatively small. The small amount of CaS might not cause the problem of nozzle clogging in one heat casting but could block the nozzle by accumulating in the nozzle after several castings in practice. The liquid inclusions region is the target for successful modification of inclusions but not the whole region is the target because either decrease in [Al] content less than 150 ppm will yield other oxides, specially FeO and MnO, or decrease in [Ca] content less than 10 ppm will not completely modify the alumina inclusions but form (high temperature stable) $CaAl_4O_7$ inclusions which may create problems during casting. The modification window of Ca-treatment is quite narrow and should be controlled precisely.

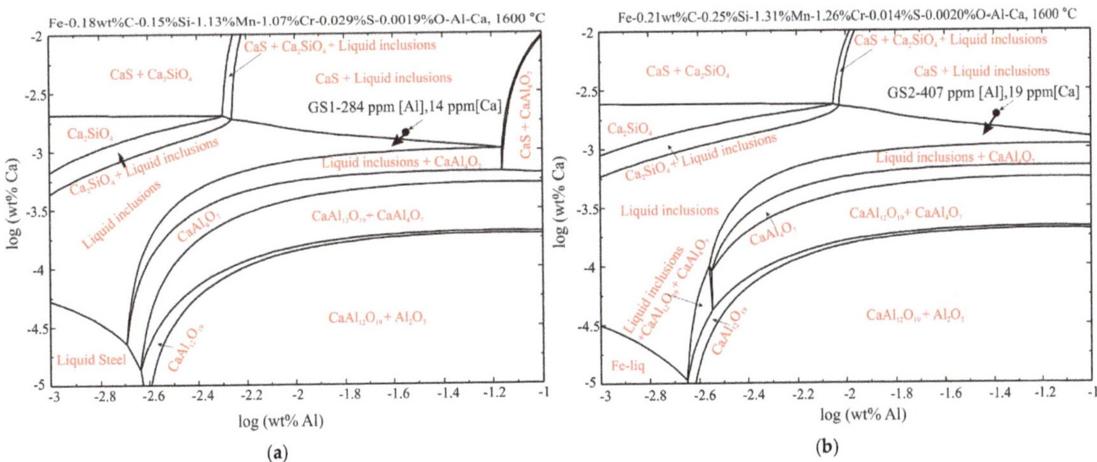

Figure 10. Stability diagram of inclusions formed in the Fe-C-Si-Mn-Cr-S-O-Al-Ca system at 1600 °C. (**a**) GS1 sample, (**b**) GS2 sample.

6. Conclusions

In the present study two gear steel samples with different sulphur and calcium contents were analyzed using an electron probe micro-analysis (EPMA) technique coupled with thermodynamic studies. The main findings are:

1. A significant number of pure MnS inclusions were observed in high sulphur steel as compared to low sulphur steel. Duplex inclusions and some spinel cores encapsulated by sulphides were also observed. (Ca,Mn)S encapsulated most of the oxide inclusions in both samples and no transient (CaO-Al₂O₃-CaS) inclusions were precipitated in the solidified steel despite it having a high sulphur content, i.e., 290 ppm and 140 ppm, respectively.

2. Calcium aluminates with a low melting temperature were formed in low sulphur steel, which is considered desirable for continuous casting. The main factors that influence the size, morphology and distribution of inclusions are the Al, S, O and Ca contents in the steel and its temperature.

3. The thermodynamic stability diagram of inclusions in Fe-C-Si-Mn-Cr-S-O-Al-Ca systems at 1600 °C agrees well with the inclusions of both sulphur-level Ca-treated samples, which shows inclusions can be completely modified into liquid ones by decreasing the content of Ca up to 10 ppm. Further decreasing the Ca content will result in incomplete modification of oxides and sulphides, and the formation of (high temperature stable) aluminates (CA2) will occur.

Metals **2021**, 11, 2051

Author Contributions: Conceptualization, S.L. and H.A.; methodology, S.Z., Y.X. and Z.H.; software, H.A.; validation, S.L., X.M. and B.Z.; formal analysis, S.L.; investigation, H.A.; resources, H.A.; data curation, S.L.; writing—original draft preparation, H.A.; writing—review and editing, X.M.; supervision, X.M.; project administration, B.Z.; funding acquisition, B.Z. All authors have read and agreed to the published version of the manuscript.

Funding: This research and APC were funded by the Australian Government through the Australian Research Council Research Hub for Computational Particle Technology, grant number IH140100035.

Data Availability Statement: The manuscript does not contain any data that has already been reported or archived elsewhere.

Acknowledgments: This research was supported by the Australian Government through the Australian Research Council Research Hub for Computational Particle Technology (IH140100035) and Baosteel through Baosteel-Australia Joint Research and Development Centre. The Australian Microscopy & Microanalysis Research Facility is acknowledged for providing characterization facilities. Technical support for the EPMA facility from Ron Rasch and Ying Yu of the Centre for Microscopy and Microanalysis at the University of Queensland is gratefully acknowledged.

Conflicts of Interest: The authors declare no conflict of interest.

References

1. Tönshoff, H.; Stanske, C. High productivity machining: Materials and processes. In Proceedings of International Conference on High Productivity Machining, Materials and Processing, New Orleans, LA, USA, 7–9 May 1985; pp. 7–9.
2. Kirsch-Racine, A.; Bomont-Arzur, A.; Confente, M. Calcium treatment of medium carbon steel grades for machinability enhancement: From the theory to industrial practice. *Metall. Res. Technol.* **2007**, *104*, 591–597. [CrossRef]
3. Vasconcellos da Costa e Silva, A.L. The effects of non-metallic inclusions on properties relevant to the performance of steel in structural and mechanical applications. *J. Mater. Res. Technol.* **2019**, *8*, 2408–2422. [CrossRef]
4. Maciejewski, J. The Effects of Sulfide Inclusions on Mechanical Properties and Failures of Steel Components. *J. Fail. Anal. Prev.* **2015**, *15*, 169–178. [CrossRef]
5. Cyril, N.; Fatemi, A. Experimental evaluation and modeling of sulfur content and anisotropy of sulfide inclusions on fatigue behavior of steels. *Int. J. Fatigue* **2009**, *31*, 526–537. [CrossRef]
6. Murty, Y.; Morral, J.; Kattamis, T.; Mehrabian, R. Initial coarsening of manganese sulfide inclusions in rolled steel during homogenization. *Metall. Trans. A* **1975**, *6*, 2031–2035. [CrossRef]
7. Ånmark, N.; Karasev, A.; Jönsson, P.G. The Effect of Different Non-Metallic Inclusions on the Machinability of Steels. *Materials* **2015**, *8*, 751–783. [CrossRef]
8. Li, M.-L.; Wang, F.-M.; Li, C.-R.; Yang, Z.-B.; Meng, Q.-Y.; Tao, S.-F. Effects of cooling rate and Al on MnS formation in me-dium-carbon non-quenched and tempered steels. *Int. J. Miner. Metall. Mater.* **2015**, *22*, 589–597. [CrossRef]
9. Huang, F.; Su, Y.F.; Kuo, J.-C. Microstructure and Deformation Characteristics of MnS in 1215MS Steel at 1050 °C. *Int. Conf. Min. Mater. Metall. Eng.* **2017**, *191*, 1333–1345.
10. Liu, Y.; Zhang, L. Relationship between Dissolved Calcium and Total Calcium in Al-Killed Steels after Calcium Treatment. *Met. Mater. Trans. A* **2018**, *49*, 1624–1631. [CrossRef]
11. Jeon, S.-H.; Kim, S.-T.; Lee, J.-S.; Lee, I.-S.; Park, Y.-S. Effects of Sulfur Addition on the Formation of Inclusions and the Corrosion Behavior of Super Duplex Stainless Steels in Chloride Solutions of Different pH. *Mater. Trans.* **2012**, *53*, 1617–1626. [CrossRef]
12. Xin, X.; Yang, J.; Wang, Y.; Wang, R.; Wang, W.; Zheng, H.; Hu, H. Effects of Al content on non-metallic inclusion evolution in Fe-16Mn-x Al-0.6 C high Mn TWIP steel. *Ironmak. Steelmak.* **2016**, *43*, 234–242. [CrossRef]
13. Guo, Y.-T.; He, S.-P.; Chen, G.-J.; Wang, Q. Thermodynamics of Complex Sulfide Inclusion Formation in Ca-Treated Al-Killed Structural Steel. *Met. Mater. Trans. A* **2016**, *47*, 2549–2557. [CrossRef]
14. Diederichs, R.; Bleck, W. Modelling of Manganese Sulphide Formation during Solidification, Part I: Description of MnS Formation Parameters. *Steel Res. Int.* **2006**, *77*, 202–209. [CrossRef]
15. Blais, C.; L'Espérance, G.; LeHuy, H.; Forget, C. Development of an integrated method for fully characterizing multi-phase inclusions and its application to calcium-treated steels. *Mater. Charact.* **1997**, *38*, 25–37. [CrossRef]
16. Higuchi, Y.; Numata, M.; Fukagawa, S.; Shinme, K. Inclusion Modification by Calcium Treatment. *ISIJ Int.* **1996**, *36*, S151–S154. [CrossRef]
17. Tiekink, W.; Santillana, B.; Kooter, R.; Mensonides, F.; Deo, B.; Boom, R. Calcium: Toy, tool or trouble. *AIST Trans.* **2008**, *5*, 395–403.
18. Verma, N.; Pistorius, P.; Fruehan, R.J.; Potter, M.; Lind, M.; Story, S. Transient Inclusion Evolution during Modification of Alumina Inclusions by Calcium in Liquid Steel: Part, I. Background, Experimental Techniques and Analysis Methods. *Met. Mater. Trans. A* **2011**, *42*, 711–719. [CrossRef]
19. Verma, N.; Pistorius, P.C.; Fruehan, R.J.; Potter, M.; Lind, M.; Story, S.R. Transient inclusion evolution during modification of alumina inclusions by calcium in liquid steel: Part II. Results and discussion. *Metall. Mater. Trans. B* **2011**, *42*, 720–729. [CrossRef]

20. Wang, Y.; Sridhar, S.; Valdez, M. Formation of CaS on Al_2O_3-CaO inclusions during solidification of steels. *Met. Mater. Trans. A* **2002**, *33*, 625–632. [CrossRef]

21. Shin, J.H.; Park, J.H. Formation Mechanism of Oxide-Sulfide Complex Inclusions in High-Sulfur-Containing Steel Melts. *Met. Mater. Trans. A* **2017**, *49*, 311–324. [CrossRef]

22. Wakoh, M.; Sawai, T.; Mizoguchi, S. Effect of S Content on the MnS Precipitation in Steel with Oxide Nuclei. *ISIJ Int.* **1996**, *36*, 1014–1021. [CrossRef]

23. Fukaya, H.; Miki, T. Phase Equilibrium between CaO· Al_2O_3 Saturated Molten $CaO–Al_2O_3–MnO$ and (Ca, Mn) S Solid Solution. *ISIJ Int.* **2011**, *51*, 2007–2011. [CrossRef]

24. Park, J.H.; Todoroki, H. Control of MgO· Al_2O_3 spinel inclusions in stainless steels. *ISIJ Int.* **2010**, *50*, 1333–1346. [CrossRef]

25. Harada, A.; Matsui, A.; Nabeshima, S.; Kikuchi, N.; Miki, Y. Effect of slag composition on MgO· Al2O3 spinel-type inclusions in molten steel. *ISIJ Int.* **2017**, *57*, 1546–1552. [CrossRef]

26. Shin, J.H.; Park, J.H. Modification of inclusions in molten steel by Mg-Ca transfer from top slag: Experimental confirma-tion of the 'refractory-slag-metal-inclusion (ReSMI)'multiphase reaction model. *Metall. Mater. Trans. B* **2017**, *48*, 2820–2825. [CrossRef]

27. Piao, R.; Lee, H.-G.; Kang, Y.-B. Experimental investigation of phase equilibria and thermodynamic modeling of the CaO-Al_2O_3-CaS and the CaO-SiO_2-CaS oxysulfide systems. *Acta Mater.* **2013**, *61*, 683–696. [CrossRef]

28. Bale, C.; Chartrand, P.; Degterov, S.; Eriksson, G.; Hack, K.; Ben Mahfoud, R.; Melançon, J.; Pelton, A.; Petersen, S. FactSage thermochemical software and databases. *Calphad* **2002**, *26*, 189–228. [CrossRef]

29. Dogan, Ö.N.; Michal, G.; Kwon, H.-W. Pinning of austenite grain boundaries by AlN precipitates and abnormal grain growth. *Metall. Trans. A* **1992**, *23*, 2121–2129. [CrossRef]

30. Kor, J.; Glaws, P. The Making, Shaping and Treating of Steel. Ladle refining and vacuum degassing. *Pittsburg* **1998**, *10*, 661–692.

31. Tabatabaei, Y.; Coley, K.S.; Irons, G.A.; Sun, S. Model of inclusion evolution during calcium treatment in the ladle furnace. *Metall. Mater. Trans. B* **2018**, *49*, 2022–2037. [CrossRef]

32. Liu, Y.; Zhang, L.; Zhang, Y.; Duan, H.; Ren, Y.; Yang, W. Effect of Sulfur in Steel on Transient Evolution of Inclusions During Calcium Treatment. *Met. Mater. Trans. A* **2018**, *49*, 610–626. [CrossRef]

33. Yang, W.; Zhang, L.; Wang, X.; Ren, Y.; Liu, X.; Shan, Q. Characteristics of Inclusions in Low Carbon Al-Killed Steel during Ladle Furnace Refining and Calcium Treatment. *ISIJ Int.* **2013**, *53*, 1401–1410. [CrossRef]

34. Zhao, D.; Li, H.; Bao, C.; Yang, J. Inclusion Evolution during Modification of Alumina Inclusions by Calcium in Liquid Steel and Deformation during Hot Rolling Process. *ISIJ Int.* **2015**, *55*, 2115–2124. [CrossRef]

35. Gollapalli, V.; Rao, M.V.; Karamched, P.S.; Borra, C.R.; Roy, G.G.; Srirangam, P. Modification of oxide inclusions in calcium-treated Al-killed high sulphur steels. *Ironmak. Steelmak.* **2019**, *46*, 663–670. [CrossRef]

metals

Article

Numerical Simulation and Application of Tundish Cover Argon Blowing for a Two-Strand Slab Continuous Casting Machine

Yang Li [1,*], Chenhui Wu [1], Xin Xie [1], Lian Chen [1], Jun Chen [1], Xiaodong Yang [2] and Xiaodong Ma [3,*]

[1] State Key Laboratory of Vanadium and Titanium Resources Comprehensive Utilization, Panzhihua 617000, China
[2] Pangang Group Xichang Steel and Vanadium Co., Ltd., Steelmaking Plant, Xichang 615000, China
[3] Sustainable Minerals Institute, The University of Queensland, 80 Meiers Road, Indooroopilly, QLD 4068, Australia
* Correspondence: leemuqiu@gmail.com (Y.L.); x.ma@uq.edu.au (X.M.)

Abstract: During continuous casting, argon blowing from tundish cover (ABTC) can greatly prevent the flow of the remaining air and decrease the reoxidation of molten steel in tundish. In the current study, a numerical model based on a tundish of a two-strand slab continuous casting machine was established to investigate the feasibility and evaluate the protective casting effect of the ABTC process. The influence of operation parameters, including sealing schemes of tundish cover holes and the argon flow rate of the remaining oxygen content, were studied in tundish. Then, industrial trials based on the operation parameters from the numerical model were carried out to evaluate the protective effect of ABTC. The results indicate that the ABTC process has a great protective effect in avoiding increasing levels of nitrogen and losing titanium and aluminum. With the ABTC process applied, the average increment of nitrogen ($\triangle w_{[N]}$) in steel from the end of RH to tundish decreases by 90% from 10×10^{-6} to 1×10^{-6}, the average loss of titanium ($\triangle w_{[Ti]}$) by 12.7% from 63×10^{-6} to 55×10^{-6}, and the amount of aluminum ($\triangle w_{[Al]}$) decreases by 7.1% from 70×10^{-6} to 55×10^{-6}. The injecting hole and baking holes should be sealed during the period of empty tundish to efficiently discharge the air. In order to ensure that the oxygen volume fraction in tundish is less than 1%, the argon flow rate should be $\geq 220 \, \mathrm{Nm^3/h}$ during the period of empty tundish and $\geq 80 \, \mathrm{Nm^3/h}$ during the period of normal casting.

Keywords: continuous casting; tundish; argon blowing; protective casting; numerical simulation

Citation: Li, Y.; Wu, C.; Xie, X.; Chen, L.; Chen, J.; Yang, X.; Ma, X. Numerical Simulation and Application of Tundish Cover Argon Blowing for a Two-Strand Slab Continuous Casting Machine. *Metals* **2022**, *12*, 1801. https://doi.org/10.3390/met12111801

Academic Editor: Noé Cheung

Received: 19 September 2022
Accepted: 19 October 2022
Published: 24 October 2022

Publisher's Note: MDPI stays neutral with regard to jurisdictional claims in published maps and institutional affiliations.

1. Introduction

High-cleanliness steel has been developed from advanced steel materials with the rapid expansion of transportation, national defense and marine engineering [1,2]. The number of oxide inclusions in molten steel is directly proportional to the total oxygen content of molten steel, which is usually treated as an indicator to evaluate the number of oxide inclusions. Although the purity of molten steel can be greatly improved by the refining of ladle [3–5], abundant oxide inclusions will produce and deteriorate the purity of molten steel if the protective casting is poor.

In the field of steel production, argon as a protective gas has been widely used in the process of continuous casting. The application of argon in ladle stirring [6–10], RH [11,12], the argon bubbling curtain [13,14] in tundish, and submerged entry nozzle [15–18] has provided a great protective effect on removing inclusions and alleviating the clogging of SEN (Submerged Entry Nozzle). However, the secondary oxidation of molten steel cannot be avoided in the early casting stage of the first ladle after baking tundish due to the untimely melt of protective slag. Argon blown into tundish from pipes installed on the tundish cover, as a kind of important protective casting process, plays an important role in decreasing the secondary oxidation of molten steel and improving its cleanliness. During the initial casting period after baking tundish, blowing argon into tundish and discharging air may prevent the molten steel from being reoxidized before the protective slag is fully melted. Wang [19] conducted a plant trial in Tang steel and found that the protective casting effect can be greatly improved when the residual oxygen volume fraction in tundish is less than 1%. Story [20] found that the loss of aluminum in molten steel may be decreased by 87.5% with argon blowing into tundish during the whole period of continuous casting. Considering the advantages of the process, some plants in Europe and Japan have been applied in industrial production and shown good effects [21]. For example, the defect rate of IF steel in Corus decreased by 38%, and the quantity of inclusions along the casting direction in POSCO significantly decreased after applying this process. Some plants in China have also adopted this process. Han Steel reported the average content of inclusions in low-carbon steel decreased from 0.47% to 0.32%. Gao [22] found the surface defect production rate of deep-drawing steel decreased from 7.78% to 2.62% as the diameter of argon blowing pipe increased from 20 mm to 34 mm.

Although ABTC has been applied in some plants, the numerical investigation and application assessment of ABTC has rarely been reported in the literature. Therefore, the current study aims to investigate the feasibility and evaluate the protective casting effect of ABTC during a continuous casting period. Firstly, a three-dimensional numerical model based on the practical two-strand slab tundish of the Pangang Group was built to investigate the behavior of oxygen volume fraction under different sealing schemes and argon flow rates during a period of empty tundish and normal casting. Then, plant trials based on the numerical simulation results and steel grade of M3A35 were carried out, where the increased nitrogen content and the loss of titanium and aluminum at the end of the RH process and tundish were tested to evaluate the protective casting effect.

2. Model Description

2.1. Governing Equations and Boundary Conditions

In order to simplify the numerical model and reduce its calculation time, the following assumptions were made:

1. The argon blown into the tundish and the air remaining in the tundish were regarded as incompressible fluids.
2. The whole process of protective casting was regarded as isothermal, and the energy loss of tundish was ignored due to the high temperature in tundish after baking and the low specific heat capacity of argon and air.
3. The temperature of argon blown into tundish was assumed to be the same as that of tundish, and the natural convection in tundish was ignored.
4. The effect of argon flow on molten steel and slag was ignored.

The flow of gas in tundish during the process of protective casting can be described by the following governing Equations:

$$\frac{\partial \rho}{\partial t} + \nabla \cdot (\rho \boldsymbol{u}) = 0 \tag{1}$$

$$\frac{\partial}{\partial t}(\rho \boldsymbol{u}) + \nabla \cdot (\rho \boldsymbol{u}\boldsymbol{u}) = -\nabla p + \nabla \cdot \left[(\mu + \mu_t)\left(\nabla \boldsymbol{u} + \nabla \boldsymbol{u}^T \right) \right] + \rho \boldsymbol{g} \tag{2}$$

where ρ is the mixture gas density, kg/m^3; t is time, s; $\boldsymbol{u}$ is the velocity vector of mixture gas, m/s; p is the local pressure, Pa; μ is the mixture viscosity, $kg \cdot m^{-1} \cdot s^{-1}$; μ_t is the turbulent viscosity, $kg \cdot m^{-1} \cdot s^{-1}$; and $\boldsymbol{g}$ is the gravitational acceleration, $m \cdot s^{-2}$.

The standard two-equation k-ε model with scalable wall function is applied to describe the turbulent behavior of flow field, where the equation of turbulent kinetic energy transport and its dissipation rate transport can be solved to obtain the turbulent viscosity:

$$\mu_t = \rho C_\mu \frac{k^2}{\varepsilon} \tag{3}$$

Turbulent kinetic energy, k, $\text{m}^2 \cdot \text{s}^{-2}$:

$$\frac{\partial(\rho k)}{\partial t} + \nabla \cdot (\rho u k) = \nabla \cdot \left[\left(\mu + \frac{\mu}{\sigma_k} \right) \nabla k \right] + G_k - \rho \varepsilon \tag{4}$$

where G is the generation of turbulent kinetic energy due to the mean velocity gradients and can be written as:

$$G_k = \mu_t \frac{\partial u_j}{\partial x_i} \left(\frac{\partial u_i}{\partial x_j} + \frac{\partial u_j}{\partial x_i} \right) \tag{5}$$

The dissipation rate of turbulent kinetic energy, ε, $\text{m}^2 \cdot \text{s}^{-3}$:

$$\frac{\partial(\rho \varepsilon)}{\partial t} + \nabla \cdot (\rho u \varepsilon) = \nabla \cdot \left[\left(\mu + \frac{\mu_t}{\sigma_\varepsilon} \right) \nabla \varepsilon \right] + \frac{\varepsilon}{k}(C_1 G - C_2 \rho \varepsilon) \tag{6}$$

where C_1, C_2, C_μ, σ_k, σ_ε are the empirical constants, whose values, as recommended by Launder and Spalding [23], are 1.38, 1.92, 0.09, 1.0, and 1.3, respectively.

The species transport equation was solved to obtain the volume fraction of air and argon, and then the volume fraction of oxygen can be obtained by the product of air volume fraction and 0.21 (the volume fraction of oxygen in air):

$$\frac{\partial}{\partial t}(\rho Y_i) + \nabla \cdot (\rho u Y_i) = -\nabla \cdot \left(-\left(\rho D_{i,m} + \frac{\mu_t}{Sc_t} \right) \nabla Y_i \right) + S_i \tag{7}$$

where Y_i is mass fraction for species i, non-dimensional; $D_{i,\,\text{m}}$ is the mass diffusion coefficient for species i, $\text{m}^2 \cdot \text{s}^{-1}$; Sc_t is the turbulent Schmidt number, 0.7; S_i is the source term for species i, $\text{kg} \cdot (\text{m}^{-3} \cdot \text{s}^{-1})$; and in the current study, i represents argon or air in tundish.

2.2. Experimental Facility and Numerical Model

Figure 1 shows the two-strand slab tundish schematic diagram. A tundish with a size of $7.77 \times 1.68 \times 1.32$ m and a taper of 1.1 between the top face and bottom face has an injecting chamber and two symmetrical casting chambers. The injecting chamber has two baking holes, and each casting chamber has a baking hole and a stopper hole. The tundish cover is composed of an outer shell made of steel plates and filler made of refractory material. The argon pipe is buried inside the refractory material. Figure 2 shows the arrangement of the argon blowing pipe in the tundish cover. Fourteen argon blowing pipes with diameters of 30 mm were installed on the tundish cover, including six for the injecting chamber and four for each casting chamber. Before a tundish is used, the blast furnace gas is first adopted to bake the tundish through those baking holes for more than 30 min to completely eliminate water vapor. After baking the tundish, its temperature was greater than 1000 K. The process of ABTC will be conducted after baking tundish. Because the covering flux on the molten steel surface cannot be fully melted immediately, argon blown into tundish through those pipes may remove air and reduce molten steel secondary oxidation. During the period of empty tundish and normal casting, it is necessary to maintain the volume fraction of oxygen in tundish at a low level (<1%) to avoid the secondary oxidation of molten steel and improve its cleanliness.

During the period of empty tundish, the gas can be freely exchanged between the injecting chamber and casting chamber. However, during the period of normal casting, the injecting chamber and casting chamber are isolated by molten steel, which forms a separate injecting chamber and two separate casting chambers, in which the gas cannot exchange. Therefore, the mesh for the empty tundish and normal casting must be built separately. The mesh for empty tundish and normal casting is shown in Figure 3. Octahedron grids are adopted to a mesh in the calculating domain, and the numbers of meshes are 508,484 for empty tundish, 41,018 for isolated casting chamber, and 52,493 for isolated injecting chamber.

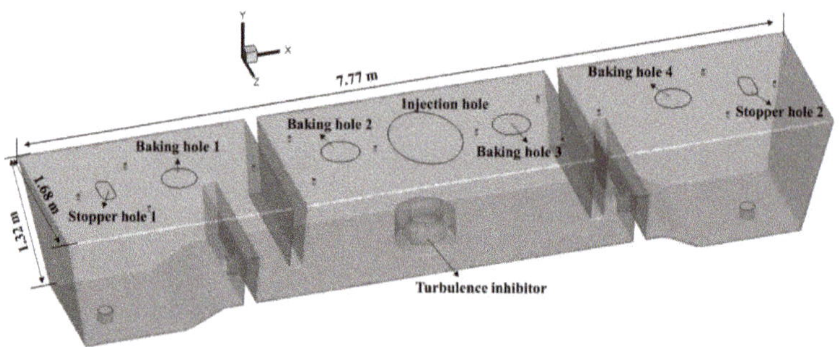

Figure 1. Structure diagram of a two-strand slab tundish.

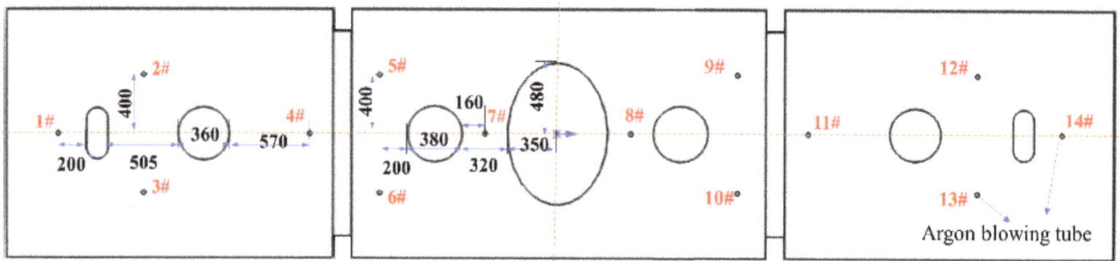

Figure 2. Arrangement of the argon blowing pipe on the two-strand slab tundish cover (mm).

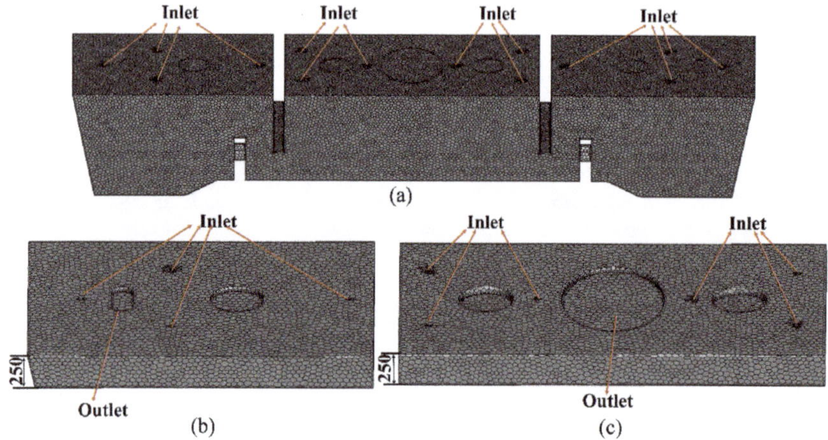

Figure 3. Mesh for empty tundish (**a**), isolated casting chamber (**b**), and isolated injecting chamber (**c**) (mm).

During the calculation, the velocity inlet boundary condition was used for those argon pipes, where the argon flow rate of each pipe is 1/14 of the total flow rate. The pressure outlet with a relative static pressure of zero was adopted for the opening hole, which is decided by the sealing scheme in the stage of empty tundish and the injecting hole and stopper hole in the normal casting according to the practical process. The non-slip wall boundary was used for all inner faces of tundish, which assumes that the gas velocity on the wall is zero. The initial condition of air volume fraction and temperature is assumed as 1 and 1000 K, respectively. The SIMPLEC algorithm was adopted to solve the transient problem. The timestep was set to 0.05 s for the current study. Calculation convergence was achieved when all residuals were lower than 10^{-4}. The average volume fraction of oxygen is monitored throughout the calculation process. The tundish parameters and other related continuous casting process parameters are summarized in Table 1.

Table 1. Parameters of tundish and the casting process.

Item	Value
Slab section size, mm × mm	230×1300
Casting speed, m·min^{-1}	0.9–1.2
Depth of molten steel during normal casting period, mm	1000
Diameter of baking hole, mm	380, 360
Diameter of argon pipe, mm	30
Distance between molten slag surface and tundish cover, mm	250
Density of argon at 288 K, kg·m^{-3}	1.6228
Density of air at 288 K, kg·m^{-3}	1.225
Viscosity of argon, kg·m^{-1}·s^{-1}	2.125×10^{-5}
Viscosity of air, kg·m^{-1}·s^{-1}	1.789×10^{-5}
Mass fraction of oxygen in air	0.21
Mass diffusivity of mixture, m^2·s^{-1}	2.88×10^{-5}
Total flow rate of argon blowing, Nm^3·h^{-1}	60, 100, 120, 140, 200, 220, 240, 260

2.3. Model Validation

In order to verify the reliability of the numerical model, an industrial trial with an argon flow rate of 220 Nm^3/h was conducted, and the injecting hole, baking holes, and stopper holes were kept open after tundish baking. The VF of oxygen at different times was measured using an Optima 7 gas analyzer. The measurement in the industrial trial was carried out 3 min after tundish baking due to the rapid pace of industrial production. A numerical simulation based on the process was also conducted. The data from the industrial test and numerical calculation are shown in Figure 4, demonstrating a better agreement between the experimental values and the simulation results.

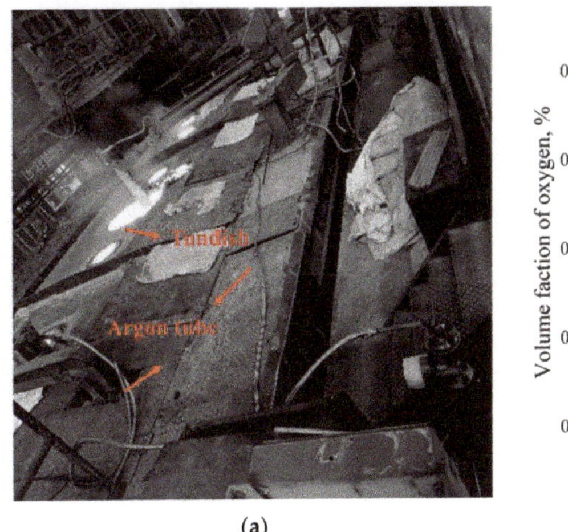

(a)

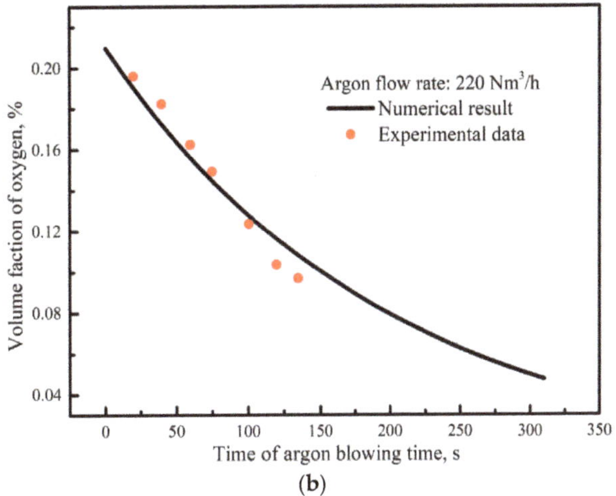

(b)

Figure 4. Snapshot of the industrial test (**a**) and comparison of the industrial test data and numerical calculation (**b**).

A verification of grid independence was also performed in the current study. The operating parameter of the argon flow rate of 220 Nm^3/h, with sealing the injecting hole, baking holes and an open stopper hole, was chosen. The average VF of the oxygen in the tundish at 300 s was discussed and used as the verification criterion. Table 2 shows the statistical results for different meshes. As shown in Table 2, the errors decrease with an increased number of cells. For the case of M3, the error relative to the finest mesh M4 is less than 5%, which is within the allowable error range. Therefore, considering the computational cost and efficiency, the mesh of M3 was adopted for the numerical calculation.

Table 2. Error statistics of different meshes.

Mesh	M1	M2	M3	M4
Total cell number	192,236	325,618	508,483	612,348
Total node number	998,050	1,878,166	2,932,932	3,532,026
Average volume faction of oxygen	0.04846	0.04765	0.04591	0.04511
$\delta_{oxygen} = |VF_{oxygen\text{-}i} - VF_{oxygen\text{-}4}|/VF_{oxygen\text{-}4}$	0.0743	0.0563	0.0179	–

3. Results and Discussion

3.1. Numerical Simulation

3.1.1. Calculation of the Sealing Scheme of Tundish Cover Holes during the Period of Empty Tundish

During practical production, there are about 5~10 min for blowing argon into tundish from the end of tundish baking to the start of casting. In order to improve the efficiency of protective casting and fully discharge oxygen, asbestos is usually used to seal some holes, such as the baking hole, injecting hole, or stopper hole. To determine the best sealing scheme during the period of empty tundish, three schemes were designed and described, as shown in Table 3. The average volume fractions of oxygen with argon flow rates of 200 Nm^3/h for different schemes were calculated in 10 min. The variations of average argon volume fraction with respect to time for different schemes are shown in Figure 5. As shown in Figure 5, the difference in average oxygen volume fraction at 10 min between the Schemes 1, 3 and 4 is small, which is 0.0174, 0.0167, and 0.0170 respectively. Scheme 2, sealing the injecting hole and all baking holes, is the best, and its average oxygen volume fraction is 0.0132 at 10 min.

Table 3. Sealing schemes of tundish cover holes.

Scheme No.	Seal of Tundish Cover Hole
1	Stopper hole 1–2, baking hole 1–4
2	Injecting hole, baking hole 1–4
3	Injecting hole, stopper hole 2, baking hole 1–4
4	None

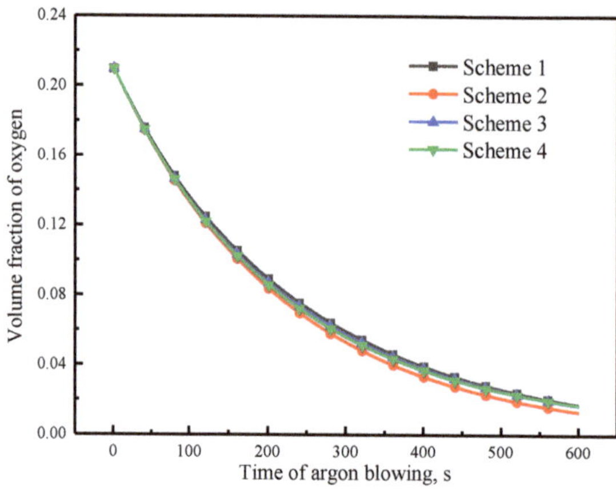

Figure 5. Variation of oxygen volume fraction in tundish for different schemes.

Figure 6 shows the contour of oxygen volume fraction at 10 min. For Scheme 1, sealing all stopper holes and all baking holes, the oxygen volume fraction in the casting chamber is higher than that in the injecting chamber. The dam and wall of tundish isolate the casting chamber and injecting chamber, which results in the oxygen in the casting chambers not being smoothly discharged and forming a dead zone. For Scheme 2, sealing the injecting hole and all baking holes, the oxygen volume fraction in the injecting chamber and casting chamber is relatively uniform, which indicates a better effect of discharging oxygen. For Scheme 3—sealing the injecting hole, stopper hole 2, and all baking holes—the oxygen volume fraction in casting chamber 1 is lower than that in chamber 2, which indicates the negative effects of discharging oxygen due to the long distance from casting chamber 2 to the outlet (stopper hole 1). As for Scheme 4, opening all holes, the oxygen content in the injecting chamber is higher than that in the casting chamber due to the stronger backflow of air. Overall, Scheme 2 is the best and used for the analysis of argon flow rate.

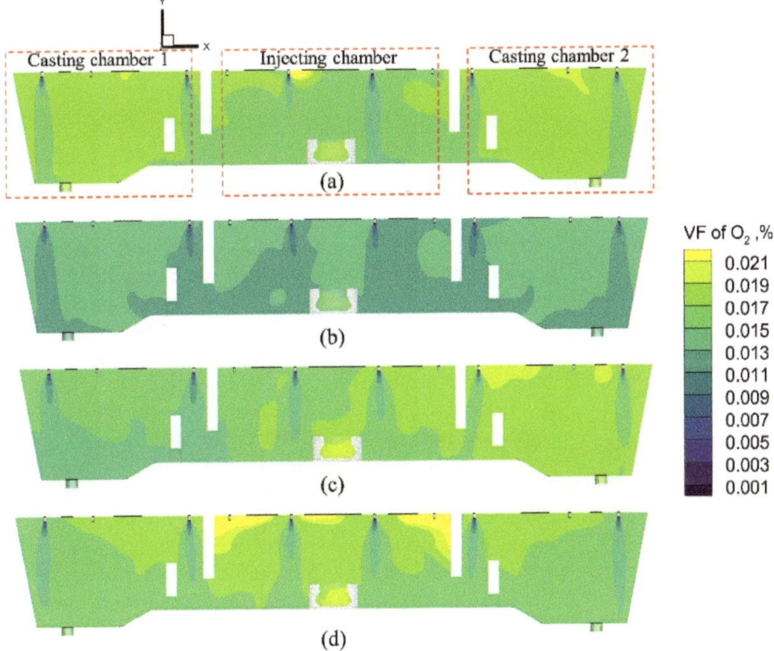

Figure 6. Contour of oxygen volume fraction at $z = 0$ m section, (**a**) for Scheme 1, (**b**) Scheme 2, (**c**) Scheme 3, (**d**) and Scheme 4.

3.1.2. Calculation of Argon Flow Rate during a Period of Empty Tundish

Figure 7a shows the variation of average oxygen volume fraction with time under different argon flow rates. The volume fraction of oxygen shows similar tendencies for different argon flow rates, but the increase in argon flow rate accelerates the reduction of oxygen volume fraction, which shows that increasing the argon flow rate is conducive to discharging oxygen in tundish. Figure 7b shows the effect of argon flow rate on average volume fraction of oxygen at different times. The average volume fraction of oxygen linearly decreases with the increase in argon flow rate. It can be reduced to 1% after 10 min when the argon flow rate is 220 Nm^3/h, which indicates that the argon flow rate should be greater than 220 Nm^3/h to ensure that the remaining volume fraction of oxygen is less than 1% and achieves a better protective casting effect during a period of empty tundish.

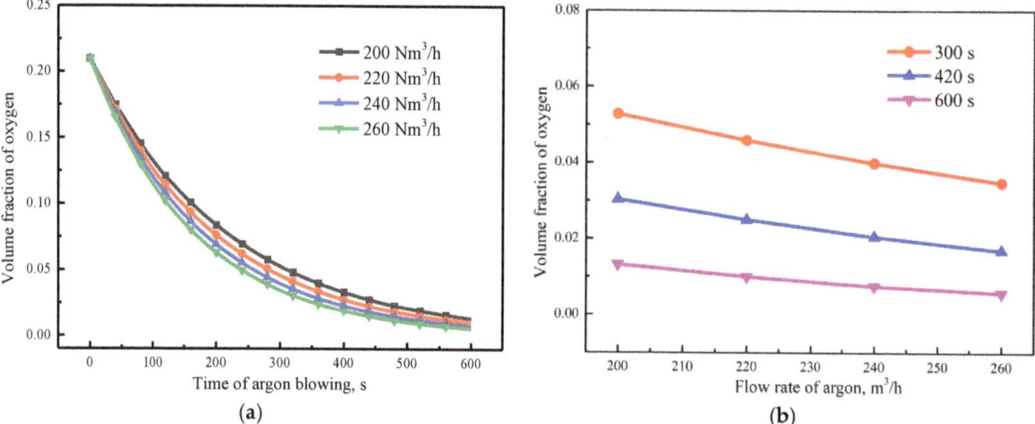

Figure 7. Variation of volume fraction of oxygen with time for different argon flow rates (**a**), and effect of argon flow rate on the volume fraction of oxygen at different times (**b**).

3.1.3. Calculation of Argon Flow Rate during a Period of Normal Casting

During a period of normal casting, the injecting chamber and casting chambers are isolated into independent chambers by molten steel. There is no gas exchange between those independent chambers. Compared with a period of empty tundish, the holding capacity of gas in the tundish largely decreases, and the argon flow rate should be adjusted during a period of normal casting. In order to determine the optimal argon flow rate, the variation of average oxygen volume fraction in an isolated injecting chamber and casting chamber was investigated, respectively. During the calculation, the oxygen volume fraction of Scheme 2 with an argon flow rate of 260 Nm^3/h at 5 min was taken as the initial condition. During a period of normal casting, the baking holes are usually sealed except for the sampling operation, and the stopper holes and injecting hole remain open.

Figure 8a shows the variation of oxygen volume fraction with time in the isolated injecting chamber. The volume fraction of oxygen decreases with increasing argon blowing time, but the downtrend shows a tendency of increasing first and then decreasing with the increase in argon flow rate. The oxygen volume fractions after argon blowing for 5 min are 0.03, 0.0171, and 0.0259 when the argon flow rates are 60 Nm^3/h, 80 Nm^3/h, and 120 Nm^3/h, respectively. Figure 8b shows the oxygen volume fraction at a given time under different argon flow rates. At a fixed time, the oxygen volume fraction first decreases and then increases with increase in argon flow rate, showing a minimum value at an argon flow rate of 80 Nm^3/h.

In order to explain this phenomenon, the contours of oxygen volume fraction and velocity vector at $z = 0$ m section are abstracted and shown in Figures 9 and 10. As shown in Figure 9, when the argon flow rate is 80 Nm^3/h, the oxygen volume fraction in the injecting chamber is uniform, except the region near the injecting holes, where it is slightly higher than that in the other region. With the argon flow rate increasing to 140 Nm^3/h, the region of oxygen volume fraction of more than 0.027 in injecting chamber significantly increases, and even extends into the vicinity of baking holes. This phenomenon can be explained by the velocity vector, as shown in Figure 10. Argon jets into the injecting chamber by the argon pipe installed on both sides of the injecting hole. After the argon jet impacts the molten steel, two horizontal streams occur and collide at the center of the injecting hole, and then discharge from the injecting hole. With the discharge in argon, a large amount of air is drawn into the injecting chamber and results in an increase in oxygen volume fraction. The phenomenon of air entrainment increases with the increase in argon flow rate. Therefore, the best argon flow rate in the injecting chamber is 80 Nm^3/h.

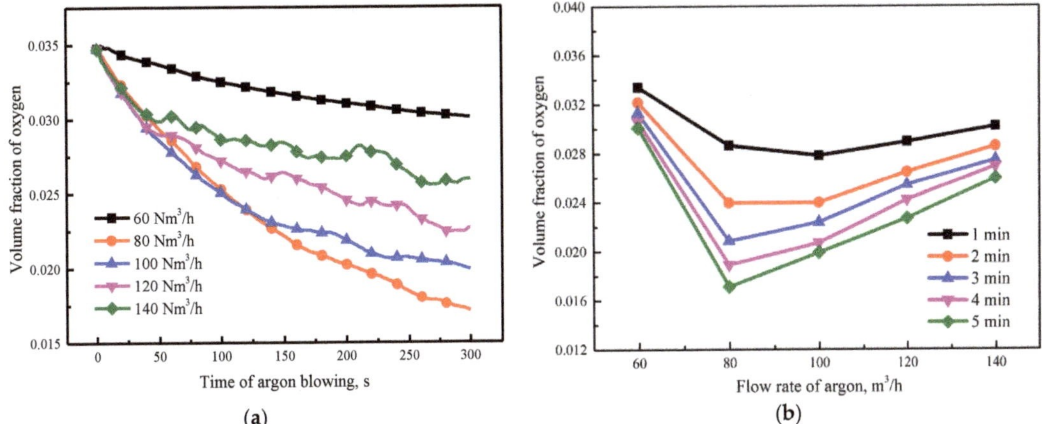

Figure 8. Variation of oxygen volume fraction with time for different argon flow rates (**a**) and effect of argon flow rate on volume fraction of oxygen for different times (**b**) in an isolated injecting chamber.

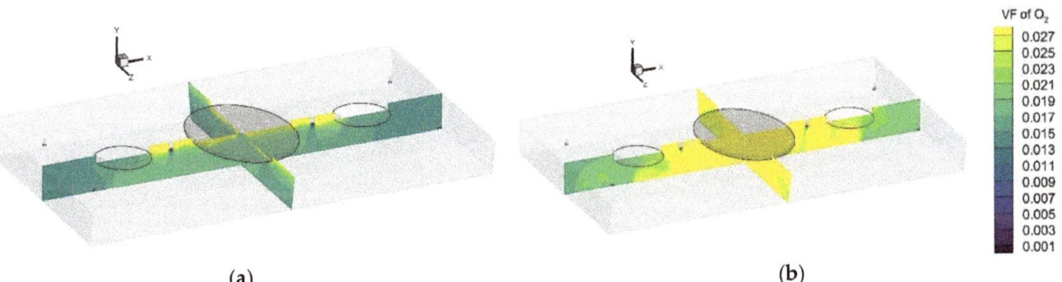

Figure 9. Contour of argon volume fraction at the $z = 0$ m section for different argon flow rates in the injecting chamber: (**a**) 80 Nm^3/h, and (**b**) 140 Nm^3/h.

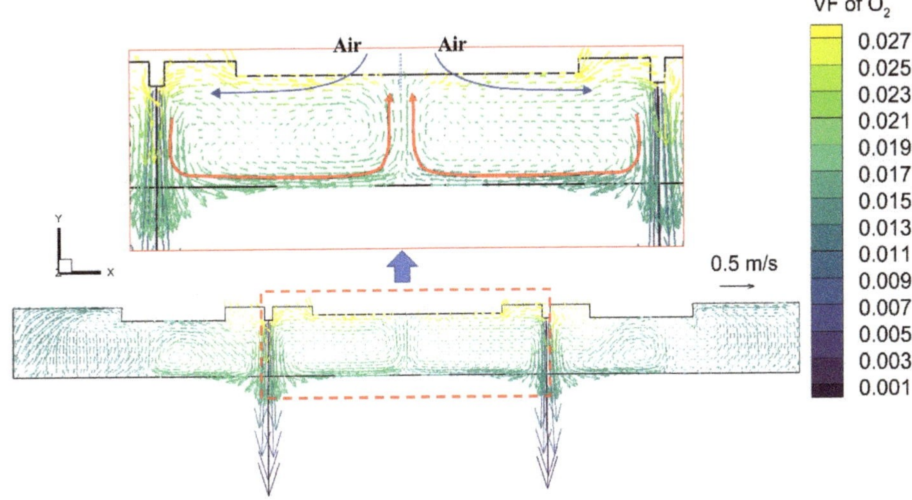

Figure 10. Vector of argon velocity at $z = 0$ m section in the injecting chamber.

The variation behavior of oxygen volume fraction in the isolated casting chamber was also investigated and is shown in Figure 11. The volume fraction of oxygen in the isolated casting chamber continuously decreases with the increase in argon blowing time, and the downtrend shows a positive correlation with argon flow rate. Figure 11b shows the oxygen volume fraction at a given time under different argon flow rates. The oxygen volume fraction decreases with the increase in argon flow rate at a given time and achieves 1% at 5 min when the argon flow rate is 80 Nm^3/h, which indicates the argon flow rate in the isolated casting room should be greater than 80 Nm^3/h. However, combined with the variation behavior of oxygen volume fraction in an isolated injecting chamber, the argon flow rate should be kept as 80 Nm^3/h for the period of normal casting.

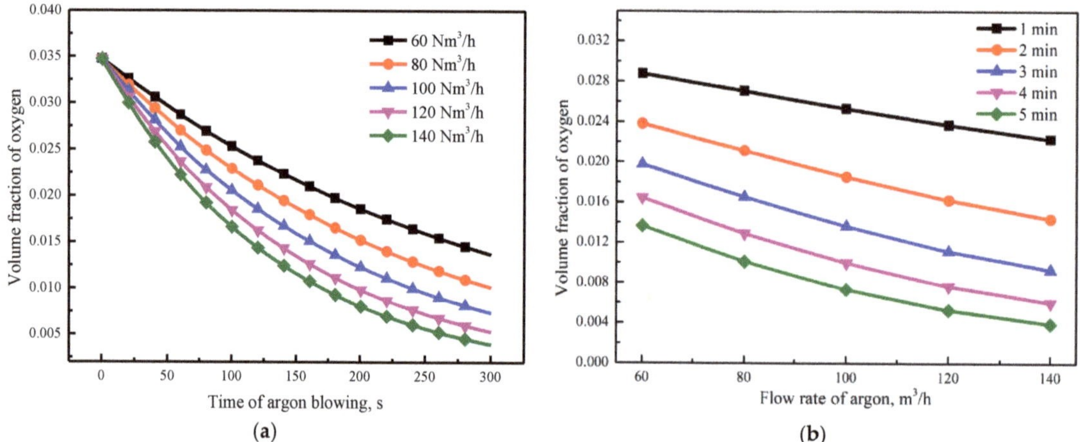

Figure 11. Variation of oxygen volume fraction with time for different argon flow rate (**a**), and effect of argon flow rate on volume fraction of oxygen at different times (**b**) in an isolated casting chamber.

3.2. Industrial Application

In order to evaluate the process effect of ABTC, plant trials of two consecutive tundishes for continuous casting were conducted, and each tundish was used for six ladles. The first tundish with ABTC, under Scheme 2, was taken as the trial group, and the second tundish without ABTC was taken as the control group. The process parameters of ABTC and continuous casting are listed in Table 4.

Table 4. Process parameters of argon blowing tundish cover and continuous casting.

Item	Value
ABTC during a period of empty tundish	Argon flow rate: 220 Nm^3/h; Time of argon blowing: 3 min
ABTC during a period of normal casting	Argon flow rate: 100 Nm^3/h; Time of argon blowing: all the other time
Steel grade	M3A35
Steel composition, wt%	C $\leq$ 0.002, Si $\leq$ 0.005, Mn: 0.05~0.08, Ti: 0.065~0.085, Al: 0.02~0.035, Cr $\leq$ 0.05, Ni $\leq$ 0.05, Mo $\leq$ 0.01
Casting speed, m/min	1.12
Section size of slab, mm $\times$ mm	1300 $\times$ 230

During trials, steel samples at the end of RH and in the tundish for each ladle were taken, and the contents of nitrogen, titanium, and aluminum in the steel samples were measured and are listed in Table 5. By applying ABTC, the average nitrogen content of steel samples in tundish decreases from 22×10^{-6} to 16×10^{-6}, and the average increased nitrogen content ($\triangle w_{[N]}$) from the end of RH to tundish decreases by 90% from 10×10^{-6} to 1×10^{-6}. The average loss of titanium and aluminum decreases by 12.7%

from 63×10^{-6} to 55×10^{-6} and 7.1% from 70×10^{-6} to 55×10^{-6}, respectively. In general, the ABTC process is conducive to decreasing the increased nitrogen content of molten steel in tundish and the loss of titanium and aluminum. Therefore, the protective casting effect could be significantly improved by applying ABTC.

Table 5. Mass fractions of nitrogen, titanium, and aluminum of steel samples at the end of RH and in tundish.

| Heat No. | Trial Group with ABTC, $\times 10^{-6}$ | | | | | | | | | Control Group without ABTC, $\times 10^{-6}$ | | | | | | | | |
| | End of RH, w | | | Tundish, w | | | Increment, $\triangle w$ | | | End of RH, w | | | Tundish, w | | | Increment, $\triangle w$ | | |
	[N]	[Ti]	[Al]	[N]	[Ti]	[Al]	[N]	[Ti]	[Al]	[N]	[Ti]	[Al]	[N]	[Ti]	[Al]	[N]	[Ti]	[Al]
1	15	760	434	15	710	350	0	−50	−84	16	750	448	17	700	360	1	−50	−88
2	12	820	427	16	760	380	4	−60	−47	16	800	462	28	750	420	12	−50	−42
3	14	800	422	13	710	310	0	−90	−112	16	820	466	16	700	360	0	−120	−106
4	19	760	395	18	730	340	0	−30	−55	0	790	474	23	760	450	23	−30	−24
5	16	770	407	17	720	360	1	−50	−47	24	810	470	22	730	390	0	−80	−80
6	15	820	436	17	770	390	2	−50	−46	0	800	418	27	750	340	27	−50	−78
Average	15	788	420	16	733	355	1	−55	−65	12	795	456	22	732	387	10	−63	−70

4. Conclusions

In this study, the feasibility and application effect of ABTC during a continuous casting period were investigated and evaluated through simulations and plant trials. The following conclusions can be drawn:

(1) The process of ABTC shows a great protective casting effect in avoiding increasing nitrogen and losing titanium and aluminum. As for the steel grade of M3A35, the average increment of nitrogen content ($\triangle w_{[N]}$) decreases by 90%, and the average losses of titanium $\triangle w_{[Ti]}$ and aluminum $\triangle w_{[Al]}$ decrease by 12.7% and 7.1%, respectively.

(2) During ABTC, the injecting hole and baking holes should be sealed, and the argon flow rate should be ≥ 220 Nm^3/h to obtain better protective effects during a period of empty tundish.

(3) During a period of normal casting, the rational argon flow rate is 80 Nm^3/h due to the strong air entrainment in the isolated injecting chamber.

Author Contributions: Conceptualization, Y.L. and X.Y.; Data curation, C.W.; Investigation, Y.L.; Methodology, X.X.; Software, Y.L.; Supervision, X.M.; Validation, J.C.; Writing—original draft, Y.L.; Writing—review and editing, C.W., X.X. and L.C. All authors have read and agreed to the published version of the manuscript.

Funding: This research is supported by the Pangang Group Research Institute Co., Ltd.

Institutional Review Board Statement: Not applicable.

Informed Consent Statement: Not applicable.

Data Availability Statement: Not applicable.

Conflicts of Interest: The authors declare no conflict of interest that are relevant to the content of this article.

References

1. Chu, J.; Bao, Y. Study on the relationship between vacuum denitrification and manganese evaporation behaviours of manganese steel melts. *Vacuum* **2021**, *192*, 110420. [CrossRef]
2. Kawakami, K.; Taniguchi, T.; Nakashima, K. Generation Mechanisms of Non-metallic Inclusions in High-cleanliness Steel. *Tetsu–Hagane* **2007**, *93*, 743–752. [CrossRef]
3. Ling, H.; Zhang, L.; Liu, C. Effect of snorkel shape on the fluid flow during RH degassing process: Mathematical modelling. *Ironmak Steelmak.* **2018**, *45*, 145–156. [CrossRef]
4. Ling, H.T.; Zhang, L.F. Investigation on the Fluid Flow and Decarburization Process in the RH Process. *Metall. Mater. Trans. B* **2018**, *49*, 2709–2721. [CrossRef]

5. Cao, Q.; Nastac, L. Numerical modelling of the transport and removal of inclusions in an industrial gas-stirred ladle. *Ironmak Steelmak.* **2018**, *45*, 984–991. [CrossRef]

6. Liu, Y.; Ersson, M.; Liu, H.; Jonsson, P.G.; Gan, Y. Reply to the Discussion on "A Review of Physical and Numerical Approaches for the Study of Gas Stirring in Ladle Metallurgy". *Metall. Mater. Trans. B* **2020**, *51*, 1847–1850. [CrossRef]

7. Jardon-Perez, L.E.; Gonzalez-Rivera, C.; Ramirez-Argaez, M.A.; Dutta, A. Numerical Modeling of Equal and Differentiated Gas Injection in Ladles: Effect on Mixing Time and Slag Eye. *Processes* **2020**, *8*, 917. [CrossRef]

8. Liu, W.; Lee, J.; Guo, X.; Silaen, A.K.; Zhou, C.Q. Argon Bubble Coalescence and Breakup in a Steel Ladle with Bottom Plugs. *Steel Res. Int.* **2019**, *90*, 1800396. [CrossRef]

9. Li, L.; Liu, Z.; Cao, M.; Li, B. Large Eddy Simulation of Bubbly Flow and Slag Layer Behavior in Ladle with Discrete Phase Model (DPM)–Volume of Fluid (VOF) Coupled Model. *JOM* **2015**, *67*, 1459–1467. [CrossRef]

10. Li, Y.; Deng, A.Y.; Li, H.; Yang, B.; Wang, E.G. Numerical Study on Flow, Temperature, and Concentration Distribution Features of Combined Gas and Bottom-Electromagnetic Stirring in a Ladle. *Metals* **2018**, *8*, 19. [CrossRef]

11. Chen, S.; Lei, H.; Wang, M.; Yang, B.; Dou, W.; Chang, L.; Zhang, H. Ar-CO-liquid steel flow with decarburization chemical reaction in single snorkel refining furnace. *Int. J. Heat Mass Tran.* **2020**, *146*, 118857. [CrossRef]

12. Ling, H.; Xu, R.; Wang, H.; Chang, L.; Qiu, S. Multiphase Flow Behavior in a Single-Strand Continuous Casting Tundish during Ladle Change. *ISIJ Int.* **2020**, *60*, 499–508. [CrossRef]

13. Cwudzinski, A. Physical and mathematical modeling of bubbles plume behaviour in one strand tundish. *Metall. Res. Technol.* **2018**, *115*, 101. [CrossRef]

14. Chatterjee, S.; Chattopadhyay, K. Physical Modeling of Slag 'Eye' in an Inert Gas-Shrouded Tundish Using Dimensional Analysis. *Metall. Mater. Trans. B* **2016**, *47*, 508–521. [CrossRef]

15. Yang, W.; Luo, Z.; Gu, Y.; Liu, Z.; Zou, Z. Simulation of bubbles behavior in steel continuous casting mold using an Euler-Lagrange framework with modified bubble coalescence and breakup models. *Powder Technol.* **2020**, *361*, 769–781. [CrossRef]

16. Wang, Y.; Fang, Q.; Zhang, H.; Zhou, J.; Liu, C.; Ni, H. Effect of argon blowing rate on multiphase flow and initial solidification in a slab mold. *Metall. Mater. Trans. B* **2020**, *51*, 1088–1100. [CrossRef]

17. Zhang, H.; Fang, Q.; Xiao, T.P.; Ni, H.W.; Liu, C.S. Optimization of the flow in a slab mold with argon blowing by divergent bifurcated SEN. *ISIJ Int.* **2019**, *59*, 86–92. [CrossRef]

18. Yang, W.D.; Luo, Z.G.; Lai, Q.R.; Zou, Z.S. Study on bubble coalescence and bouncing behaviors upon off-center collision in quiescent water. *Exp. Therm. Fluid Sci.* **2019**, *104*, 199–208. [CrossRef]

19. Feng, W. *Reoxidation Phenomenon and Inhibition Mechanism of Aluminum-Deoxidized Steel in Tangsteel*; University of Science and Technology Beijing: Beijing, China, 2019.

20. Story, S.R.; Goldsmith, G.E.; Fruehan, R.J.; Casuccio, G.S.; Potter, M.S.; Wlliams, D.M. A study of casting issues using rapid inclusion identification and analysis. *Iron Steel Technol.* **2006**, *3*, 52–61.

21. Sun, Y.; Cai, K.; Zhao, C. Effect of transient casting operation on cleanliness of continuously cast strands. *Iron Steel* **2008**, *43*, 22–25. [CrossRef]

22. Gao, J.; Yan, S.; Ding, Z. Analysis on surface defects of deep-drawing piece of steel DC04-LQQ for filter and process control. *Spec. Steel* **2017**, *38*, 27–31.

23. Launder, B.; Spalding, D.B. The numerical computation of turbulent flows. *Comput. Method. Appl. Mech.* **1974**, *3*, 269–289. [CrossRef]

MDPI

St. Alban-Anlage 66

4052 Basel

Switzerland

Tel. +41 61 683 77 34

Fax +41 61 302 89 18

www.mdpi.com

Metals Editorial Office

E-mail: metals@mdpi.com

www.mdpi.com/journal/metals